William Rounseville Alger

# The Solitudes of Nature and of Man

The Loneliness of Human Life

William Rounseville Alger

**The Solitudes of Nature and of Man**
*The Loneliness of Human Life*

ISBN/EAN: 9783337370947

Printed in Europe, USA, Canada, Australia, Japan

Cover: Foto ©berggeist007 / pixelio.de

More available books at **www.hansebooks.com**

THE

# SOLITUDES

OF

# NATURE AND OF MAN;

OR,

*The Loneliness of Human Life.*

BY

WILLIAM ROUNSEVILLE ALGER.

Hast du Begriff von Oed' und Einsamkeit *r*
GORTHE.

BOSTON:
ROBERTS BROTHERS.
1892.

# PREFACE.

———◆———

THOSE who have the key for interpreting the signs
of genuine thought and emotion will perceive that this
book has sprung sincerely from the inmost life of the
writer. His ambition has been to make it the Book
of Solitude, whose readers may learn from it how at
the same time to win the benefits and shun the evils
of being alone. The subject — the conditions and influences of solitude in its various forms — is so largely
concerned with disturbed feelings that it is difficult, in
treating it, to keep free from everything unhealthy, excessive, or eccentric. In view of this, great pains have
been taken to avoid every morbid extravagance, and
stay close by the standards of sanity, truth, and cheer-
fulness. For an author ought not to dishearten, but
to inspire his readers; not to exhale around them an
infecting atmosphere of hates, griefs, and despairs, but
to warm and strengthen them with his health, valor,
and contentment. We grow old to trust and joy, and
they become vapid. Doubt and sadness keep their
fresh force, we are always young to them. They should,
therefore, never be disseminated. In treating themes
pertaining to the deepest emotions, the temptations to
satire and to sentimentality are both strong; but for the
exertion of a sound influence those temptations should
be resisted. Faith, direct sincerity, undiseased tenderness, and the authority of well-mastered experience, are
the best qualities in a teacher.

In dealing with the affairs of the heart, every form
of unfeelingness is an offence. It is by drawing out

and satisfying, not by freezing or searing, the affections, that true happiness and peace are to be won.
The warm effusion of Christianity is better adapted to
human nature than the dry chill of Stoicism. Every
man obscurely feels, though scarcely any man distinctly
understands, the intimacy and vastness of his connections
with his race. It is true that the real world of the soul
is an invisible place, removed from the rush and chatter
of crowds, and that the most important portion of life
is the secret and solitary portion. Yet the most influential element even of this secluded world and this hidden
life, is the element which consists of the ideas and feelings we habitually cherish in relation to our fellow-beings.

The philosophy of solitude has been well discussed —
bating an occasional romantic vein with morbid tinges —
by Zimmermann, whose work permanently identifies his
name with the subject. I had not read that celebrated
treatise until after the completion of my own, and am indebted to it for nothing beyond the citations explicitly
made from it in the revision of these pages. Zimmermann's personal experience of solitude, and long brooding
over it, contributed much to the value of his work ; which
is comprehensive in survey, rich in learning, penetrative
in thought, vivid in sentiment, and eloquent in diction.
On the other hand it is too diffuse in style, too expanded
in form. It was translated into the chief languages of
Europe and had a vast sale. The English translation, so
widely scattered half a century ago, comprised only about
a third part of the matter in the four volumes of the
original German, which the author had elaborately rewrit-
ten with great enlargements. Besides the important work
of Zimmermann there are in the literary remains of eminent men — among whom Petrarch, Montaigne, Cowper,
and Schopenhauer deserve special mention — a multitude
of essays, poems, and letters, on solitude. These I have
carefully searched, and have endeavored to enrich my
own disquisition by appropriately quoting from them their
best things. I have not gathered these numerous quotations in a pedantic spirit, for a show of learning, nor in a
poetic spirit, for mere ornament, but in a didactic spirit.

in the belief that the collection thus made of the solemn
and weighty or beautiful and pathetic expressions of
authoritative minds would be valuable. I cannot help
believing that the best readers, so far from bringing against
me the charge of superfluity in quotation, will be grateful
for the use of these auxiliaries. It stands to reason that
no man can handle a great moral theme from his single
mind so well as he can when aided by the contributions
of the wise men who have handled it before. The form,
divisions, and method of the present work have grown out
of my own meditations. Its development in details has,
of course, been modified as well by the inspiring suggest-
iveness of the writings of others on the same subject, as
by the transplantation into it, with due credit, of many
of the most nutritious thoughts and sanative sentiments
met with in them.

Contempt and scorn, unless directed by nobler emotions,
are as pernicious as they are easy and vulgar. Pure
forms of reverence and aspiration are more rarely felt as
facts of experience, and are more difficult of attainment
as attributes of character. But it is better to lift the eye
than to curl the lip. The aphorism of Lavater is good:
Trust him little who smilingly praises all alike, him less
who sneeringly censures all alike, him least who is coldly
indifferent to all alike. Did I not believe this book
adapted to develop both a healthy dislike for what is bad
in men, and a becoming admiration and love for what is
good in them, I would fling it into the fire instead of
committing it to the press. Is there anything else so
odious as the passions of hatred and envy? What else
is so desirable as the qualities of devoutness, wisdom,
magnanimity, and peace? Unless the author is ignorant
of his own heart he has written the following pages with
the warmest pity for the victims of the ignoble traits of
human life, and with a fervent desire to remove the causes
of their sufferings. Unless he is deceived he has also
been actuated by a religious veneration for great and good
men, the heroic masters in virtue, and by a purpose to
exalt them before the multitude as ideals which shall
exert an influence to mould to their likeness those who

earnestly contemplate them.   Great men heighten the
consciousness of the human race; and it is our grateful
duty to magnify him whose genius magnifies mankind.
The roll of persons admiringly treated of in the following
leaves composes a list of names fit to be kept in the cas-
ket of a king.

The majority of men in every age are superficial in
character and brittle in purpose, and lead undedicated
lives; swarming together in buzzing crowds in all haunts
of amusement or places of low competition, caring little
for anything but gossip and pastime, the titillation of the
senses, and the gratification of conceit.   To state the
conditions and illustrate the attractions of a holier and
grander happiness, — to hold up the examples of nobler
characters and lives, lifted into something of loneliness
by their gifts and achievements, — is, accordingly, always
a timely service.   All better lives are so much redeeming
leaven kneaded into the lump of humanity.

There are many disappointed and discontented men
and women, exasperated with society, uneasy with seclusion, galled by the bonds of the world when they feel its
multitudinous emulation, unable to enjoy freedom and
repose when they retreat into solitude.   Sometimes this
state is a consequence of poor health.   Then the patient
has more need of the physician than of the divine, the
first desideratum being the restoration of the nervous
system to its normal tone.   Generally, however, there is
equal occasion for moral counsel and medicinal direction.
But when this experience is more purely a moral result,
it is, in most cases, the product of a too magnified opinion of self combined with a too acute feeling of that
opinion.   Exactingness is the bane, renunciation the
antidote.   Self-respect may be the sternest wisdom, but
self-idolatry is infatuation.   This is one of the many
questions which must be analyzed in any adequate pre-
sentation of the causes of human loneliness.

The author now dismisses his book of solitude, to find
its way and do its work among men, in the hope that it
may render many an unhappy heart that service of sympathetic guidance which he feels such a work would, at
an earlier day, have rendered himself.

# CONTENTS.

---

## PART I.

### THE SOLITUDES OF NATURE.

## PART II.

### THE SOLITUDES OF MAN.

## PART III.

### THE MORALS OF SOLITUDE.

# PART IV.

## SKETCHES OF LONELY CHARACTERS:

OR, PERSONAL ILLUSTRATIONS OF THE GOOD AND EVIL OF SOLITUDE.

# THE SOLITUDES OF NATURE.

# THE SOLITUDES OF NATURE.

*Gregariousness and Solitariness.*

AT the first glance every form of being appears to be social, all the world gregarious. The trees interlace their branches and wave their tops in multitudinous union; from the equator to the poles the waves shoulder their fellows, glistening with innumerable smiles; whole orchards of apple-blossoms blush in correspondence; in regiments the ranks of corn laugh on the slopes; ponds of lilies uncover their bosoms to the moon; meadows of grass-blades bend before the breeze; and the barley rustles millions of beards together on the lea. Shoals of herring solidify acres of the sea with moving life. Infinitudes of phosphorescent organisms, covering the surface of the deep, turn its heaving field of darkness into a sheet of fire. There are ant-hills, animated cities whose inhabitants outnumber Jeddo and Pekin. Villages of beavers build in company. Shaggy hosts of bisons shake the globe with the dull thunder of their tread. Herds of antelopes are seen crowding the entire horizon with their graceful forms. The naturalist in the tropics sometimes beholds clouds of gorgeous butterflies, miles in width, flying past him overhead all day. Captain Flinders saw, on the coast of Van Diemen's Land, a flight of sooty petrels, in a stream, which, as he calculated, contained a hundred millions. Audubon, while crossing the Kentucky Barrens, in eighteen hundred and thirteen, journeyed for three days beneath a flock of passenger-pigeons, which, according to his careful estimate, formed an oblong square a mile in breadth and a hundred and eighty miles

in length, and included more than a billion of birds.
Moving firmaments of locusts hide the heaven and
darken the earth.   And what mathematics will compute
the sum of the insects that toil in the erection of a coral
reef?

Everywhere, then, we see nature collecting her products,
—sands on the shore, leaves in the wood, fields of flowers,
aggregations of mountains, firmaments of stars, swarms
of insects, flocks of birds, herds of beasts, crowds of
persons.   Life would thus seem to be attractive, the enemy of isolation, huddling its subjects into social closeness, from heaps of mites to tribes of men.   But, after
all, these phenomena are exceptional, and the inferences
delusive.   There is more loneliness in life than there is
communion.   The solitudes of the world out-measure its
societies.   If consciousness sometimes draws, it has its
pole of repulsion as well; and much of that which looks
like fellowship is really but an amassment of separations.
What sociality is there in compact leagues of animalculæ?
Each one, shut in his incommunicative cell, might as well
have the solar system to himself.   The higher we look
on the scale of strength and individuality, the more isolated we see that the nature and habits of creatures are.
The eagle chooses his eyrie in the bleakest solitude; the
condor affects the deserted empyrean; the leopard prowls
through the jungle by himself; the lion has a lonely lair.
So with men.   While savages, like the Hottentots, gibber
in their kraals, and, among civilized nations, the dissipated and the frivolous collect in clubs and assemblies,
dreading to be left in seclusion, — the poet loves his solitary walk, the saint retreats to be closeted with God, and
the philosopher wraps himself in immensity.

Preparatory to fixing attention on the various forms of
the loneliness of human life, a contemplation of some
of the gigantic solitudes of nature may envelop the soul
in a befitting atmosphere of sentiment.

### The Solitude of the Desert.

As we advance into the solitude of the desert, not an animal, not an insect, breaks the perfect silence ; not a tree or a shrub varies the interminable monotony of sand. Over the arid and level floor you may sweep the circumference of vision with a glass, and not behold a moving speck. Only when, here and there, a bleached skeleton peers out of the drift, Solitude seems to find a speechless voice in death, and mutely to proclaim its sway. At noon, in the glaring furnace, the eye faints to see the air incessantly quiver with heat ; and night, when it comes, broods, chill and still, under the low-arched sky, sparkling with magnified stars.

### The Solitude of the Prairie.

THE solitude of the prairie is wonderful. Day after day, from morning till evening, the traveller journeys forward, wearing the horizon as a girdle, without seeming to change his spot ; for the immense circuit of which he is the centre appears to move with him. An ocean of grass around, an immitigable gulf of azure above, he feels as if he stood on the top of the world, the circular, sharp-cut level of an inverted cone, upon which the bulging dome of heaven shuts down in accurate adjust ment. He looks around the unvarying wilderness of verdure, and it seems as if the whole universe were that, and there were nothing beside.

### The Solitude of the Ocean.

THOUGH civilized man has grown more familiar with the ocean, it is none the less a solitude. How melancholy is its ceaseless wash, how lonely its perpetual swing, without a comrade in its convulsion or its calm ! How, beneath the immense stoop of naked sky, within the blue walls of air, in illimitable fluctuation, it stretches

away from the stagnation of the weedy gulf, in one direction, to where winter locks its moaning billows in silence
to the polar cliffs; in the other direction, to where its
cataracts of surf crash on the Indian coast. Everywhere,
out of sight of land, its spirit and expression are solitary,
awful, scornfully exclusive of sympathy. Perched alone
on the mast-head, gazing on the unbroken horizon, how
inexpressibly little a man feels himself to be! Whether
he contemplates the unity of the ship, frail speck on the
fearful abyss, the unity of the overarching heaven, the
unity of weltering desolation around, or the unity of
mystery enveloping all, it awakens an appalling sense of
lonesomeness.

### The Solitude of the Pole.

THE most dense and dreary of physical solitudes is
that of the polar realm. Now, with the cracking in splits
of the frozen fields, the falling of ice-cliffs, the grinding
of floes, the shrieks of the gale, one would imagine
heaven and earth were going to pieces in the uproar.
Again, the elemental strife at rest, the mariner treads his
deck, or wanders inland, where the total life of the globe
appears suspended, and a silence, oppressive as if Nature
held her breath, prevails. Occasionally a single walrus
crawls out in the cold-gleaming sunlight, and vainly looks
around the horizon for a living fellow; the dwindled fir-
mament, full of large, lustrous stars, circles, swift and
noiseless, and everywhere is one unrelieved expanse of
ice and snow. Sometimes the voyager meets a flock of
floating mountains journeying southwards, huge masses
of deathly whiteness, slow, silent, solemn as messengers
from a dead world. At another time the Aurora Borealis
suffuses the spectral world of ice; and fantastic villages,
battlements, cloisters, pinnacles and spires, with unimaginable colors, make it look like a gorgeous collection of
oriental cities. But always are found belonging there a
remoteness, a strangeness, a terror, essentially solitary.

## The Solitude of the Forest.

THERE are striking peculiarities about the solitude of the forest. These solitudes are very numerous. Vast woods of magnolias and rhododendrons, on the untraversed flanks of the Himalayan range, outspread an immeasurable wilderness of blossoms, and conceal in their fragrant solitude the mysteries of immemorial ages of nature. The great reaches of pine and fir on the Apennine and Alpine sides, of Norway spruce and Russian larch occupying the uncleared north of Europe, their billowy tops rolling in the summer breeze, their branches whistling to the icy blast, hide the unprofaned retreats of the primeval world, in whose ancient gloom man is as much alone as though transported to another planet. Maurice de Guérin describes a scene of awful loneliness he witnessed in a French wood. "A tremendous north wind roars over the forest and makes it give forth deep groans. The trees bow under the furious blows of the gale. We see through the branches the clouds which fly swiftly in black and strange masses, seeming to skim the summits of the trees. This vast, dark, swimming veil occasionally lets a ray of the sun dart through a rent into the bosom of the forest. These sudden flashes of light give to the appalling depths in the shadow something haggard and strange, like a smile on the lips of a corpse." Enormous tropical forests in Africa, superb with pomp of palms and baobabs, of rosewood, ebony, teak, tamarinds and acacias, brilliant with oriental exuberance of colored flower, fruit and vine, have never echoed stroke of axe, step or voice of humanity, in their recesses. The aboriginal woods of western North America seem as if they might harbor a million anchorites, not one of whom should be within a day's journey of any other. The traveller who pauses in the gigantic cedar-groves of Mariposa, penetrated by the spirit of unspeakable seclusion and rest that reigns there, feels as if he had reached the heart of solitude, where the genius of antiquity is enthroned on a couch of gray repose.

But all other forests are trifling, every other solitude on earth, except that of the sea, is small when compared with the tract of colossal vegetation which covers the South American basin between the Orinoco and the Amazon. In one part of this green wilderness a circle may be drawn, eleven hundred miles in diameter, the whole area of which is virgin forest, presenting impenetrable masses of interwoven climbers and flowery festoons, impenetrable walls of huge trunks in actual contact, — showing what stupendous room, yet unimproved, God has made for the multiplication of men and their homes; showing that Malthus and his theory were born in undue time. At night, when the beasts of prey are abroad, the noise is as though hell were holding carnival there. But at noon, the sultry stillness, almost palpable, is broken only when some hoary giant, undermined by age, crashes in columnar death. The grandeur and solemnity of this verdant temple fill the mind with awe. The gloom and loneliness are so depressing that it is a relief to emerge from under the sombre roof, once more to see the blue sky, once more to feel the clear sunshine.

### The Solitude of the Mountain.

A far different solitude is found on the summits of mountains, in the upper veins of air. We leave the warm valley below, with its snugly shielded villages and the busy stir of labor and merriment. Over many a weary height we climb, at each stage of ascent leaving more of the domestic world behind. Few hearts or eyes follow our progress. As our diminishing forms are traced from beneath gradually ascending, we find everything stunted and bleak; we pass the line of perpetual snow; we reach the zone of shrubless desolation, where not a leaf nor a bird is to be seen: the shepherds and their flocks have disappeared in the dim deeps, and a little cloud is our only comrade, as, still mounting, we finally pause on the summit and gaze. The world lies unfurled below; its forests, patches of carpeting; its rivers, silver threads, its inhabitants, annihilated ; its noise inaudible. We are alone on the top. Mists, dismal and heavy with loads of darkness and hail, drift by. The moon, rapidly hidden and shown by the clouds, hangs in the empty air and glares at us on a level with our eyes. Uplifted thus amidst the uncompanionable concave, a crushing sense of loneliness, of orphanage and want, possesses the soul, and makes it sigh for a humbler station, hedged with the works of society, warm with the embrace of love, brightened by the smiles of friends.

## The Solitude of the Ruin.

FINALLY we come to the solitude of ruins, — relics of the past, the dolorous dials Time in his passage has raised to count his triumphs and measure his progress by. A ruin is forlorn and pathetic wherever seen, — in an isle of African Nilus, or in a forest of American Yucatan. The traveller falls into a pensive mood, as, leaning against the stony masses of Meroe, whose glory the barbarian overthrew and the ,,ands buried, he scans the fading marks of the life that once flourished on that now silent plain. The same experience comes over him when his steed wearily penetrates the rank grass among the mounds of Copan and Palenque, the riddle of whose forgotten civilization baffles every guesser who inspects its remains, where the luxuriant vegetation has overgrown tombs and temples, here and there a palm, in its resistless upshoot, cleaving altar and image, column and skull. The Sphinx, that strange emblematic creature, half beast, half humanity, sixty-two feet in height, a hundred and forty feet long, still tarries amidst the mute desolation whence the whole race and civilization that set it there have vanished. Between its protruded paws originally stood a temple in which sacrifices were offered. The temple has crumbled in pieces. The sands have drifted over the feet and high up the sides of the mysterious monster, on all whose solemn features decay has laid its fingers. Yet the pilgrim is awed as he looks on the colossal repose, the patient

2

majesty of those features, and feels the pathetic insignifi-
cance of his own duration, in contrast with the unknown
ages and events that have sped by that postured enigma.

Yes, a ruin, whether mantled rich with ivy or swept
bare by the blast, — a feudal castle, crumbling on the cliff,
the snake in its keep and the owl in its turret; or a tri-
umphal pillar, thrown down and broken, its inscription
obliterated, its history in the maw of oblivion, — wears
the mien of solitude, breathes the sentiment of decay, and
is a touching thing to see. Ruins symbolize the wishes
and fate of man; the weakness of his works, the fleeting-
ness of his existence. Who can visit Thebes, in whose
crowded crypts, as he enters, a flight of bats chokes him
with the dust of disintegrating priests and kings, see the
sheep nibbling herbage between the fallen cromlechs of
Stonehenge, or confront a dilapidated stronghold of the
Middle Age, where the fox looks out of the window and
the thistle nods on the wall, without thinking of these
things? They feelingly persuade him what he is.

And how thickly these gray preachers are scattered
over the world, preaching their silent sermons of evanescence, wisdom, peace! Tyre was situated of old at the
entry of the sea, the beautiful mistress of the earth,
haughty in her purple garments, the tiara of commerce on
her brow. Now the dust has been scraped from her till
she has become a blistered rock, whereon the solitary
fisher spreads his nets; and along all her coasts, to Sidon
and Tarshish, the booming billows, as freightless they rise
and fall, seem to ask, "Where are the ships of Tyre?
Where are the ships of Tyre?" A few tattered huts
stand among shapeless masses of masonry where glorious
Carthage stood; the houses of a few husbandmen, where
voluptuous Corinth once lifted her splendid array of marble palaces and golden towers. Many a nation, proud
and populous in the elder days of history, like Elephanta
or Memphis, is now merely a tomb and a shadowy name.
Pompeii and Herculaneum are empty sepulchres, which
that fatal flight before the storm of ashes and lava
cheated of their occupants: the traveller sees poppies
blooming in the streets where the chariots once flashed;

unbidden tears come as he lingers where the veil has
been ripped from the statue of Isis, or pauses where the
fire is extinct on the altar of Vesta. Etruria is one stupendous grave, teeming with an empire's dust. The muffled abode of millions of men mingled for ages on ages
with the mould of the globe, it yields no admonishing
reverberation as we tread over it, unless we meditate and
listen ; then, indeed, the mystic soil, borrowing the
tongues of time and destiny, makes every particle of air
in the solitude vocal with pathetic tidings. As the mild
effulgence of lunar light mitigates the ruinous austerity
of the Coliseum, look up and recall the time when the
buzz of a hundred nations ran round those mighty
walls ; and, by contrast, how vacant and how dreary the
desolation is ! Roaming among the remnants of Moorish
grandeur in widowed Granada, strolling through the
chambers of Alhambra, admiring the delicious proportions with enjoyment subdued by pity, the air seems
charged with tearful sighs, and along the lonely halls the
spirit of tradition and sympathy wails in the tones of an
Æolian harp, "Ah, woe is me, Alhama, for a thousand
years ! " When we reflect that tigers foray in the palace-
yard of Persepolis, and camels browse in Babylon on the
site of Belshazzar's throne,—when, in imagination, at Baal-
bec, we march down majestic avenues littered with decay,
see lizards overrunning the altars of the Temple of the
Sun, and in the sculptured friezes, here the nests of obscene
birds, there the webs of spiders, — when we survey the
extent and noble forms of the ruins of Pæstum-amongst-
the-Roses, or of Tadmor-in-the-Wilderness, forming a
scene more exquisitely mournful than earth otherwhere
affords, what heart of man will not fill with regret and
presages, and own the unfathomable power of that natural solitude which crumbling art fills with the lost history
of our race?

# THE SOLITUDES OF MAN.

# THE SOLITUDES OF MAN.

*Physical Solitude and Spiritual Loneliness.*

AFTER every description of the monotonous wastes
and wilds of outward nature, we receive a heightened
impression of what true lonesomeness is, by turning to
the intenser inner deserts of mental and moral being.
Bleak and monstrous and unvaried as the sterile and
gloomy steppes of Mongolia and Tartary are, the feeling
of vastness and terror they impart is weak in comparison
with that obtained from contemplating the character of
Tamerlane, —

> Timour — he
> Whom the astonished people saw
> Striding o'er empires haughtily —
> A diademed outlaw.

The spirit and career of Attila, the cosmogonic dreams
of Swedenborg, the schemes revolving in the mighty
brain of Mirabeau, the Titanic aloofness and misanthropy
of Schopenhauer, the oceanic soul of Spinoza ringed only
by the All, are more appalling, more suggestive of the
infinite, than any material bulks or abysses. Is it a terrible chasm in which the sprinkled ranks of the galaxy are
hung? What, then, is the lonely mystery of the mind in
whose meditations the spectral infinitude of astronomy
lies like a filmy dot?

The physical solitudes of nature are without any feeling
of their own incommunicable separation and dreariness :
but the spiritual solitudes of man are conscious, and either
pine under the burden of isolation or groan for relief.
The sea, as its murmuring lip caresses the shore, or its

mountainous surges shatter against the cliff, seems not to
feel lonely, is company enough for itself, until deserted
yearning man approaches to give it contrast and interpretation. When shipwrecked man lies tossed on the strand,
thoughts and fears of home, love, death, eternity, thun-
dering at the base of reason, then first the sympathizing
phenomena without form a scene of genuine solitude, and
loneliness becomes an experience of anguish. Obviously
there can be no external expanse so deserted, so sublime,
as that night-scene of the soul when it muses, alone, with
faith and wonder, overhung by a still immensity of starry
thoughts.

Physical solitude and spiritual loneliness suggest, but
do not imply, each other. Either may blend with the
other to heighten it, or to relieve it. Either may in-
clude or exclude the other. On a morning of May,
long ago, a young man rode across an Illinois prairie,
with a friend. They passed, on the boundless expanse,
far out of sight of any human habitation, thousands of
crab-apple-trees in full blossom, their beauty and fra-
grance surpassing all that he had ever dreamed of vegeta-
ble loveliness and perfume. It seemed as if the whole
world had been converted into green grass, blue sky,
apple-blossoms, odor, golden sunrise, and two men on
horseback. Yet loneliness was an impossible feeling.
Every capacity of the soul was crowded by the complex
and strange exhilaration of that hour. Compare with such
a scene and experience those presented by a convict un-
dergoing execution in front of a hundred thousand spectators. While the officer adjusts the cap and rope, the
most awful interests of man are brought to bear on the
soul of the unhappy victim. Eternity seems condensed
in the dropping moments. There is no solitude here, but
how dread a loneliness! There is also often a profound
loneliness, full of pain, in the upper rooms of those high
houses in great cities, in which the poor single occupants
hearken to the almost inaudible murmur of the streets
below, and look up at the stars. Countless thousands of
men close around each wretched garreteer, yet he as
bleakly alone as though drifting on a plank in mid ocean!

To sit on rocks, to muse o'er flood and fell,
To slowly trace the forest's shady scene,
Where things that own not man's dominion dwell,
And mortal foot hath ne'er or rarely been ;
To climb the trackless mountain all unseen,
With the wild flock that never needs a fold ;
Alone o'er steeps and foaming falls to lean ;
This is not solitude ; 't is but to hold
Converse with Nature's charms, and view her stores unrolled.

But 'midst the crowd, the hum, the shock of men,
To hear, to see, to feel, and to possess,
And roam along, the world's tired denizen,
With none who bless us, none whom we can bless,
Minions of splendor shrinking from distress !
None that, with kindred consciousness endued,
If we were not, would seem to smile the less
Of all that flattered, followed, sought and sued :
This is to be alone ; this, this is solitude.

Epictetus, in his fine and brave little essay on Solitude, gives this as his definition of it. "To be friendless is Solitude." The more sharply we meditate on it, the more thoroughly we test it, the more deeply to the root of the matter we shall find this word of the cheerful Phrygian Stoic to go. Zimmermann says, " Solitude is that state in which the soul freely resigns itself to its own reflections." This is really no definition, it is a partial and superficial description, of solitude. More strictly it is a statement of one of the effects of being unoccupied from without. Obviously solitude is the deprivation of companionship ; but our own reflections are often the bestowers of a vivid companionship. The true definition is this. Solitude is *the reaction of the soul without an object and without a product.* If our activity has objects, those objects serve as comrades : if it is creative, the results serve as comrades. But if our activity is the overflow of unemployed powers with no object to meet and return it, with no product to embody and reflect it, we are conscious of an unrelieved loneliness. Solitude, therefore, is the reaction of the soul without an object and without a product.

*The Solitude of Individuality.*

THE first specification to be made of the loneliness of human life is that which results from the fatal separateness and hiddenness of each individuality. The innermost secret of the self-hood of any being can never be communicated, can never be shown, to another. Only little superficial fragments of our life are revealed, in comparison with the portion which moves on in unguessed concealment. That marvellous something which makes us ourselves, constitutes in us an impenetrable adytum where only the Power that created us can be or look

> Vainly strives the soul to mingle
> With a being of our kind :
> Since the deepest still is single,
> Vainly hearts with hearts are twined.

It is a well-known fact in physics that no two particles of matter ever truly touch ; their contact is but virtual. An ultimate sphere of force surrounds each atom with a repulsion absolutely invincible. Were the total universe made a press and brought to converge on two atoms, that dynamic investiture could not be broken through and an actual meeting effected. So with souls. Alas, how widely yawns the moat that girds a human soul ! Each one knows its own bitterness, its own joy, its own terrors and hopes ; and no foreigner can ever really touch, but only more or less nearly approach, and exchange signals, like distant ships in a storm.

> O the bitter thought, to scan
> All the loneliness of man !
> Nature by magnetic laws
> Circle unto circle draws :
> But they only touch when met,
> Never mingle, — strangers yet.

> Will it evermore be thus —
> Spirits still impervious ?
> Shall we never fairly stand
> Soul to soul as hand to hand ?
> Are the bounds eternal set
> To retain us, strangers yet ?

Every man wrestles with his fate not in the public am-
phitheatre, but in the profoundest secrecy. The world
sees him only as he comes forth from the concealed conflict, a blooming victor or a haggard victim. We hate or
pity, we strive or sleep, we laugh or bleed, we sigh and
yearn ; but still in impassable separation, like unvisiting
isles here and there dotting the sea of life, with sounding
straits between us. It is a solemn truth that, in spite of
his manifold intercourses, and after all his gossip is done,
every man, in what is most himself, and in what is deepest in his spiritual relationships, lives alone. So thoroughly
immersed is the veritable heart behind the triple thick-
ness of individual destiny, insulating unlikeness and suspicion, that only the fewest genuine communications pass
and repass ; rarely in unreserved confidence is the draw-
bridge lowered, and the portcullis raised. Frequently
the most intimate comrades of a life, when the whole tale
of days is told, know little or nothing of each other ; so
successfully are our disguises worn, so closely are these
impervious masks of sense and time and fortune fitted to
the being we are. Occasionally, urged by overstress of curiosity and tenderness, taking the dearest ones we know by
the hand, we gaze beseechingly into their eyes, sounding
those limpid depths, if haply, reading the inmost soul, we
may discern there a mysterious thought and fondness,
answering to those so unspeakably felt in our own. But
again and again we turn away, at last, with a long-drawn
breath, sighing, alas, alas ! No solicitation can woo, no
power can force, admission to that final inviolate sanctuary of being where the personality dwells in irreparable
solitude.

Were this all, however, only the fewest persons would
be troubled by their isolation. There is another experience, more open to view, and more oppressive to bear,
that in its sharpness aches for companionship. What is
it ? And what are its conditions ? The solitude necessarily belonging to the inmost essence, structure, and contents of every personality we accept as a law of our
being and circumstances. But to have a *peculiar* personality is to know a special loneliness which is a trial.

Peculiarities, in the degree in which they mark a soul, make that soul unintelligible to others. And the more unlike a soul is to the souls around it, as a general thing the greater desire it feels to see itself reflected in them, understood by them, sympathized with and cherished by them. Chamisso's unique tale of Peter Schlemihl or the Man without a Shadow, powerfully illustrates this. Wherever poor Peter goes, his lack of a shadow insulates him in wretched singularity. Every Jew, curmudgeon, hunchback, roguish school-boy, spies out his fatal defect; and the mob pelt him with mud. He wears away days and nights in his chamber in solitary sorrow. He wanders on the heath alone with his misery, and at last betakes himself to a cave in the Thebais.

It is not simply for one to be by himself that makes him feel lonely. In the quaint phrasing of Sir Thomas Browne, we must confess that "they whose thoughts are in a fair and hurry within, are sometimes fain to retire into company to be out of the crowd of themselves." When our noisy task is done, and fellow-laborers retire, and outer tools and cares are dropped, and leisure ushers an inner world of congenial pursuits, we may truly say we are never more completely occupied than when idle. So a man, as Scipio said of himself, is really never less solitary than when physically alone, if his solitude be filled with spiritual presences that give employment to his mind, keep the currents of consciousness flowing.

> Who contemplates, aspires, or dreams, is not
> Alone : he peoples with rich thoughts the spot.
> The only loneliness — how dark and blind ! —
> Is that where fancy cannot dupe the mind ;
> Where the heart, sick, despondent, tired with all,
> Looks joyless round, and sees the dungeon wall.

So long as the fluent and refluent tides of thought and feeling freely rise and fall, we need not companions to make us happy : when that condition fails, no society can prevent the painful longings of our lonesomeness. The fruition of a blessed communion is, in essence, simply a harmonized action and reaction of the soul and what surrounds it. Be this realized, and there is fellowship everywhere;

silence is melodious, and desertion itself social. Then out of the tender exuberance of his heart one may exclaim,

> There is a pleasure in the pathless woods,
> There is a rapture on the lonely shore,
> There is society where none intrudes,
> By the deep sea, and music in its roar :
> I love not man the less but nature more,
> From these our interviews, in which I steal
> From all I may be, or have been before,
> To mingle with the universe, and feel
> What I can ne'er express, yet cannot all conceal.

True desertedness and its pangs are experienced when we want the appropriate nutriment and stimulus for our faculties and affections, fit dischargers and outlets for their fulness. It is to miss loved objects, the wonted excitants and channels of our souls, and to have no sufficing new ones in their stead, and to feel that none of the people around understand us and feel with us. The exiled Switzer pines in a foreign clime for his native mountains, the sublime prospect, the familiar legendary spots, the upland breeze, the stimulant variety, the boundless freedom : and as he remembers, he weeps till his heart breaks. The soul, too, has its own deeper homesickness. An unappropriated enthusiasm ; a full heart aching for a vent and a return, and finding none ; a spirit thwarted of its proper action and reaction : — this is the painful essence of solitude, the live vacuum of lonesomeness.

Not the mere presence of numbers can heal this spiritual pain. There is no solitude in the world so heavy as that of a great city to the sensitive stranger who stands in its streets, and sees the endless tides drift by, till he turns away, feeling, Of all these multitudes hurrying past, not one, not one, cares anything for me ! Appropriate objects of thought and affection, if present in imagination, may furnish satisfying employment for the activities of the soul, however far they are removed in fact. The wild bird whose little heart throbs instinctively towards her nest and broodlets, is happy, as, all alone, she cuts the desert air towards home with a worm in her mouth. Galileo,

gazing at the constellations through the grating of his
cell, and feeling the fellowship of the illustrious conquerors of science in all ages, was less alone than when he
knelt amid the scowling throng of inquisitors to retract
the truth.   Not visible approximation, but conscious affinity, is the chief condition of inter-communication.   What
good is it that prison wards are in juxtaposition, and that
the stars are thick ?   As well for each other not to exist,
as to exist hopelessly sundered from knowledge and sympathy.   The king and the footman may consort as the
lion and the jackal : but bodily presence is not friendship ;
exchange of command and obsequiousness between superior and inferior is not the satisfaction of the natures
of both in common communion.   Unlike souls, though
crowded together in ranks, may all the while be as lonely
as the rows of funeral urns in a columbarium.   John
Foster writes in his journal, " Relapsed into the *solitaire*
feeling of being a monad ; a self-originating, sad and
*retiring* sentiment which seems to say, ' No heart will
receive me, no heart needs me.' "   Again he writes, in
the same journal, " Feel this insuperable individuality.
Something seems to say, ' Come away ; I am but a
gloomy ghost among the living and the happy.   There is
no need of me ; I shall never be loved as I wish to be
loved, and as I could love.'   I will converse with my
friends in solitude ; then they seem to be within my soul ;
when I am with them, they seem to be without it."

The grave-digger, wholly by himself, shovelling up the
skull of poor Yorick, was in a jovial entertainment of
merry thoughts.   Hamlet, isolated by his sad endowments, shaking his disposition with thoughts beyond the
reaches of his soul, moved about in the busy press of
ladies and courtiers, appallingly alone.   To a great nature, deeply in earnest, frivolous and shallow company
makes desertion twice desolate, as certain sounds serve
but to make stillness seem doubly still.   The tenacious
tenants of holy moods and mighty tasks have little in
common with the fugitive hoverers who flutter in and out
of every whim that rises.   Any exceptional deprivation,
gift or experience, either in kind or degree, in proportion

to its distinctive intensity, separates, — emphasizes its subject with solitariness. The loss of any sense by man, as that of hearing, lifts a sad, dark barrier between him and his fellows. The solitude of blindness is pre-eminently deep and oppressive. And it is pathetic to think how many great men have, like Homer and Milton, had the windows of their souls thus closed. Galileo, in his seventy-third year, wrote to one of his correspondents : "Alas! your dear friend has become irreparably blind. These heavens, this earth, this universe, which by wonderful observation I had enlarged a thousand times beyond the belief of past ages, are henceforth shrunk into the narrow space which I myself occupy. So it pleases God ; it shall, therefore, please me also." Handel passed the last seven years of his life in total blindness, in the gloom of the porch of death. How he and the spectators must have felt when the great composer, in seventeen hundred and fifty-three, "stood, pale and tremulous, with his sightless eyeballs turned towards a tearful concourse of people, while his sad song from Samson, ' Total eclipse, no sun, no moon !' was delivered."

Nothing can be more lonely than the chief characters in literary fiction, with exceptional endowments, aims and achievements, such as Prometheus, Faust, St. Leon, Zanoni. Hawthorne has expressed a kindred thought, with his usual vigorous felicity. "The perception of an infinite shivering solitude, amid which we cannot come close enough to human beings to be warmed by them, and where they turn to cold, chilly shapes of mist, is one of the most forlorn results of any accident, misfortune, crime, or peculiarity of character, that puts an individual ajar with the world." Hawthorne was himself a lonely man afflicted with a morbid shyness. He had a preternatural insight into the secrets, especially the pathological secrets, of human nature. That high idea of himself, intensely emotional, which with his genius he could not fail to have, was associated with a feeling of inability to impress it properly and see it reflected in others. In such an example, extreme shyness, with all its miserable torture, is no proof of pride or egotism in its subject. It simply proves the sharp power with which a sub-conscious occupation with his reflection in others possesses him. It is that he has extraordinary sympathy, not extraordinary selfishness. But it is, unfortunately, a viscid and attached, not a sparkling and free, sympathy. And it is one of the most fatal barriers to surrender, fusion, and joy in company.

### The Solitude of Grief.

THE most common and obvious of the secluding experiences of man is grief. Bereavement, in its essence, is always the loss of some object accustomed to draw forth the soothing or cheering reactions of the soul. The activity thus deprived of its wonted vent becomes a source of pain. Turned back upon itself, it aches with baffled yearning; or, forced upon objects unfitted to the fine habit of its affection, it feels desecrated and agonized. A necessary sense of loneliness is therefore associated with every deep form of grief. Amidst all its changing elements a feeling of desertion is the steady characteristic. Those who have stood by the death-bed of a beloved being whose departure from the earth seemed to leave the earth poor and cold, can never forget the desolating sense of solitude that came when the parting breath went. The soul of the dying seems borne away from us on the long-drawn sigh of his last fondly whispered farewell, as on a wave sweeping him far up the heavenly beach, but leaving us behind to struggle alone in the dark flood.

The removal of customary objects of love, hope, and care, — the blasting of a cherished enterprise, the decay of a once inspiring faith, — around which our thoughts danced in melodious measure and the currents of our emotions ran merrily, — causes a revulsion, leaves behind a wretched emptiness and a more wretched fulness, with no joining channel between, which compose the very substance and anguish of lonesome sorrow. Such an experience is a natural consequence of a great defeat, flinging the deep permanent shadow of disappointment athwart

the landscape of after-life. It is difficult to conceive of
a denser internal solitude than that which might enwrap
a defeated general, a captive king or queen, borne in triumph over the Appian Way, through a fluctuating ocean
of Romans. Paulus Æmilius, five days before enjoying
the most brilliant triumph Rome had ever seen, lost one
of his two younger sons; and three days after the triumph he lost the other. He was borne like a god in his
car, through miles of glittering and shouting humanity,
amidst endless throngs of captives and chariots loaded
with spoils, — his heart breaking within him. Past
glory and bliss set in an exile's memory against present
shame and woe, personal loss and sorrow contrasted with
public gain and exultation, is the very separation of separation. Such an image of loneliness, hard to surpass,
is presented in Keats's picture of old Kronos, the father
of the gods, dethroned and banished by the rebellious
young Zeus ; —

> Deep in the shady sadness of a vale
> Sate gray-haired Saturn, quiet as a stone.

What an eloquent image of grief, and what a tragic picture of loneliness, is the prophet Jeremiah, with white
beard and broken harp, sitting on the ruins of Jerusalem,
when the heathen had ravaged the city of his idolatry,
and the darling hope of his life was blasted ! When out
of this bereft and forsaken lot a voice of lamentation is
heard, whose articulations are sobs, no wonder they sound
to the vulgar revellers of the world as the accents of a
strange tongue which they cannot understand. No wonder, either, that the delicate and profound children of
misfortune and sadness should shrink from exposing their
afflictions to the superficial heirs of success and gayety,
but should rather flee into retirement, there to ease their
pangs with tears, and with exercises of trust and prayer
charm their souls into the embrace of nature and God.
Here we reach the loneliness of the closet, where no
echo of the roistering crew or the toiling crowd penetrates ; a retreat sacred to sad memory, healing thought,
and pious rites. I have seen, in an Alpine pass, a slow

deep-tinted mist wind itself around the cruel crags, splin-
tered peaks which stood like so many horrid tusks goring
the sky, — wind itself around them till they seemed
couches soft enough for angels to furl their wings to repose on.   So, in its patient religious loneliness, does the
rich sensibility of genius gather beautiful associations
around the lacerating points and passes of grief, robbing
them of their harshness.

But besides the solitude shed around the afflicted by
their inward grief, they seek seclusion on account of the
exquisite state of their sensibilities, freshly torn and unable to encounter the miscellaneous exposures of society.
The grieved heart, like the wounded deer, retreats into
solitude to bleed.   Sometimes it is cruel, ah! sometimes
it is kind, to leave the unhappy alone with his unhappiness.   The subject of a severe sorrow, the fibres of his
spirit rent from their habituated clingings, shrinking in
self-defence from every coarse contact, courts the secrecy
of his chamber, of lonely walks, or wraps himself in the
protection of an unnoticing absorption.   The mind
bruised by the blows of calamity, the tendrils of its
affections hanging lacerated, is so susceptible in its soreness that it cannot bear, even in thought, the collisions
of the careless.   To its exacerbated tenderness every
breath is a shock, every touch torture.   Under these circumstances the instinctive safeguard is the shelter of
silent loneliness.   It has well been said, "Solitude is the
country of the unhappy."   On the death of his darling
daughter Cicero fled from Rome to a still retreat whence
he wrote to his friend Atticus : "Nothing can be more
delightful than this solitude, nothing more charming
than this country place, the neighboring shore, and the
view of the sea.   In the lonely island of Astura, on the
shore of the Tyrrhenian Sea, no human being disturbs
me ; and when early in the morning I retire to the leafy
recesses of some thick and wild wood, I do not leave it
till the evening.   Next to my Atticus nothing is so dear
to me as solitude, in which I hold communion with philosophy, although often interrupted by my tears."   That
social usage which gives the afflicted an investiture em

blematic of their grief, happily seconds this instinct by enabling them, in a degree, to carry their solitude with them, even through street and market, in a mute appeal to all considerate observers for a softened behavior in their presence. Mourners, sphered by their dark garb in a sacred and touching solitude, glide through the crowd, shielded from whatever is unseemly, no sharp sound reaching them save as muffled and blunted, everything frolicsome or boisterous growing reverently sober as they approach. Before the van of the army of grief the rude cold waves of the world of mirth and harshness divide, as if invisibly struck by a wand, and let the silent ranks pass, untouched, between.

Not only, however, is the afflicted sufferer made lonely by the deprivation of the cherished objects of fruition whose loss grieves him. And not only does the shrinking of his hurt feelings from every frequented scene to court the curative calm and balm of silence and repose make him feel solitary. There is still another element of desertedness and pain in the case ; namely this. Any kind or degree of experience which others cannot enter into with us leaves us alone. And it is the very characteristic of a keen and massive grief that it takes its subjects into depths where the untried are unwilling to follow. Thus is it that

> Misery doth part
> The flux of company.

The mother whose only child has been borne across the shadowed threshold into the mysterious Nevermore, missing with unspeakable woe the darling nestler, as she bethinks her that the bright eyes are darkened, the sweet voice hushed, the soft lips closed and cold forever, in her convulsive desolateness feels as if all were gone : and, since nobody else can quite feel so, she is alone.

> Alone ! — that worn out word,
> So idly spoken, and so coldly heard :
> Yet all that poets sing, and grief hath known
> Of hope laid waste, knells in that word, ALONE !

In the glowing imagination of one person rises a goal

decked with a fragrant garland : in the pensive memory
of another stands a perpetual bier.   How can there be
any assuasive communion between those whose states and
impulses are so opposed?   The widow of Herder, on the
anniversary of his death wrote to their dear common
friend, Jean Paul Richter : — " I am alone to-day, and in
the other world.   It is the second of May."   Could any
Babylonian be expected to enter into the feelings of the
expatriated Israelites when, weeping, they hung their
harps on the willows by the banks of Babel, and declared
themselves unable to sing the songs of the Lord in a
strange land?   The Siberian exile journeying over the
homeless plateau, wrapped in sables, on his huge snow
shoes plodding wearily through the petrifying desolation,
thinking of home, encounters no one capable of repro-
ducing his emotions and reflecting them to him in relieving sympathy.   Nor is the patriot, banished to fairer
lands, welcomed into the hospitable circles of a gentler
clime than his own, much less lonesome.   For it is not
the absolute, but it is the *relative*, what is possessed or
missed *in his consciousness*, that makes man blithe or melancholy.   In meeting the great agonies of experience
there must be solitude.   For the commonalty of mankind
and the average hours of life are alien from the transcend·
ent touches of woe, incompetent to feel fitly for their
victims.   There every soul that reaches knows its own
bitterness, and strangers intermeddle not.   The wife,
robbed of the intimate companion of her existence, on
whom she leaned, with whom she associated every joy and
sorrow, hope and fear, — clad in black, bending in monumental memory by the thought of him with tears and
sighs, lonely as the weeping willow that droops over a
tomb, her voice mournful as the breeze that complains in
its branches at midnight, — feels the whisper of frivolous
tongues as an intolerable burden, the presence of a throng
as a degradation and an insult.   The tender Robertson,
on the death of a young daughter, wrote to one of his
friends : — " I have just returned from putting my little
beautiful one myself into her grave, after a last look at her
placid countenance lying in her coffin.   It was by starlight,

with only the sexton present ; but it was more congenial to my heart to bury her so than in the midst of a crowd, in the glaring daylight, with a service gabbled over her." He does not use the word *gabbled* in any disrespect to the service, but as a vent for his intense feeling of the sacrilege publicity would be to the holiness of his grief.

With the exception of rapturous love, there is no sympathy in the world so intense and profound as that between those who have known the same griefs. One of the chief services the highest souls render to lower ones is to reveal to these their own sorrows, not experienced in bleak and bitter nakedness, but associated with nobler strengths, comforts, and supports. What a divinely solacing music is mixed with the griefs of humanity, as their echoes come back from the solitary heart of Jesus! There is a Church of Grief whose members deeply and tenderly know each other. A great element of power in the bond between the afflicted Christian and the Saviour is their fellowship in suffering. "He was tried in all points like as we are." The penitent, in grief and guilt, filled with devotional awe, lonely amidst the worshippers that crowd the cathedral, sees the lonely face of Christ lifted on the crucifix, and his heart leaps towards it with a spasm of worshipping sympathy.

A person of great gifts but with a wearied mind and a wounded heart, profoundly convinced of the vanity of worldly honors and pleasures, and of the calming efficacy of divine contemplations, resolved to enter on a monastic life by joining the Port Royalists, and on doing so, laid the following poem on the altar of their retreat.

O ye dark forests, in whose sombre shades
    Night finds a noonday lair,
Silence a sacred refuge ! to your glades
    A stranger worn with care
And weary of life's jostle, would repair.
He asks no medicine for his fond heart's pain;
He breaks your stillness with no piercing cry ;
    He comes not to complain,
    He only comes to die.

To die among the busy haunts of men
    Were to betray his woe ;

But here these woods and this sequestered glen
    No trace of suffering show.
Here would he die that none his love may know.
Ye need not dread his weeping, — tears are vain ;
Here let him perish and unheeded lie ;
    He comes not to complain,
    He only comes to die.

Many poor hearts, long struggling in bitter forlornness
between thwarted aspirations and dissatisfying realities, at last break, and go down, unpitied, unnoticed.
Sometimes, borne out into the wilderness of grief without
a comrade to stanch our wounds, we are tempted to say,
There is no other gulf so wide and cold as that which
flows between heart and heart.   Would you gain deliverance from this grievous loneliness?   Seek it not, as so
many do, by lulling those importunate longings to rest in
the dismal shade of oblivion.   Neither seek to escape by
a wilful avoidance of them in the rush and clamor of ex-
ternals.   These are artificial expedients, temporary and
violent, and must ever prove, at last, in the deepest truth,
unwholesome and calamitous.   The normal and divine
procedure is not to suppress nor to elude grief, but to
confront, cure, and improve it ; transforming it into something higher, and passing on to purer substance and
sweeter blessing in its assimilated and transfigured might.
So shall we hereafter retrace in our successive sorrows
the seasonal stages of our growth, and look back on our
wounds converted into ornaments.   So shall we cast off
sufferings, hardships, misfortunes, and suspend them
along the ascending way of life, to be used as rounds and
steps for climbing into more magnanimous comprehension, firmer tenderness, loftier being.   Then the blows of
time and fate will leave on our souls not disfiguring scars
but inserted buds, inoculating us to bear diviner fruit.
The loneliness of an overwhelmed grief, as destitute of
religious faith as of human sympathy, is like that which

                    Broods
In winter nights o'er frost-bound solitudes, —
Darkness, and ice, and stillness, all in one ;
A silence without life, a withering without sun.
But o'er that silence, when, at night's full noon,
Through breathless cloud, shimmers the glorious moon,—

the pale illumination of the frozen forest is like the beau-
tifying light shed on sorrow by religious belief and resolve.

## The Solitude of Love.

VERY different from the forlorn retirement of grief, but
sometimes almost as exclusive in its kind, is the solitude
of love. That contrasts with this as the loneliness of
the closet with the loneliness of the grove. There, is
the oppression of an imprisoning limit; here, the freedom of abounding impulse; but in both alike, an isolating quality, a consecrating intensity, an insuperable
repugnance to the indiscriminate intercourse of the
world. In both, when at their height, the desire to be
alone is so keen that the subject of the experience feels
the presence of a single person to be equivalent to the
presence of a multitude. Then, as Ovid said, *Nos duo
turba sumus.*

> Shall I whisper aloud,
> That we two make a crowd?

With all unwontedly earnest love mingles an obscure fore-
boding of wreck and loss, bereavement and agony, to
come. On its upper surface affection admits acquaintances, to see their smiles, and to hear their words re-
echoed; but in this lower deep, where the wonderful
omens move, it excludes curiosity, and even sympathy,
and broods alone with its unsharable bliss or its strange
presentiments of ill. Whether so rich a boon is felt to
be unsuited to the conditions of earth, too fair to last,
sure to provoke some envious power to blast it, I know
not. But truly so it is, that the finer any experience of
love becomes in our human relations, the more surely it
is haunted by a formless fear, dispensing, where it prevails, an air of solitude, a lonesome misgiving, like that
derived from the undefined Fate which fills the background of a Greek tragedy.

There is a bitter loneliness resulting from the absence
and need of love, as well as a sweet loneliness resulting

from the presence of it. Few have felt this more sharply than Charlotte Brontè; and she has described it : — "Sometimes when I wake in the morning, and know that Solitude, Remembrance, and Longing are to be almost my sole companions all day through, — that at night I shall go to bed with them, — that they will long keep me sleep-less, — that next morning I shall wake to them again, — I have a heavy heart of it." Charles Lamb, the exquis-ite affectionateness of whose nature, with his poverty and many bitter trials, made him especially susceptible of such an experience, shows us a glimpse of his sufferings from it in the poem he addressed to his friend Lloyd, when the latter sought him out in London, "alone, ob-scure, without a friend, a cheerless, solitary thing." He says : —

> For this a gleam of random joy
>     Hath flushed my unaccustomed cheek ;
> And, with an o'ercharged bursting heart,
>     I feel the thoughts I cannot speak.
>
> O, sweet are all the Muses' lays,
>     And sweet the charm of matin bird ; —
> 'T was long since these estrangéd ears
>     The sweeter voice of friend had heard.

George MacDonald, referring to an English traveller among the Swiss mountains, who snobbishly regarded all but himself as intruders, well says : — "Was there not plenty of room upon those wastes for him and them? Love will provide a solitude in the crowd ; and dislike will fill the desert itself with unpleasant forms."

Jesus is the supreme example of that loneliness which is felt as a consequence of the greatness of the love with-in and the smallness of the love without. "The foxes," he sighed, "have holes, and the birds of the air have nests, but the Son of Man hath not where to lay his head." And when the Pharisee, at whose table he dined, complained of the toleration he showed for the sinful woman, what a world of lonely and sorrowing tenderness is unveiled in his reply, — "Simon, thou gavest me no kiss ; but this woman hath not ceased to kiss my feet. Wherefore I say unto thee, her sins, which are many, are forgiven : for she loved much."

Vivid and profound love shrinks from communicating its confidences, lest injury be done them, lest their hallowedness be profaned. Their delicacy is too ethereal for a rough hand; their vestal bloom is too holy for unconsecrated spectators. Grief, in its soreness, allows only the tender themes that are wonted and soothing to touch its hurt fibres; love, in its scrupulous sacredness, permits only the trusted and adored object to come near and read its confessions. The priest alone can be admitted to the shrine. Numa felt not lonely in his cave, but when he returned among the citizens. Interviews with any sacred Egeria tend to unfit us for ordinary fellowships. The dedicated privacy of a pure and modest heart cannot expose its shaded secrets to vulgar lookers. The more pertinaciously they explore, the more bashfully it shrinks and veils. It can calmly brook no eye save that of God and the elected one. Therefore, around this most choice and sensitive experience there ever spreads a kind of solitude. It is true that the experience resulting from an access of fervent affection often has another aspect. Its expansiveness makes it many times seem emphatically social. "The heart, enlarged by its new sympathy with one, grows bountiful to all." Nevertheless the phase of the experience here insisted on is true too. Love affects not the dusty highway, but the woodland path. It retires to brood over its thick-clustering and honeyed thoughts. The maiden, with the picture of her lover, runs not into the crowd to gaze on it, but wanders into some umbrageous nook, where imagination may feed on itself, nor fear rebuke from the ring-dove balancing on yonder bough, or betrayal from that brook, the babbling tongue of the glen.

Solitude is not only the sanctuary, it is also the nursery, of sentiment; where, brooding over itself in quiet, and sympathetically brooding over whatsoever is friendly to it, it grows deeper, and draws around itself an ever-enlarging mass of nutritious associations. Petrarch, the high Laureate of this feeling, sings:

> From hill to hill I roam, from thought to thought,
> With Love my guide; the beaten path I fly,

3                                                        D

For there in vain the tranquil life is sought.
If 'mid the waste well forth a lonely rill,
Or deep embosomed a low valley lie,
In its calm shade my trembling heart is still ;
And there, if Love so will,
I smile, or weep, or fondly hope, or fear.

That which is true of sentiment in general is true of a just and genuine piety in particular. It is the shallow that is garrulous, — the deep is silent. The name of Christ, the idea of Deity, the sense of eternity, the anticipation of heaven, the mysteries of sin and regeneration, are things too solemn and sublime to be bandied from mouth to mouth in technical debates and conventional conferences. Cleaving to the marrow of life, to the dividing asunder of spirit and flesh, they fitly appropriate to themselves the most select and awful moments of meditation, the most secret and sanctified moods of affection, when not a taint of passion befouls the heart, and the fewest vestiges of earth linger on the mind. It is an impressive fact that the subject of a religious conversion, in the freshness of his experience, instinctively shrinks from the world, seeks seclusion on some pilgrimage, or in some convent. The ideas of God, purity, judgment, the feelings of remorse, sanctification, joy, which have come into his soul with such revolutionizing power, are too stupendous for gossip. They withdraw him. He knows by instinct that he can maintain himself at their height only in solitude. The Christian convert flies to the monastery, to feed and hedge his faith with a guardian ritual ; the Buddhist devotee betakes him to a sanctuary of the contemplative Buddha, to muse and aspire ; the Brahmanical ascetic journeys on, over hill and plain, his alms dish in one hand, his staff in the other, alone, silent, buried in a thought. Who that has any appreciation of divine things, of what is becoming, can bear to drag these innermost sanctities into the light, where a thousand discordant scrutinizers are gazing and listening, eager to handle and to criticise ? No soul, save a hard and narrow one, can be otherwise than full of lonely awe when confronting that thought of God before which the globe

ts but a bubble, and the sky a shadow. Plotinus defined
the fruition of piety as "a flight of the alone to the
Alone." We always think of the oracles of the gods as
dropping in grove and grotto, not in street and stadium.
Lenau wrote to his friend, Anastasius Grün, from the
summit of the Alps, — "Solitude is the mother of God in
man." When Jesus would pray he "went apart into a
mountain." Even to his dearest disciples, and that at a
time when the need of sympathy was sorest, he said,
"Tarry ye here while I go and pray yonder."

Few persons have genius and soul enough to experience the highest religious emotions at first hand. The
most can but poorly simulate and echo them, or copy
their forms and attitudes at a distance. Thus the gigantic
personalities, in whose tremendous powers and passions
the chief religious experiences and rites now current first
originated, come to be reproduced in dwarfed proportions
and faded hues, as though the motions of a figure whose
colossal bulk filled the space between earth and heaven
were seen reflected in diminishing mirrors as the postur-
ings of a puppet. To mirror livingly and originally the
transcendent realities and relations from whose corre-
spondences in consciousness the primal religious emotions are born, requires a depth and translucency of
sensibility not possessed by one man out of a million.
The mighty objects and truths which create religion by
surpassing and baffling our powers, which engender in
our ignorance and weakness the dread sense of dependence, wonder, and aspiration, refuse to reflect themselves in the shallow and turbid pools which poorer souls
furnish. Religion, therefore, is essentially lonely and not
social. The common notion to the contrary is a vulgar
fallacy ; a fallacy, however, almost unavoidable from the
intimate association of sociality with religious phenomena. The true and pure religious emotions are essentially solitary, and love only loneliness ; but the awe,
mystery, helplessness, connected with them terrify us and
force us to seek fellowship in our experience of them, as
a relief and reassurance. It will always be found that
for the exercise of their ultimate religious feelings the

highest, greatest, deepest souls irresistibly seek solitude,
unspeakably enjoy it, and shrink from society at such
times with insuperable repugnance.  But to the multitude the direct and solitary contemplation of their relations with the unknown and the infinite is too awful ; it
must be shared, diluted, relieved by organic fellowships
and poetic associations.

There are topics appropriate for speech, which naturally find utterance in address or conversation ; there
are other topics meet but for private contemplation and
ordering, which find fit expression only in soliloquy.
This subject has been treated with admirable precision
and grace in two discourses on "The Sphere of Silence,"
by an English divine.  They are to be found in that series
of wonderful discourses by James Martineau, entitled
"Endeavours After the Christian Life." *  The naked
verities of religion dwell in the last penetralia of our
being where no mortal communion can reach.  The
knowledge and love of them must ever be a recluse experience, because their grandeur is so great as to monopolize the attention it secures, and because their modesty
is such that they die away at the first proposal of exhibition or flattery.  They will bestow their fellowship and
reveal their forms in the dark mirror of the mental holy
of holies, only when every wind of the world is whist,
and a silence as of the primordial solitude reigns throughout the spaces of the soul.  For experiences celestially
fine and sensitive as these, public comparison, giddy
talk, any sort of notoriety, is desecration.  To strew
pearls before the unclean who will turn and rend you for
it, is an outrage on all that is fit ; those of swinish character, having no taste for adorning themselves, but only a
greed for coarse food, must be expected to turn angrily
on the inconsiderate man who disappoints with indigest-
ible jewelry their appetite for corn.  A drunkard dis-
paraging or eulogizing temperance, — a harlot descanting

* Each one of these most beautiful and most valuable discourses
is a key to some important compartment of human experience.  He
who really masters them carries thenceforth a precious bunch of the
keys of life.

on the nature of virtue, — or an epicure discussing the
worth of denial and heroism, — is an odious spectacle.
The highest instincts of the soul demand moral con-
gruity. Who could endure to pour the weird strains of
Mendelssohn's Dream amidst the rattling of the square
and the mart? Who would not rather hide the pictures
of Perugino forever than display them on the walls of a
slaughter-house? There are pure and holy women who
never expose their charms or share their delights with
the world, as there are lakes that, on the untrodden tops
of mountains, like eyes of the earth, look only up to
heaven. Every virgin solitude is perfumed with the
Divine presence, and balsamic for mental bruises. Divinely drawn, the soul flees thither to be the guest of
God, and Silence is the sentinel of their interview. A
retired and self-guarded life of devotion to nature is like
a priestly life of temple-worship; as a German woman
of genius has said, " When the boy Ion steps before the
portals, and signs to the flying storks not to defile the
roof, when he sprinkles the threshold with sparkling water,
and cleans and decorates the halls, I feel in this solitary
occupation a lofty mission which I must reverence. Ah,
I too would be a youth, to fetch water in the fresh morning, while all yet slumber, to polish the marble pillars
and bathe the statues, to cleanse everything from dust
until it glistens in the gloaming; and then, when the
work is done, to rest my hot brow on the cool marble,
rest the bosom that palpitates with emotion at the beauty
which breaks into the temple with the dawn."

There is a suitableness of person, of scene and season,
required for the unveiling of the secrets, and the contemplation of the treasures, of affection. Refined and
thoughtful must be the person, not harsh and reckless ;
the scene and season, not obtrusive and noisy, but retired and still. Whatever reeks and roars with the rushing world, shocks and defiles. Pure and pensive solitude
is the setting that woos the living pictures. Nor is there
any one wholly destitute of this lonely companionship
of love, this saddening wealth of joy. The veriest wretch
in the world has some dear memory, some beautiful long-

ing, some guarded ideal, so fondly prized that he loves
to set apart secret moments for pilgrimages to its inner
sanctuary, there to worship, perhaps to weep, where no
eye can see and no ear can hear.   So even the most su
perficial votary of fashion, the most inconsiderate retailer
of petty scandals, has her times of uncompanionable reflection, unfathomable emotion and desire.   Occasionally
this is found to be true in cases where it would have been
least  suspected,  so carefully had  it  been concealed.
Reckless critics often make the cruelest misjudgments
here.   Not unfrequently those thirsting most for love
shrink most from notice.   Obscurity is their shield.

What can be so melancholy as to have sacred experiences, which ache for expression and sympathy, and not
dare to expose them for fear of repulse or ridicule?   It
is more melancholy not to have them.   The glorious, sad
solitude of one devoted to the highest ends, who can find
no comrades, who roams the streets at night, weeping,
longing for some one to walk and talk with him, to aspire
and work with him, — is more to be admired than to be
pitied.   The weeping is indeed a weakness, but it expresses a strength.   To call such an one an egotist or a
sentimental fool, to laugh or sneer at his pain, is a wicked
heartlessness, however often it is done.   The wealth of a
soul is measured by how much it can feel ; its poverty, by
how little.   God hands gifts to some, whispers them to
others.   In the former the divine charm is followed by
immediate popular recognition : in the latter it is usually
hidden for a long time from all except the deep-souled
and deep-seeing few.   It is not improbable that the truest
saints have never been heard of : —

> Too divinely great
> For Fame to sully them with state,

they have modestly offered themselves up to the Universal in seclusion and silence.

There is an hour, the transition between day and night,
celebrated by the poets, with Dante at their head, which
fine souls in all ages have felt as the votive season of
sentiment, — pensive twilight, the dim-tinted habitation

of solitude and sacredness, hailed with mountain-horns and hymns, bells and prayers, while Nature herself, half steeped in roseate hues, half mantled in shadow, seems to be tenderly musing.

> Soft hour which wakes the wish and melts the heart
> Of those who sail the seas, on the first day
> When they from their sweet friends are torn apart;
> Or fills wi*h love the pilgrim on his way,
> As the far bell of vesper makes him start,
> Seeming to weep the dying day's decay.

It is the favorite hour of all poetic lovers who have ever consecrated it to their beloved, love they what they may; when they retreat by themselves from " the thick solitudes called social," to indulge and nourish their master-sentiment; when sensitive genius keeps tryst with its idolized ideal, the betrothed keep tryst together, and saints keep tryst with the spirit of devotion and their God.

### The Solitude of Occupation.

PURSUING our subject a step further, we come to a separated experience, resulting neither from the injured sensibility of grief, nor from the enshrined devotedness of love, but from lack of room for forms of extra fellowship. It is the solitude of an absorbing occupation. Whatever fills the capacity of the soul, of course, for the time, excludes everything else; and there thus results an apparent singleness and separation. Augustine, struggling in the crisis of his conversion, in the chamber of his friend Alypius, says, — "I was alone even in his presence.' This principle is the key to one of the marked varieties of the isolation in human life. A man with a great mission, an intense passion for some definite object, is thereby set apart from the common crowd of associates whose free impulses are ready to respond to every random appeal. He has no loose energies to spare in reaction on stray chances or incoherent claims : his whole soul is g'ven to the one aim and its accompaniments. Sometimes an illusion, fastening in the mind, appropriates the thoughts and passions as its food, and makes the man its servant.   Others laugh at his absurdity, or turn carelessly from him as an oddity.   Elated with his error, fondling his idol, he heeds not their scorn or their neglect.   Lost in his idiosyncratic joy or anxiety, hugging his peculiar purpose to his breast, he drifts through the frigid wilderness of society, as essentially alone as a sailor lashed to a spar on the ocean.

Dante was made lean for many years by the exactions of his supreme poëm.   Devouring his time, thought, feeling, soul, in his wanderings and poverty, it made him passing solitary among men, and kept him stern, sad, and serene on a wondrous fund of tenderness and vehemence. Ceaselessly quarrying at the rock of eternal flame and fame, he conquered daily peace.   If not thus absorbed how his mighty heart must have gnawed itself, and the insect swarm of care, hate, and sorrow have stung him mad !

Who could be more distinctively by himself than Columbus, made a lonely visionary by a sublime dream which he had determined to embody in a visible demonstration of fact before the world.   Equally solitary in his soul and in his design, whether pacing the strand, buried in thought, or reasoning with the monks of Salamanca, his scheme absorbed him, his originality set him on a pedestal above the heads of living men, among the illustrious pioneers of history, of whom he claimed lineage, with whom he felt his place and sympathy to be.

Every first-rate mechanician or inventor who has created astonishing machines, has been remarkable for his absolute abstraction from outward things, and his intense interior absorption during the incubation of his projects. All discoverers or schemers of the highest order, all intense idealists and workers, are in this manner taken possession of by their destined vocation.   And thenceforth they know nothing else.   Conversing with their thoughts, toiling at their plans, devising methods, or imagining the results of success, they walk up and down, deaf to every foreign solicitation and to every impediment.   Come what will their task engrosses them, their late cries out, and all else must give way. Such men are essentially alone ; though it is an unresting, contentful isolation, unlike the vacant, asking isolation of unabsorbed men. Its proper type is the loneliness of a waterfall in the bosom of unreclaimed nature ; or the loneliness of a beehive in a hollow oak in the heart of the untrodden forest.

We must not overlook, however, the wide difference between a solitude felt as such in pain and pining, which implies unappropriated powers, and is a condition of misery, and the solitude which is unconscious, wherein the soul is self-sufficing, its occupation leaving nothing unsupplied for the time, no wish for external sympathy or help. The latter is one of the happiest forms of life, in spite of its somewhat withdrawn and melancholy aspect. Apart from social interchanges, it may appear dreary and monotonous ; but it is not so. Mendelssohn was repeatedly known to wander through crowds, with abstracted face, soliloquizing strains of music to himself, lost, in this improvisation, to all about him. On writing the last sentence of his "Decline and Fall of the Roman Empire," Gibbon looked up at Mont Blanc, and drew a deep sigh. "The sudden departure of his cherished and accustomed toil, left him, as the death of a dear friend would, sad and solitary."

In fact, for solid happiness and peace, there are none more favored than those blessed with a master-passion and a monopolizing work. In the congenial employment thus secured, the earnestness of their faculties is called out and dedicated. They thus find for themselves and in themselves an independent interest, dignity, and content, together with exemption from most of the vexatious temptations by which those are beset whose enjoyment rests on precarious contingencies beyond their own power. An enthusiastic ornithologist, like Audubon or Wilson, roaming through trackless forests and prairies beyond the outermost haunts of civilization, busy now with rifle and knife, now with brush and palette, lover of nature, lover of beauty, lover of solitude, lover of his chosen pursuits, — what matchless health and cheer and

3 *

delight and peace are his! Palissy the potter, clad in
rags, starving, burning his last chair as fuel for his experiments, — his haggard wife and children almost fancying
him insane, — was by no means the unhappiest of men.
Inspired by a splendid hope, already clutching the prize,
he wist not of hunger or of sneers ; thrills of rare bliss
visited his breast, and bankers and cardinals might well
have envied him. When we think of the astronomer in
his secluded tower, in the gloom, hour by hour turning
his glass on the unbreathing heaven, peering into the
nebulous oceans, or following the solemn wanderers ; —
when we notice the lamp of some poor student, burning
in his window, his shadow falling on the tattered curtain
where he sits with book and pen, night after night, "out-
watching the Bear and Thrice-great Hermes," — we may
fancy that he leads a tedious and depressing life. Ah,
no. The august fellowship of eternal laws, the thought
of God, the spirits of the great dead, kindling ideas and
hopes, the lineaments of supersensual beauty, glorious
plans of human improvement, — dispel his weariness,
cheer every drooping faculty, illumine the bleak chamber, and make it populous with presences of grandeur
and joy. The solitude is unreal, for he is absorbingly
busy. He is alone, but not lonely.

When with a great company one listens to fascinating music, gradually the spell begins to work ; little by
little the soft wild melody penetrates the affections, —
the subtle harmony steals into the inmost cells of the
brain, winds in honeyed coils around every thought, until
consciousness is saturated with the charm. We forget
all. Distraction ceases, variety is gone. Spectators,
chandeliers, theatre, disappear. The world recedes and
vanishes. The soul is ravished away, captive to a strain,
lost in bewilderment of bliss, its entire being concentrated in a listening act ; and we are able to believe the old
legend of the saint who, caught up into paradise by overhearing the song of the Blest, on awakening from his
entrancement found that a thousand years had passed
while he was hearkening. Such is the solitude of absorption, when it touches its climax. He is wise who

endeavors to know something of its elevation and blessedness by giving his soul to those supernal realities
which are worthy to take his absolute allegiance, and
swallow him up.  Though such an one lives in solitude,
the solitude itself is inexpressibly sociable.

## The Solitude of Selfishness.

TURNING to still another province of the subject, we
find a less congenial topic awaiting us.  There is a repulsive species of loneliness very different in its origin
and nature from the forms thus far portrayed.  It may
be designated the solitude of meanness or guilt.  Of all
the unfellowshipping styles of life this is the bleakest
and the most unamiable.  In fact, the other moods of
segregate experience, however sad or painful they may
be, are not ignoble nor pernicious.  But the persons who
here come under notice, with their ominous habits of
aloofness, are marked by gloomy or narrow and despicable qualities which cause them to be disliked and
shunned.  To enjoy company we must be able to trust
each other, frankly unbosom ourselves, think similar
thoughts, feel accordant emotions, blend hearts in unreserved surrender to common influences.  The action and
reaction of souls in the same manner and on the same
objects, is the fruition of friendship, — the experience
of harmonized states of consciousness sympathetically
awakened and sympathetically changing.  But this is
comparatively the prerogative of the virtuous, the tender,
the disinterested.  In proportion as any one is morose
and hateful his cold or jealous vileness cuts him off from
the happiness of genuine fellowship.  Wherever he may
be he is alone.  To be destitute of sympathy is the very
solitude of solitude, no matter what the circumstances, —
whether from the window of a diligence you look with
aching heart on a village merry-making, or pause, risen
above the clouds, a solitary wanderer, amid the glacial
sea, gazing in horror on its dumb desolation.  And if
absence of sympathy be the essence of loneliness, who

so lonely as the cold earthlings who form the various embodiments of selfishness, who take no interest in others except to make use of them, giving no impulsive love, asking none. The heartless, it is certain, cannot perform the functions nor enjoy the satisfactions of heart. They may not know the difference themselves, their very impoverishment securing them immunity from the pangs of baffled affection, so that they do not suffer from conscious and painful isolation. Only the loving pine for love. The most unsympathetic, obviously, will care least for society. But the repulsive solitude in which the incapacity of their mutilated natures imprisons them, preserves one of its aspects of penalty in undiminished reality. If they are not aware of the negative, to languish under its deprivations, neither can they possess the positive, to thrill with its bestowments. Such an one dwells unconsciously chained in a movable prison which he carries around him wherever he goes, which hopelessly shuts the sweetest boons of existence from access to his soul ; and though that prison is invisible to him, every other eye discerns it. Thus the miser, whose sordid love of money receives all other feelings into its sea-like passion, who withdraws every fibre of his soul from friend and foe, from truth and beauty, to cling exclusively around his yellow heaps, isolated within his squalid show of rags and penury, when he retires to gloat secretly over his hoards, does not himself feel lonely ; but to those who regard him he seems profoundly so. They see him, the abject outcast, as an unclean waif tossed into the sewer of society from the gutter of civilization. They give him a glance of contemptuous pity in passing, somewhat as they would fling a bone to a starving dog. Is not such a life a horrible loneliness ? Outwardly viewed, it is a fearful solitude ; although inwardly it may swarm with an obscene activity of greed and complacency.

There is, then, an experience carried on within itself, quite aloof from the joyous companionship of life, not for lack of time and space for social interchange, but from want of the personal material and conditions. This is the solitude of a heartless or wicked breast. A man locked up in a shrivelled and frigid self-hood, with no living currents of faith and love between him and his fellow-creatures, is as much alone amidst a Parisian holiday, surrounded by a bedecked and huzzaing world of humanity, as the traveller who loses his way, benighted in the centre of a Polish forest, and, in the drifted snow, leans against a tree, starving and freezing, while the dis-- tant yell of wolves is borne to his ears.

A Greek philosopher, referring to two opposite kinds of loneliness, experienced from antithetical causes, said that he who loved solitude must be either a god or a beast. He only stated the truth a little extravagantly. Man is made for society and brotherhood. He who is content to dwell alone, then, without society or brotherhood, is on a plane of endowment and desire either superior or inferior to that of common humanity, approximates the level either of a divinity or of a brute. In other words, solitude may be approached by ascent or by descent. There is the separation of the throne, and there is the separation of the sty. Man may soar into experiences too exalted and complex for easy communication with comrades of the earth, too sublime and holy to be vulgarized in plebeian speech, — the solitude of a god. Man may sink into experiences too poor and base to bear articulation, too secret and selfish to be capable of sympathy, — the solitude of a beast. Thus one may be alone because he is above, or because he is beneath, the conditions of satisfactory companionship with his neighbors. While these two are alike in being isolated, the distinction between them traverses the entire distance from the august to the despicable. The sentiment of the lonely which invests the self-seeker in his plot differs from that which surrounds the poet in his dream as the solitude of the buzzard, picking his prey in the glen, differs from the solitude of the sun, burning in the zenith.

The legitimate effect of sin, of everything that serves private interest to the injury of the universal interest, is to sunder and segregate. Evil bristles with negative polarity, and would disintegrate the society in which it prevails. On the other hand, the power of virtue leads its subjects to commune, clasp, coalesce. The fox, th. hawk, the leopard, from their selfish dispositions, are solitary; they shun a company that they may the better pounce on their prey, and glut their appetite. But the bees live in swarms, the friendly swallows fly in flocks and build their nests in contiguity. Brave impulses and magnanimous sentiments, every moral or religious affection, all qualities loyally allied to principles that subordinate individual whims to the general good, are attractive, have a public regard, yearn spontaneously outward to love and be loved, to bless and be blessed. They draw men into groups, set the nerves of relationship vibrating, fill the channels of mutual life with invitation and energy. This is the instinctive tendency of all rich and gentle hearts, unless, as sometimes unhappily occurs, tragic rebuffs, failures and sufferings teach them to act otherwise in self-defence. But ignoble passions, cruel indulgence, all the suspicious and hateful characteristics of selfishness, which would gratify the lawless craving of the individual at the expense of the solemn and permanent weal of the whole, naturally creep into secrecy, and, repulsively electrized with fear and malignity, walk apart there.

An intense feeling of solitude is produced in a man of dark designs when his confederates turn against him and desert him. In the revulsion from busy associates and elated hopes to isolation, overthrow and despair, he must feel a fearful loneliness. Wallenstein, betrayed by Gallas, Piccolomini, Altringer, and nearly all the rest in whom he had confided, standing solitary, overwhelmed, yet upright, amidst the ruins of his guilty projects, furnishes an impressive instance. The superstitious dreams with which he had linked his destiny to the stars, nursing his vast and sombre ambition with astrological prognostics, only served then to make his solitude more gigantic.

The fittest emblem of the solitude of a completely selfish man moving about in society, is the loneliness of an iceberg drifting amidst the crowds of waves, now feebly glimmering with moonbeams, now shattering the tempest on its breast, finally, honeycombed with rottenness, toppling over and swallowed in the maw of the maelstrom its own plunging makes. There are several classes of persons who, as exiled from the open and genial fellowship of life, are alone, even when, from their egotistic absorption or their hardened indifference, they are not lonesome. The cynic, who admires and enjoys nothing, despises and censures everything, eager, morose, his milk of human kindness turned sour; the misanthrope, whose blood has been turned into gall by deception or disease, a malevolent villain, whose first impulse is to hate and avoid every one he meets, or to blast them with his scorn; the proud, haughtily holding their heads aloft, snuffing the incense of their own conceit, unable to stoop to the sweet offices of meek humanity, fancying the earth too base for their feet, and other men only good enough to be their servants; the mean, all whose experiences sneak in dark by-ways, too cowardly to face the sun and the loving eyes of men, — unable to rise to the level of a generous sentiment, a noble enthusiasm, a momentary self-forgetfulness; — such as these, destitute of the essential conditions of friendship, must be deprived of all the best fruitions of human society. The world and life must be to them comparatively what they were to the leper of the Middle Age. This abominated outcast, clad in a coarse gray gown reaching from head to foot, with the hood drawn over the face, went about carrying in his hand an enormous rattle, called Saint Lazarus's rattle, whose frightful sound warned every human being to keep at a distance; he was thus banished from his fellow men by a cordon of disease and horror drawn around him, which drove every one before it as he advanced.

The tiger, in his awful strength and voracity, when he forays in the trembling haunt of antelopes, is not more alone than the tyrant, wrapped in the pomps of power as in robes of ice, shaking a nation with his murderous nod, having a taster for every dish lest it be poisoned, wearing a secret shirt of mail lest some assassin reach him. The abandoned devotee of debauchery, giving full swing to depraved propensities, now rioting in excommunicate gratifications with sickening gusto, now shuddering with nameless horrors and anguish, lurking in hidden retreats, like Tiberius at Capri, exists in a hideous solitude. The criminal is drearily alone ; temptation, struggle, guilt, remorse, despair, are the loneliest of experiences. Evil seduces and assails, is embraced or vanquished, singly and in private. Secretly and alone are we all led up into the wilderness to be tempted of the devil.

No one can be so unspeakably alone as the possessor of a foul and dreadful secret which turns nature into a listening confessional, and the disclosure of which he feels would instantly discharge the thunderbolts of doom. The prisoner of a guilty past, aching for communication, yet shrinking from it with terror, dying for sympathy, yet not daring to seek it, dwells in the most terrible of all solitudes. The memory of his crime, charged with dire bodements, fastened inextricably to his soul, — he feels as a victim bound to the stake, a distant girdle of faggots burning towards him. Though agonizing for deliverance, he fears to accept it : for, appalling as his loneliness is, how can he bear society, when he knows that, at any instant, the fatal secret sunk in the depths of his consciousness may slip the shot from its shroud and bolt on his horrified gaze !

There are things, as we thus see, too mean and bad to be voluntarily disclosed, too wicked and terrible for a trustful communion ; the fears they engender, the shocks they would impart, and the dangers they threaten, keep their subjects apart and taciturn in the suspicious and sinister seclusion of an inner secrecy. The wickedest man in the world is the most completely alone, in the etymological sense of the word, that is, *all one,* — sundered from these virtuous and blessed junctions with others which properly make each man a part of the whole of humanity.

## The Solitude of Genius.

THE extreme of experience just described is the loneliness of the leprous.  On the other hand there are souls
occupied with matters so exaltedly noble and sensitive as
to be generally incommunicable.  This extreme is the
loneliness of the laurelled.  This class of men are lonely
not because they do not dare, or cannot bear, or do not
wish, the most intimate companionship.  They are lonely
because their states of consciousness are so swift and fine,
their height of soul and range of life so vast and arduous, that their associates are unable to appreciate them.
This brings us to the saddest and sublimest part of our
theme, the solitude of genius.  The lark rises against
the rosy ceiling of day, far beyond the emulation of
ground-birds ; and genius soars into heaven in its worshipping joyousness until no earth-bound spirit can follow.  The scale of its experience, in both directions
equally, joy and sorrow, surpasses that of common
persons.

> Chords that vibrate sweetest pleasure
> Thrill the deepest notes of woe.

All men of unusual mass and height of character wear
a sombre hue of purpose which repels familiarity.  "The
love of retirement," Johnson impressively remarks, "has
in all ages adhered closely to those minds which have
been most enlarged by knowledge or elevated by wisdom.  They have found themselves unable to pursue the
race of life without frequent respirations of intermediate
solitude.  There is scarcely any man, eminent for extent
of capacity or greatness of exploits, that has not left behind him some memorials of lonely wisdom and silent
dignity."

Every one conspicuous above his fellows in endowments is made solitary in that degree, unless his gifts, by
ministering to their gratification, bring him into social
relations with them and win him their applause.  Even
then the solace he finds is usually obtained by turning

the ordinary side of his nature into view and action, veiling or suspending the peculiar endowment in which he so far surpasses others as to be an insulated unique. Mediocrity need not search for sympathizers; they swarm. Originality may seek widely and long, but in vain, for the equal love it desires. Originality is understood slowly and with difficulty, easily gains notice, less easily commands disciples, but most easily provokes dislike and creates foes, then itself revolts into disguise and seclusion, and only with the utmost labor and infrequency succeeds in discovering or making an adequate friendship. Extraordinary minds are painfully alone in the world because their actions cannot elicit harmonized reactions from the ordinary minds by which they are surrounded. And the latter are trained into satisfying conformity with the former only in such rare instances and with such pains, because that educational process requires a tenacity, a patient affectionateness, which the ordinary mind is not supplied with. The soul touched by God is separate. Prophets are lonely; Elijah, fed by ravens beside his secret cave and stream, — fed with meat in whose strength he travelled forty days unto Mount Horeb, we cannot think of as a social man. Paul, after his miraculous conversion and commission, says, "I conferred not with flesh and blood"; he withdrew into Arabia for a long season of meditation and spiritual training. It is reported of Jesus himself, that he oft "withdrew into the wilderness and prayed." What a lonely and strengthening time of it Luther had in Wartburg castle on the edge of the dark Thuringian forest; and Loyola in the sepulchral cavern of Manresa, on the banks of the limpid Cardinero! Great teachers too, as well as prophets, are lonely; there are so few prepared to understand them and give them welcoming response. "The light shineth in darkness, and the darkness comprehendeth it not."

Genius is alone both as to the world it constitutes and as to the world in which it moves. Souls of coarse fibre and mean store cannot responsively reproduce the delicacy and wealth of its inner experiences; neither car they see the supernatural glory of its outer visions. For genius beholds without, the wonders it first feels within. To its perception, in imaginative grief, the ocean is a universe of tears murmuring human woes. In its moods of abounding love and serenity every material object is an emblematic voice, a window of spirit, a divinized hieroglyph. When the two friends, Beaumont and De Tocqueville, were floating together at evening in a boat on one of the great lakes of the western continent, the latter says the moon stood in the edge of the sky "like a transparent door opening into another world." Such an expression would be unmeaning or distressing to a mere proser. Soft, rich, capacious genius, looking with eyes of inquiring tenderness into every soul it meets, and seeing nothing there correspondent with what is deepest and dearest in itself, is repelled into solitude. Then in pathetic disappointment, with rebounding and ebullient faith, it laves the void with the copious overflow of its emotions, until that void, filled with immortal spirits, with heaven and God, reflects upon the yearning giver and recipient wonderful answers of beauty and love. And so a divine peace is won, and solitude becomes more sufficing than society. When the young Michael Angelo went to Rome and began to study and labor there, he wrote home, — "I have no friends ; I need none." The huge "confusion of the life of the metropolis only penetrated like a distant murmur" the solitude in which he dwelt and toiled, with little sympathy from other men, though with much admiration. His chief happiness was in absorbing work, and in the visions of that ideal realm where he walked as king.

The famous platonizing English divine, Henry More, was lonely among the earthlings and partisans of his time. His ideality, learning, and earnest love removed him in spirit to a planetary distance from his worldly minded neighbors, whom he characterized as "parrot-like prattlers boasting their wonderfull insight to holy truth, when as they have indeed scarce licked the outside of the glasse wherein it lies." He was wont to think "the angels looked on this troubled stream of the perish-

ing generations of men to as little purpose almost as idle
boys do on dancing blebs and bubbles in the water."
Knowing how truly catholic and genial he was, we recognize with a personal regret what the experience must
have been which caused him to sing, —

> Cut off from men and all this world,
> In Lethe's lonesome ditch I 'm hurled ;
> Sad solitude 's my irksome bliss.

There belongs to such natures as that of poor David
Gray, the Scottish poet, at least for a time, the experience
of a piteous, half-frightened loneliness.   Intensely conscious of his own difference from those around him, but
with his feeling of superiority not sufficiently powerful
and pronounced to give him peace, he hungers for love
and admiration from them to assure him that this difference is not a weakness or a delusion, but is the stamp
of genius.   This eagerness to be noticed and praised in
order that he may not fall into despair and betray his
mission, is often repulsive to poor observers, and wins
him aversion, perhaps hate and ridicule.   He should in
such a case not droop with distrust and grief, but gird
himself with noble convictions ; comfort himself not with
disdain but with benignity, perceiving that the truly great
can be appreciated only by their mental kindred.   Had
an ignorant shepherd, and a Plato, in climbing the Cau-
casus, come upon Prometheus, what different estimates
and emotions would have arisen in their respective souls
as they saw there the worn form of the august sufferer
nailed to the wintry mountain wall !   Each man can judge
of other men only in the light and with the aid of the
data he carries in himself.

The chamois browses by himself on the blue cliffs of the
sky because his food is in that high haunt, and because he
is so shy of his foes.   Is it not something the same with
the rare specimens of humanity ?   The most exalted contemplations are the nourishment of their life, and they are
wonderfully sensitive to the hostile influences that threat-
en them on the low level of the crowd.   Accordingly they
shrink from the elbows and sneers of the vulgar, climb

out of the stifling vapors of the valley, feel the exhila-
rating attraction of the free empyrean.  The higher they
ascend the fewer are able to accompany them, and the
bleaker grows the desolation, until at last all are left
behind, and the glorious isolation that invests them is
like the cold loneliness that surrounds the sunset-head
of Monte Rosa.  The penalty affixed to supremely
equipped souls is that they must often be thus left alone
on the cloudy eminence of their greatness, amidst the
lightnings and the stars of the canopy, commanding the
sovereign prospects indeed, but sighing for the warm
breath of the vale and the friendly embraces of men.  A
naturalist, caught in a terrible tempest on Mount Etna,
at the height of ten thousand feet, spent twenty-four
hours there in a cavern, amidst the awful uproar, feeling
quite certain that in all Europe not another human being
passed that night in the same stratum of air.  Many a
deep and bold thinker often deems himself the exclusive
occupant of some stratum of new ideas and emotions.
Though frequently a mistake, the supposition is undoubtedly sometimes well founded.  When Amerigo Vespucci
saw the Southern Cross it was a baseless boast he made,
declaring that he now "looked on the four stars never
seen till then by any save the first human pair."  In the
boundless regions of speculative thought there still are
innumerable solitudes, but very few virgin solitudes.  It
is usually but a vain conceit that prompts us to believe
that we are standing in view of a mental prospect no
mortal imagination has before seen.  Few, indeed, are
the positions in the intelligible universe open to man,
which have not been occupied and commanded by the
minds of Plato, Dante, Shakespeare, Leibnitz, Kant,
and Goethe.  When Belzoni, with great labor, had pene
trated the rocky sepulchre of Setei-Menephthah to its
inmost secret, he found he was not the first who had
made a violent entrance thither ; for the sarcophagus
was broken and the mummy gone.  Exceedingly rare
are the discoverers of solitudes, either in the material or
the spiritual world, of whom it can be truly said, —

> They were the first that ever burst
> Into that silent sea.

Those grandeurs of the material and of the intellectual universe which overpower our self-sufficiency are the favorite subjects of contemplation with the grandest souls. And whatever object of nature or of thought is so vast as to impress us with a sense of our own littleness and evanescence, makes us feel lonely. What a desolate sense of isolation comes over a stranger in a strange land when he feels the poor atom, self, sheerly contrasted with the vast cold mass of all humanity beside! The nightly illumination of the houses and streets of London consumes fifty million cubic feet of gas, representing three thousand tons of coal. Yet this light, sufficing as it does for two millions of persons, is in the awful cone of night but as a glowworm flying in the valley of the Mississippi. The solitude of the sky when not a bird flies, not a cloud floats, through its eternal dome, is not deeper or sublimer than that of the mind of a Copernicus or a Malebranche. How can they come down to mix in the conflict of jabbering mediocrities?

The fine deep soul of Weber, — out of which came Der Freischutz, Oberon, and many another weird and tender strain, — often felt a dismal loneliness in the crowd. His insulating unlikeness from the average of men, in his truest moments, was dreary. When, after fourteen careful rehearsals, he had brought out Beethoven's Fidelio, at Berlin, and it was received with cold indifference, he exclaimed in indignation, "They could not understand the greatness of this music. Vulgar folly would suit them far better. It is enough to drive one mad." Over and over after the mention in his diary of the fashionable parties he attended in the aristocratic mansions of Prague, he adds the despairing exclamations: "Alas! Ah me! O God!" He writes: — "Bohemia has become for me a mere hospital of all intellect. There are so many miserable souls in the world! I cannot but feel that I unconsciously withdraw myself more and more from my fellow men." On parting with his beloved friends, Alexander Von Dusch and Gottfried Weber, at twenty-four, he wrote in his diary: — "Shall I ever again find in the world friends so dear and men so true?" Sixteen years later, when near his death, he wrote on the same page, "No!"

Almost every great man addicted to contemplation, and of literary habit, has left on record some expression of his loneliness. Erasmus, while residing in the University of Cambridge as a lecturer on Greek and Theology, writes to his friend Ammonius, under date November 28th, 1515, "Here is one unbroken solitude. Many have left for fear of the plague; and yet when they are all here the solitude is much worse." Shakespeare, whose unparalleled sensitiveness and vastness of sensibility seem to have enabled him to embrace the conscious substance of almost every form of experience ever presented to man, — who has so livingly painted the imaginative solitude of Prospero, the metaphysical solitude of Hamlet, the piteous solitude of Timon, the savage solitude of Apemantus, and the loathsome solitude of Caliban, — in one of his Sonnets speaks in his own person of a time —

> When in disgrace with fortune and men's eyes
> I all alone beweep my outcast state,
> And trouble deaf Heaven with my bootless cries.

Lessing, after the death of his wife, wrote to Claudius, — "I must begin once more to go on my way alone. I have not a single friend to whom I can confide my whole being. I am too proud to own that I am unhappy; I shut my teeth, and let the bark drift. Enough that I do not turn it over with my own hands."

The separate conditions of mental loneliness are joined and concentrated in the case of genius. A personality exceptionally emphasized, sensibility chronically as exquisite as that of others is temporarily made by bereaving afflictions or blissful boons, an absorbing activity in the line of its special vocation, — all these belong to genius; and therefore it must be largely solitary. Genius is average humanity raised to a higher power, and is distinguished from its neighbors as the king is distinguished from his courtiers by the dais and the crown. Every great passion, sublime purpose, singular pursuit, or un-

equalled susceptibility, naturally tends to isolate its subject and make him pine with baffled longings.

Furthermore, the fact that genius, by its realizing imagination and appropriating sympathy, naturally shares in
all the events and experiences of which the signs are
brought to its knowledge, as keenly as the ordinary soul
feels its own personal concerns, makes it liable to extreme
distress in the wrongs and woes of the world.   Hence
often arises a strong temptation to retreat into some remote solitude to escape the harassing pressure of this
ideal contact with the great miseries of the public battle
of life.   Cowper expresses the feeling well : —

> O for a lodge in some vast wilderness,
> Some boundless contiguity of shade,
> Where rumor of oppression and deceit,
> Of unsuccessful or successful war,
> Might never reach me more !   My ear is pained,
> My soul is sick, with every day's report
> Of wrong and outrage with which earth is filled.

There are, concealed in the undesecrate shrine of innocence, a thousand matters too modest and too holy to
suffer themselves to be laid bare to the gaze of hardened
men.   There are, in and about the virgin soul of genius,
a thousand delicious fragilities of thought and sentiment,
which, like the dewy gossamer shown on a rosebush at
sunrise, if you try to lift and convey them, are torn, dissolve, and vanish from your grasp.   Such a soul must
crave seclusion from the jar and friction of life, sweet
opportunities for musing and aspiration.   "Quiet is the
element of wisdom ; the calmest man is the wisest.   For
the mind is a coral-stone, around which thoughts cluster
silently in stillness, but are scared away by tumult."
Some persons are so crude and heavy that it requires
ponderous masses of power to disturb the stolid poise of
their attractions ; others are as alive to imponderable influences as electrometers.   Between such a great gulf is
fixed.   A fine interior nature, exuberant with affection
and fancy, set in a world of capricious external hurriers,
frigid mockers, ever eluding his embrace, is as lonely as
an Alpine flower nestled in the crevice of a crag and
blooming there on the edge of the glacier.

The man whose heart is such a sensitive plant that every cloud which floats remotely above it causes its petals to close, — what adequate communion can he have with the herds of jokers, the noise of whose mirth intrudes on the silence of his prayers? He feels more at home on the margin of a lonely stream than in the thoroughfares of the metropolis. The bell of a sequestered convent is much more congenial to him than the hum of a reception-room. No wonder rich and delicate natures protect themselves by retreating; they suffer less cruelly from their melancholy desertedness than from the lacerations of ungenial society. An awkward, coarse companion disturbs the reveries that hang in live suspense on the altitudes of their minds, as rudely as when, floating in a canoe at midnight on a forest-girt pond, the idiotic laugh of the loon suddenly breaks the spell, dispersing the solemn hush of wood and lake. It is natural enough, that, after such an experience, loneliness should be, for a while, preferred to company. Solitude is the refuge of the sensitive.

It is a necessity for genius to feel, in a certain sense, a complacent aloofness and superiority to the herd of the world, in order to sustain itself at its own proper height. Among the two hundred thousand men who rose up when Virgil entered the Roman Theatre, there was but one Varius competent to correct the Æneid. Knowing the thoughtlessness and fickleness of the folly-swayed mass of the people, if the great man did not cherish a keen conviction of his own greater elevation, insight, and nobleness, he would soon cease to be a great man. Thus Goethe wrote in his old age, " I was first uncomfortable to men by my error, then by my earnestness : so, do what I would, I was alone." So Adam Smith said, " The mob of mankind are admirers and worshippers of wealth and station." So Bishop Butler said, " Whole communities may be insane as well as individuals." So Spinoza, pitching his tent as on an Ararat in the desert of disdain, from the incomparable loftiness and scope of his intellectual horizon, looked down on the undiscriminating and incompetent multitudes of men with a quiet and pitying contempt.

This was full of solace and strength for him.  Without it
he would have died of heart-break and despair.  His distance from the grovelling victims of ignorance, delusion,
and hate, measured his nearness to God ; and he was supported.  There was no unkindness in his mood ; it is removed by a whole moral world from everything like vin-
dictive spleen.  Madame Swetchine was very free from
pride and the spirit of contempt ; yet she writes from
Paris to a friend : " My God, the pitiable thing the conversation of these assemblies is !  It was the first of the
year ; nonsense, silliness, gossip, frivolity, were in all their
freshness.  It is indeed well to repose through the summer, away from what is called the grand world, a taste for
which is the greatest misfortune that can happen to mind
and heart."  A complacent reaction from the vices and
pettiness of the crowd upon the superior nobleness of their
own loyalty, powers, and pursuits, is the unfailing internal
support of the truly great.

God forbid that the highest should hate or insult the
lowest.  And it is not their true nature to do so.  They
yearn pityingly over their farthest inferiors.  Yet it is vain
to attempt to hide the prodigious disparity between them.
And when those beneath force this disparity on the notice
of those above, by assuming superiority, it is not to be
wondered at if the latter experience a shock of revulsion.
The accusers of Socrates arrogated to themselves a highe*
virtue and wisdom than his.  Undoubtedly his consciousness of the relative moral height between them and himself was a godlike consolation to him.  The sublime
courage and calmness with which he claimed from his
judges, instead of death, a support by the city as a public
benefactor, show that he was perfectly aware of the immense moral distance between Socrates and Anytus, Me-
litus, and Lycon.  In every case of martyrdom, perhaps
the cruellest feature is the self-assumed superiority implied
by the judges in the very fact of condemning their victim :
his greatest support, on the contrary, must come from the
conviction of their injustice in putting him to death, and
of his own worth in standing loyally by his duty.  There
is a surpassing heroism, there must be a deep pain, and

there certainly is a terrible loneliness in singly confronting, as so many noble men have done, an infuriated mob,
to stem its wrath, stay its folly, avert its vengeance, even
at the cost of falling a prey to its headless and horrid
passion.  Who can dwell on such an example without a
pang of pity and a thrill of grateful admiration.  Surely
no one can recall, without profound and indignant pain,
how the horde of soldiery, inflamed with hatred for the
blameless Ulpian, the immortal jurisconsult and states-
man of Rome, broke into the palace of Severus and killed
the great unspotted lawyer before the faces of the emperor and his mother.  Who can read of the good Priestley, driven " by the madness of riot from the town which
he adorned by his virtues, his philosophy, and his fame,"
without a mingling of sorrow for the confused crowd and
of homage for the clear individual?  Coleridge paints the
scene : —

> Patriot and saint and sage,
> Him full of years from his loved native land,
> Statesmen blood-stained and priests idolatrous,
> By dark lies maddening the blind multitude,
> Drove with vain hate.  Calm, pitying, he retired,
> And mused expectant on the coming years.

When we think of Alexander Hamilton, hooted and
stoned in the streets of New York, in what relief his
beautiful form stands out against the howling mass of ignorance and ferocity below!  But should we undertake
to make a list of the wronged and hated benefactors of
the world, the exiled or martyred guides and exemplars
of our race, up to the crucifixion of the Saviour, there
would be no end to the tearful tale.  The crowning moral
of the narrative would be the inspired sentiment sighed
from the summit of Calvary, " Father, forgive them, for
they know not what they do."

But since man was made for society, it is not good for
him, no matter how great he is, to be always alone.  If
he is doomed to be so his lot must be full of sad wishes.
There are wounds the world cannot balm, wants no outer
success can satisfy, though to poor and cold natures these
sharpest of griefs are never known.  It is the soft-hearted

who are heavy-hearted.   The loftiest mind may shelter
the most, but it must be the least sheltered.   There is no
desertion like that of a soul sublimely incongruous with
its mates and with the conditions of its time and place.
The choicest hearts are the ones most likely to know the
experiences of disappointment and cruelty in all their
wasting bitterness.   Such hearts there are, which, once
misunderstood and aggrieved, never dare to confide again.
In the mournful isolatedness of their balked yet unap-
peasable longings, well may they exclaim, —

> Come, Death, and match thy quiet gloom
>    With being's darkling strife ;
> Come, set beside the lonely tomb,
>    The solitude of life !

Generally speaking, the man of genius is a lonely man,
not only from the greatness of his endowments, the
height at which he lives, and the absorbing action of
his faculties, but also from his scorn of conventionalities.   Sneers at conventionality, — sneers rising from
failure to see its inevitableness and use, — are cheap.
Conventionality is the unavoidable expression of social
averages.   But it must be naturally irksome to the man
of genius, who belongs outside of the average.   How
can he be otherwise than solitary when he sits on the
great white throne of imagination, gazing at the panoram-
ic phenomena of the creation in the light of transcendental philosophies, till from before his face earth and heaven
flee away, and no place is found for them ?   Impatient of
custom, contemptuous of fashionable decrees, he must
frequently be a banished man.   The epicures, the butterflies, the selfish plotters, and all such, cannot understand
him ; and to be mentally baffled is painful.   He sets an
example they cannot follow ; and to feel inferiority is
painful.   His ideas and beliefs are strange to them, —
apparently inconsistent with the familiar ideas and beliefs
with which they identify their welfare, perhaps their salvation ; and what is unintelligible and is supposed dangerous, is feared.   Accordingly they desire to rid themselves
of his presence.   The great man acts from spontaneity ;

society acts from habit, and is intolerant of original action, because it makes such exorbitant demands. To act
conventionally costs little ; to act from fresh impulse requires a large supply of power. Fashion always aims to
live with the least expenditure of force ; genius is always
seeking outlets for its overflowing force. Consequently
luxurious society is the natural enemy of genius, and, as
far as it can, exiles it into solitude.

The most ignoble men — still more than average men
— hate the superiors whom they are unable to appreciate.
Their thwarted mental reactions generate spite and wrath.
Disappointed of the husks for which they look, they
furiously trample the pearls they know not what to do
with, and bite at the odious hands that flung them.
True, this is only one phase, the darker side, of the
facts. Multitudes of men are full of reverential devotion for their superiors. Nevertheless the reality of this
darker side is fearful. The treatment of great men by
the world in all ages exemplifies the mysterious law of
vicarious redemption, confirms the words which Jesus
spoke out of his own experience : " Behold I send unto
you prophets and wise men and scribes ; and some of
them ye shall kill and crucify, and some of them ye shall
scourge in your synagogues, and persecute from city to
city."

Columbus writes, in the letter to Ferdinand and Isa-
bella describing his fourth voyage, — " For seven years
was I at your royal court, where every one to whom the
enterprise was mentioned treated it as ridiculous ; but
now there is not a man, down to the very tailors, who
does not beg to be allowed to become a discoverer. It
is right to give God his due, and to receive that which
belongs to one's self. This is a just sentiment and proceeds from just feelings. The lands in this part of the
world, which, by the Divine Will, I have placed under
your royal sovereignty, are richer and more extensive
than those of any other Christian power ; and yet, while
I was waiting for ships to convey me in safety, and with
a heart full of joy, to your royal presence, victoriously to
announce the news of the gold that I had discovered, I

was arrested and thrown, with my two brothers, loaded
with irons, into a ship, stripped, and very ill treated,
without being allowed any appeal to justice. I was
twenty-eight years old when I came into the service of
your Highnesses, and now I have not a hair upon me
that is not gray ; my body is infirm, and all that was left
to me, as well as to my brothers, has been taken away
and sold, even to the frock that I wore.    The honest devotedness that I have ever shown to the service of your
Majesties, and the so unmerited outrage with which it
has been repaid, will not allow my soul to keep silence,
however much I may wish it.    I implore your Highnesses
to forgive my complaints.    Hitherto I have wept over
others ; may Heaven now have mercy upon me, and may
the earth weep for me !    Solitary in my trouble, sick,
in daily expectation of death, surrounded by millions
of hostile savages full of cruelty, and thus separated
from the blessed sacraments of our Holy Church, how
will my soul be forgotten if it be separated from the body
in this foreign land ?    Weep for me, whoever has charity,
truth and justice !    I humbly beseech your Highnesses,
that, if it please God to rescue me from this place, you
will graciously sanction my pilgrimage to Rome and
other holy places."

A majority of the noblest geniuses who have conferred
the greatest benefits on mankind, have been spit upon or
gnashed at and banned by the dominant class of their
contemporaries.    Prophets, discoverers, inventors, martyrs, illustrious company gathered from many times and
countries, and associated in one fellowship of sublime
genius, heroic devotion, and tragic fate, — history has
nothing left of equal pathos to reveal when it has shown us
these men, dreaded, despised, persecuted, outcast, dying,
appealing to after generations to do them the justice so
cruelly denied in their own.    Nor has posterity proved
recreant to the holy trust.    They are revered and celebrated now with an enthusiasm in strange contrast with
the obloquy they suffered when alive.    And to enter into
sympathy with them is an inexpressible comfort to those
who in later times are called to a kindred experience.  As

Heine says, " An equally great man sees his predecessors far more significantly than others can. From a single spark of the traces of their earthly glory he recognizes their most secret act; from a single word left behind he penetrates every fold of their hearts ; and thus the great men of all times live in a mystical brotherhood. Across long centuries they bow to each other, and gaze on each other with significant glances, and their eyes meet over the graves of buried races whom they have thrust aside between, and they understand and love each other." It is delightful to notice the geniality with which, in his Cosmos, the grand old Humboldt recognizes his great predecessors in the enterprise of surveying the universe as a whole, — Strabo, Eratosthenes, Ptolemy, Hipparchus, Galen, Aristotle, Lucretius, the elder Pliny, Albertus, Roger Bacon, Galileo, Copernicus, Newton, and the rest, — with what joy and piety he signalizes, from a height like their own, these intellectual peaks looming in clouds and stars athwart the historic table-land of science. The picture, in the New Testament, of Jesus on the Mount of Transfiguration in converse with Moses and Elias, is a beautiful symbol of the fellowship of the highest kindred spirits in all ages.

The consciousness of thinking and feeling in unison with a multitude, of believing doctrines and observing rites in common with the great majority of our brethren, yields to sympathetic genius an invisible, peace-giving fellowship which causes an indescribable pleasantness to breathe in the air, an infinite friendliness to saturate the landscape. To abandon all the dear familiar beliefs and associations in which one grew up, in allegiance to reason to go exploringly forwards into the obscure future to find some better substitutes, more divinely real and solid, is to be, at least temporarily, like one who advances into a cave in a mountain side ; the sight of the green fields, the light of the sun, the sound of the waterfall, the bleat of the goats, and the songs of the herdsmen, all becoming fainter and fainter, until he is lost in darkness and silence. It is impossible that severe pangs should not be involved when conscience sternly orders a sensitive and clinging

soul to renounce prevalent creeds, to cast off current
prejudices and usages, to leave popular favor, estranged,
behind, and accept newly revealed and persecuted truth
with its austere duties.  It is to undergo a coronation of
hate and agony, and, carrying a crucifix *within* the bosom,
journey on a lonesome way of dolor, publicly shrouded
in scorn, secretly transfigured with the smile of God.
The loneliest of all mortals are the pioneers of new principles and policies, new faiths and feelings; for they
alone have none on earth with whom they can hold
brotherhood of soul.  Having emerged from the beliefs
in which they were educated, thrown away habituated
reliances, trusting themselves to original perception as
they advance into the unknown, out of which new revelations are breaking on them, their solitude is sometimes
as appalling as the experience of one who for the first
time rides on a locomotive across a midnight prairie,
where, through the level gloom, he seems just plunging
off the world into banks of stars.

The bigotry of those whose opinions he rejected has
succeeded in attaching an unjust odium to the name of
David Hume, who was a man of remarkable goodness
of heart and life.  He was endowed with a mind of
wonderful acuteness and strength, exceedingly suggestive
and stimulative in its working on other minds.  His place
in the history of philosophy is of epochal importance.
Kant ascribes his own original work, of such immense
moment, to the impulse directly imparted to him by
Hume.  One of the results of his unsettling inquiries,
his idealistic speculations, has been thus impressively
depicted by himself.  "I am affrighted and confounded
with that forlorn solitude in which I am placed in my
philosophy, and fancy myself some strange uncouth monster, who, not being able to mingle and unite in society,
has been expelled all human commerce, and left utterly
abandoned and disconsolate.  Fain would I run into the
crowd for shelter and warmth, but I cannot prevail with
myself to mix with such deformity.  I call upon others
to join me, in order to make a company apart, but no
one will hearken to me.  Every one keeps at a distance,

and dreads that storm which beats upon me from every side. I have exposed myself to the enmity of all meta-physicians, logicians, mathematicians, and even theo-logians; and can I wonder at the insults I must suffer? When I look abroad, I foresee on every side dispute, contradiction, anger, calumny and detraction. When I turn my eye inward, I find nothing but doubt and igno-rance. All the world conspires to oppose and contradict me; though such is my weakness that I feel all my opin-ions loosen and fall of themselves, when unsupported by the approbation of others.''

What other experience can be so forsaken and grand as the loneliness of the man who has outgrown the opin-ions of his age, surveyed all the realms of knowledge and theory thus far achieved, traversed the constellated wastes of spiritual space to the outermost verge of thought, where he confronts the scintillating abyss of mystery, leaves contemporary humanity behind, pitches his tent a hundred leagues ahead of his nearest peer, and lives there, striving to conquer fresh realms for the occupation of posterity? He may be happy even in that forlorn station if he preserves a noble heart of kindness to his kind, and a spirit of self-surrendering trust in God. Such a man needs not recognition by official diplomas. Load him with conventional honors, he would lay the trinkets aside, and retire into himself to commune with his true dignity. He is an emperor, himself his empire. He will not in his self-sufficingness forget the dependence of feebler natures, nor cease to yearn over them in their wants and sorrows. Though isolated from the people by his intellectual transcendency, he will be joined with them by his affections and services; as the snow-capped summit of Dhawalaghiri commerces with the sky in inac-cessible solitude, while his gushing streams and his slopes of bloom wed him with the plains. Should the lofty thinker lose his confidence in reason and truth, and give way to a fundamental distrust, — as the tendencies are often so terrible in him to do, — becoming a misanthrope and an atheist, — his experience may be compared with the fate of that aeronaut who ascended into the congealing space until he suffocated from the thinness of the air, and his frozen form, borne in the fragile car, floated about at the will of the atmospheric currents in the cold un-sounding vastitude, under the dark sky-vault, the earth shrunk into a great ball below.

## The Solitude of Death.

IN this attempt to describe the loneliness of human life in its various kinds and relations, one more specification remains, — the solitude of death.   However filled with the strife and the gayety of bustling throngs the life of the toiling citizen, the queen of fashion, or the popular states-man may be, there is one passage of intense isolation which none can escape.

> A lonely hour is on its way to each,
> To all ; for *death* knows no companionship.

The approach of a mortal towards the bourn of his earthly destiny is a pilgrimage in which all that composes his ex-ternal company successively falls away ; and as he reaches the brink of the mystery, the last friend shrinks back and leaves him singly to the universal Parent.   Ought we not often to be alone with God in anticipation of the hour when He alone will be with us ?

Death invests every man with a solemn sphere of soli-tude, — the patriarch amidst his tribe, the victim on the rack, the felon on the gibbet, the gladiator in the arena, the martyr in the flame, the saint on his pallet, smiling at the uplifted cross.   Yet there are different densities of loneliness in the experience, between the departures from the sobs and clasped hands of loving families, the cold isolation of suicides, and the horrible desertedness of such fates as those of the forsaken Roman emperors, — Tiberius, Caligula, Domitian, Nero and Vitellius, extin-guished in the darkness of murder and ashes.   Despite the disparities, however, there is a fundamental identity in the last moment.   In every case, to die is to break, one after another, the ties that bind us to persons and things,

and, retreating into utter seclusion, migrate, silent and
separate, to the ultimate secret of the universe. Whoever contemplatively envelopes himself in that boundless
mystery, the idea of death, feels as one who, lost on some
strange heath, is wrapped in a night without a taper or a
star. All clews lost in the gloom, his unshared individuality revolves within itself in appalled wonderment, an
atom cut loose from social laws and plunging through immensity. Think of it beforehand, or not think of it, all
must at last come to this. Noisy and crowded as our
walks are, social and garrulous as our life is, every human
being has at least three moments of incommunicable separation. As the Hindu says, " Alone man is born ; alone
he dies ; alone he goes up to judgment."

The sentiment of loneliness pervades everything associated with death. The monarch, watched by attendants,
never free from obsequious company, is touched at last
by the wand of dissolution. His palace dwindles to a
coffin, his empire narrows to a grave. How quickly
slaves and courtiers, soldiers and people, shrink away
and leave him to be forever alone ! Terrible lessons are
taught by the sight of a tyrant in the hands of death.
Who can gaze on such a spectacle and retain cruel ambition in his heart? The imagination demands a certain
isolation and solemnity as the fit accompaniment of every
picture of death. A fop, like Brummel, lying dead in his
garret, affects us with a melancholy incongruity. Meant
to flutter in the sunshine of fashion, he is a dismal sight in
the grim storm and tragedy of mortality, — a belated butterfly frozen on a leaf. Much more becoming was the
funeral environment of the old Norse sea-king. Death-
struck, he seated himself on the deck of his ship, had her
set on fire and launched before the gale. His sword in
his hand, his white hair streaming, he vanished from sight,
and, perishing in this gallant pyre, was " buried at once in
the solitude of the sea and of the sky."

Graves are solitary, however thickly they lie together.
There is no other lonesomeness in nature so deep as that
which broods over the tombs of men and nations. The
visitor who pauses in the huge catacombs of Thebes

stuffed with death, the hollowed hills so heaped with stacks of bandaged humanity that they are but thinly masked mountains of mummies, feels for a time as if he were the survivor of a world. Go to one of the skeleton kingdoms of the East. Pitch your tent beneath the palms. Gaze around on the scene of ruin where once a nation of heroes sunk into their urns, and where, in a subsequent age, the dust of those heroes, spilt from their shattered urns, was blown about the desert, — and it will only be n. 'ural if the Spirit of Desolation sighs through your soul a lament as mystic as that of the summer-breeze soughing through the pines laden with tales from a primeval antiquity. Ponder on the fate of your race from its unknown beginning till now, see the procession of the innumerable generations of the dead steadily defiling into the grave, — and the whole earth is a funeral barrow.

Not only is there a solitude in death itself as experi enced by the dying, and an air of solitude around all the places and mementos of death ; there is an unparalleled loneliness created by death in the lot and feeling of him who, enduring loss after loss, grows old in an ever-widen- ing circle of missings and estrangement. " To a man," Dr. Johnson said, " who has survived all the companions of his youth, this full-peopled world is a dismal solitude." A heart-breaking sense of desertion must be felt by the last member of a decayed family, his ancestral castle dis - mantled, the proud crest bowed, the escutcheon dimmed with poverty and shame, the familiar glories of hearth and song become a tradition. It is more impressive still to imagine the loneliness of the last representative of a once puissant race who ruled hill and glen, but whose tents and banners have faded from the landscape, and whose weap- ons moulder in the dust. The Indian chief returns from far to stand in the light of the setting sun on the burial- mound of his fathers ; he muses there in mournful taci- turnity till the white man's step is heard, then glides into the woods, adding to the twilight forest one shadow more.

There is a deep loneliness too in all the preparatory steps and approaches to death. Who can fitly describe the solitude of extreme age ? The feeling of desertedness

and separation of an old man, who has survived all his
contemporaries, survived the copiousness and fire of his
own heart, — this is loneliness indeed.   And we are all
relatively old, — have outlived many dear comrades and
dreams, out-grown many darling hopes and plans.   When
we think of our school-days ; when, even in middle age,
we recall the fair and guileless companions whose eyes,
that looked all the tenderness of romance into ours, are
dust now, whose feet, that once sprang with ours in elastic joy over hillock and stream, now lie bound and still, —
many a bolt of lonesome sorrow pierces the heart.   Byron, old while he was yet young, asks, on hearing of the
death of one of his early friends, —

> What is the worst of woes that wait on age ?
> What stamps the wrinkle deeper on the brow ?

And then he answers, with the startling emphasis that
belonged to his intense and suffering genius, —

> To view each loved one blotted from life's page,
> And be alone on earth, as I am now.

The Wandering Jew, cursed with earthly immortality, seeing generation after generation disappear from the scene
of his pilgrimage, whatever he clasped to his breast immediately dropping into dust upon it, was forever alone, his
yearning agony itself an awful solitude wherever he went.
Who lives too long in this world of evanescent things
must, perforce, in some degree, taste that dreary experience.

There is, occasionally experienced by many in their
early years, an enchanted solitude, in which ecstasy ab-
sorbs them and makes them oblivious of everything but
itself: this is rarely known after youth has ended, except
by natural poets, romantic souls, remaining ever young.

> When youth, the dream, departs,
> It takes something from our hearts,
> And it never comes again.

In dismal contrast with this there is a disenchanted solitude, in which all the genial aspects of society are hidden,

every generous illusion destroyed, and existence left a haggard waste. Such is that cynical condition in which some men find themselves in the closing period of life, tyrannical, irritable, with the temper of a hyena.

In extreme age, when the last friend has gone, and the last hope of earth ceased to charm, the old man, deserted and doleful, stands on the dull plain strewn with the wrecks of youth, like a despairing mourner in a grave-yard, where the moonshine lies on the motionless scene, and excited fancy turns every tombstone into a ghost, and takes every shadow for an omen. And he looks around in vain for a hand to clasp, or a heart to quicken his by its responsive beatings. This loneliness is so sharp and profound because of the contrast between the memory of the past and the consciousness of the present. What *was* stands in sunlight there, what *is* rests in shad-ow here; and the opposition of the pictures makes the lonesome soul doubly lonesome. It is thus with the ruins of abbeys; these are so intensely solitary from the con-trast which haunts the imagination of the pilgrim, between the former show of processions, chants, church-banners, bells, censers, and hymns, and the present scene of silence and decay, roofless walls, ivy-grown arches, great trees growing in the aisles, foxes burrowing in the refectory, rooks and daws perched on the mouldy brackets.

More appalling, however, to the spectator, than this solitude of bereft old age, or any experience of physical dissolution, is that solitude of madness sometimes exhib-ited, a death-in-life existence, the virtual destruction of the mind, the temporary suppression of the soul. There are patients in asylums of the insane, who are so shut up in one mood, so possessed by one thought, that nothing else can reach them. The convulsion of some tremen-dous moment has petrified the before flexible mechanism of the brain so as to allow the forces of consciousness to operate only in one way and to one result. They turn their faces to the wall, taking no interest in anything more, never looking up, never speaking again. In a dumb, impassive, fearful solitude they abide, till death, the great deliverer, comes. Then, at length, they go

forth, like all the rest of us, each one alone, to encounter the dark secret which both repels and invites, and is
at once unavoidable, insoluble, and eternal.

What other solitariness is conceivable so unrelieved as
that of the unhappy lunatic who was convinced that he
should never die, but as a punishment of his demerit
should be kept forever alone in the world when all other
men were dead? Ah! lonely and dread as death appears
to many, it is unspeakably sweet and welcome when it
comes at last to one weary of the hubbub, and sick of
the insufficiency, of earth. It is easy for such an one to
sympathize with what the dying Howard said when seized
by an infectious fever at Kherson, in the midst of his philanthropic labors, far from home and friends. "Lay me
quietly in the earth, place a sundial over my grave, and
let me be forgotten." Such was the trustful resignation
of the mind, such the complete weariness of the flesh,
that he shrank from the effort of the thought of fame.
This desire to cease and be forgotten is the divinely
natural preparation for our transition into futurity. The
passion for life sinking parallel with the failing force of
the organism, the two go quietly out together, so that
there is no rebellion. If the theologians, with their superstitions and artificial horrors, will but let him alone,
man is competent to death as well as to life, and dies in
peace.

Death is a new thing to every one who experiences it.
Neither is it the same thing to any two persons; for each
brings to it his own special qualities and accumulated
experience. What a different thing death is to one whose
thought includes and whose sensibility overspreads the
whole world, and to one whose consciousness is commensurate with little more than his own person and the sensible facts closely about him! To the latter it is as the
mere physical expiration of an animal; to the former it
is as the collapse of a solar system. It has been said
that murderers have met their doom on the gibbet with
more fortitude than Christ on the cross. Not with more
fortitude, but with more insensibility. The ruffian dies
like a wild beast at bay. The infinitely diffused and in-

tense sensibility of Jesus made his death like the sepa
ration of a universe.

It is an affecting peculiarity of man that he shrinks
with strong antipathy from the thought of dying alone or
among strangers.   He would have friendly eyes look on
him, feel the clasp of a familiar hand, in that silent immense passage of his being.   All things but man, when
fatally hurt or spent, retreat to die in solitude ; they are
afraid of being attacked in their weakness.   If a wolf
so much as limps, the other wolves tear him in pieces.
Instinctively, therefore, the dying animal seeks a secret
corner.   But man, with a few abnormal exceptions, never
wishes to die without some one near to count his sighs,
watch his ebbing moments, and mark his last gasp.   It
is a pathetic proof of his natural sociality.   Sympathy is
deeper than fear, and in the final failure of his own force,
in the upheaval of the bottom of his soul, he puts distrust
and hate aside, and clings to his kind with a loving expectation of help.   But no companionship of other wisdom or love can avail or endure there.   Personal insight
and trust of the truth, personal surrender to the Absolute Spirit, — these only can stay and comfort then.
Though outwardly girt by the fondest comrades, inwardly alone, each one casts his material investiture,
eludes their grasp and their gaze, and slips separately
into his curtained fate.   The loneliness of dying is like
the loneliness of the sea, whereon many ships cross and
pass without speaking.   So do many human beings die
simultaneously, but make no signals to each other as the
wonted shores recede, and the breath of the Infinite
swells the unseen sail, and the gray waste looms in the
silence of its immemorial mystery.

# THE MORALS OF SOLITUDE.

# THE MORALS OF SOLITUDE.

—◆—

## The Dangers of Solitude.

THE topic next to be treated is the perversions and dangers of solitude.  In attempting a general survey and application of the lessons of this part of the subject, scrupulous care is needed to avoid errors and exaggerations.  At the start it should be understood that there is no magic in seclusion itself to make any one strong or wise or good.  A man may keep by himself because he is a fool or a knave, and become the greater fool or knave by doing so.  The benefits of retirement are not the results of a charm, but the fruits of a law faithfully observed.  The secrets and blessings resident in solitude must be wrung from it by our energy ; they will not spontaneously drop into our laps as we approach, any more than the arrow-headed inscriptions in the desert yielded the ancient history locked up in their cipher to the caravans and armies that for so many ages ignorantly travelled by them.  Solitude works on each one and contributes to him after his own kind.  It may make a prophet or an idiot.  It excites, concentrates, and fortifies the faculties of a strong and studious soul, but bewilders and dissipates those of a weak and wandering one.  The great argument against the system of solitary confinement in penitentiaries is that it destroys the minds of those subjected to it.  Solitude has imbecility for one of its handmaids.  It was found, when the separate and silent system was introduced into the Pennsylvania prison, almost impossible to prevent the convicts from climbing up to the windows to salute each other, and from conversing through the walls of adjacent cells by signals, — so fierce was the demand of nature for sympathetic communication.  We must not let the philosophic and poetic side of the subject fascinate us, and prevent our seeing that all depends on the kind of solitude, the kind of soul, the kind of activity between them.  We should remember that there is the solitary worm as well as the solitary eagle.  One is more likely to prefer to be alone because he is too poor or too bad to furnish the conditions for agreeable company, than because he cannot find company worthy of him.  It is so much easier to get along where there is no one to thwart, contradict, or irk.  Wisdom is given to deep reflection, and lonely reflection makes wise.  Fools chatter; the gods are silent.  Though this is undeniable, there is truth on the other side too.  A sage is not unfrequently as talkative as a gossip.  Ripe experience is fondly apt to teach. Earnestness is as much akin to oratory as it is to reverie. A busy tongue may be the vehicle, as well as the substitute, for a busy brain.  If geese fly in a flock, while the condor preys alone, the moral qualities of the goose are better than those of the condor, and, undoubtedly, it is the happier bird.  The portentous gravity of the hermit owl covers not so much wisdom as the frolics of the social swallow.  There is no virtue in mere loneliness to dignify the fop or regenerate the fool, to purify a rake or make a soulless hunks a generous lover.  And when such as these affect it, the affectation is but another vent of their folly, a trick of vanity.  The solitary often occupy themselves with trivialities instead of grandeurs.  A famous pillar-saint was observed, on the top of his column, to touch his forehead to his feet twelve hundred and forty-three times without intermission.  The emperor Domitian, whose congested vanity made him ostentatious of courting sage retirement, was discovered in his seclusion stabbing flies with a bodkin.

Solitude is the retreat of the defeated as much as it is the home of the self-sufficing.  Ignatius Loyola once said to a young member of his order, who, on account of his great susceptibility to anger, was accustomed frequently

to avoid his companions and remain apart : " Irritability
and choler are not to be conquered by flight, but by combat ; solitude will not destroy them, it will only conceal
them." How these words must thrill through the ranks
of congenial souls ; a blast from the trumpet of the soldier of eternity ! He adds, " You will sacrifice more to
God, you will gain more for yourself, by acts of mortification in your intercourse with your brethren than if you
were to bury yourself in a cavern and to pass a whole
year in complete silence." One of the dangers of a retreat is the fallacy of believing that we are destroying,
while really we are only hiding, our vices. Many find it
an easier art to live alone than with their fellows, because
there they have but one to quarrel with, and that one
the most obsequious of flatterers. It is a higher accomplishment to harmonize with all than it is to melodize
alone. The former feat is to the latter as a perfect composition in counterpoint is to a good solo. There is
ground for Shenstone's rebuke, —

> In cloistered state let selfish sages dwell,
> Proud that their heart is narrow as their cell.

The truth is, man is both a gregarious and a solitary
animal, as much made for society as for solitude, and as
much for solitude as for society. His true life, in a healthy
state, is an alternation from one to the other, in due proportion. To live exclusively in either proves disease,
works ill. The office of each is to fit for the other, and
lead to it. There is something wrong in him whose
lonely interviews with nature make him dislike to meet
men ; something wrong in him whose association with
men unfits him to enjoy retirement. The one should
send him to the other with a renewed relish. As Coleridge says in his poem to Charles Lloyd,

> If this green mountain 't were most sweet to climb,
> E'en while the bosom ached with loneliness, —
> How heavenly sweet if some dear friend should bless
> The adventurous toil, and up the path sublime
> Now lead, now follow, — the glad landscape round
> Wide and more wide increasing without bound !

Society and solitude ought to rectify and supplement each other, somewhat as the states of waking and slumber do. One is comparatively for observation and comparison; the other for rumination and digestion. Among men we obtain the food of the spirit; apart from them we assimilate it. But it should never be forgotten that we are far less what solitude makes us than solitude is what we make it. Its influence on us depends on the character we carry into it, and the improvement we make of it. Its vagueness ungirds and empties an aimless soul. But a man with a mighty purpose finds room and leisure and invitations in it for his imagination to work and react until all the centres of association, the batteries of his mind, are charged with magnetic ideas. What is the true zest of life? An absorbing object. Patroclus in the tent with Achilles, Audubon with his rifle in the wilderness, Kant buried in thought among his books, Humboldt climbing the side of Chimborazo, Fenelon chastening his self-love on the way to perfection, — under the differences in all these there is an identity of joy, namely, the fruition accompanying the pursuit of an aim. If we retreat from observation for the sake of dawdling indolence or other form of self-indulgence, the retreat is pernicious; but if we withdraw to cherish a keener and deeper devotion to some noble aim, to prevent our purposes from being worn down and frittered away by the petty frictions of petty people, it is sanative, holy, inspiring. Gratification, made selfish by isolation, is degraded from the level it naturally holds when shared with others. Men, devouring their food in solitude with mere physical greed, approximate swine swallowing their swill; men, seasoning their food with conversation and affection, approximate the gods taking their ambrosia.

First, solitude is what we make it; then, we are what it makes us. To the poetic and spiritually-minded religionist, solitude is "the voluntary winding-sheet in which he wraps himself to taste the voluptuousness of being dead to earth." To the ambitious and carnally-minded worldling, solitude is a camera-obscura into which he retreats, sitting there, himself unseen, to study society, to prepare his plots and traps ere he sallies out to take it captive. A neutral mind left alone, with its squandering vagueness, legitimately ends in collapse ; a mind intent on a good aim is benignly strengthened ; a mind intent on a bad aim is perniciously strengthened. There is, therefore, in mere solitude itself, no spell to exorcise and bless its votaries. All the blessings it is capable of yielding are to be drawn from it by a faithful observance of the conditions of its improvement. Beckford, the gifted but wayward and unhappy author of Vathek, suffered continually, in his own touching phrase, from "the faint sickness of a wounded heart." In vain did he try, with every outward advantage, first the most brilliant publicity, then the most profound privacy. Neither could successfully medicate the fatal wound which was — himself.

Comparatively few can afford to do without the animating motives of fellowship and publicity. Solitude is the breeding-place of fear. Nowhere else does superstition thrive so well. Bentham observed, " Many a one who laughs at hobgoblins in company, dreads them when alone." Where one man is brave by himself, twenty are brave before a multitude ; he is a high and powerful character who is equally brave in both situations. Some when undisturbed by a foreign presence spontaneously imp their wings for a flight into the highest regions of romance and nobleness ; others sink if not incited by the consciousness of being on exhibition. He is of a royal spirit who can make the holy stimulus of duty perform the service usually rendered by the ignoble stimulus of vanity, and, at the same time, catch fresh inspiration from sympathy. There is something impressive in the fidelity with which famous public performers, great artists, in their several departments, keep themselves in training. What indefatigable pains they take to prevent any falling off in their skill or power ! With unfaltering devotion, every day, these celebrated favorites privately practise their feats, to keep every sense acute, every muscle firm, every faculty equipped. Most obvious and keen and constant, though somewhat coarse and low, is the motive that feeds their purpose and keeps their efforts from flagging ;

namely, the lavish returns of personal admiration and
pecuniary gain to be secured from the public.  The
wearisome preparatory exercises which seem so heroic,
are less impressive when we see that they are sustained
by an ever-present anticipation of golden guerdons and
intoxicating applause which will be bestowed on them
as they display their accomplishments before delighted
crowds.  The motive itself—which always decides moral
rank — is vulgar enough for the vile to feel ; it is the
power and tenacity with which they respond to the motive that are great.  But there is a spectacle of devotedness incomparably grander and more beautiful, as authoritative and sublime as anything known on this earth.  It
is afforded by those profound thinkers, exalted believers,
fervent lovers, who never make an exhibition, never receive human recognition, but toil on in secrecy, unnoticed, unthought of, set only on attaining spiritual perfection.  Winning no social appreciation, asking none,
without even a friend to look reverently and lovingly in
on their aims and struggles, they apply themselves in
their own retreats to the tasks of wisdom and piety.
With supernatural courage and energy they toil to disentangle the webs of sophistry and acquire a knowledge of
the truth, to chasten their passions, and grow pure and
magnanimous and gentle.  Though their hearts are painfully full of love and longing, with saintly renunciation
they refuse to purchase the common admiration which
they could easily obtain in over-measure if they would
but so demean themselves as to stoop for it.  They suffer
no day to go by without strenuous exercises of their
highest faculties in the rarest feats of human nature, taking scrupulous care that no sweetness of tone, no precision of touch, no delicacy of motion be lost.  And all
this they do without one public plaudit, without one sympathizing eye to see.  Here is pure heroism indeed.  It
puts all other bravery to shame.  The proudest boasts of
history are contemptible before it.  It dwarfs a Cæsar into the champion of a village brawl, glorifies a Jesus into
the hero of a universe.  These are the saints in the
church of nature, the heroes of solitude, the chivalry of

mind, the elect artists on whose curtained performance
God gazes, an invisible Spectator, distributing his applause in the functions of their characters.

To live wholly in society, or wholly in solitude, is fatal
to the best prerogatives of the soul. To be healthy and
complete we must live alternately, now with our fellows
and the world, now with ourselves and the universe.
While in each one we shall gain its best advantages by
making the most of its distinctions from the other, appreciating the contrast as vividly as possible. It is an
abuse of either to convey the unqualified conditions of
the other into it ; in company, to notice no one, see nothing, hear nothing ; alone, to be occupied with social vanities and heart-burnings. When among men, they have
claims on us, and we have no right to be self-absorbed or
absent-minded. However salutary solitude may be, it is
perverse to make a solitude of society. Yet this statement is susceptible of the modification happily afforded
in a paragraph of Schopenhauer, which as strongly rebukes his perverse practice as it exhibits his theoretic
wisdom : — " Take a little of your solitude with you when
you visit men. Self-detached, view them in a pure, objective light, with a noble freedom from prejudice. Then
society is a fire at which the wise man, from a prudent
distance, warms himself, — not plunging into it like the
fool who, after getting well blistered, rushes into the coldness of solitude and complains that the fire burns." Mixture with companions, as well as isolation from them,
has its contributions to offer towards our perfect equip-
ment. If power be born in seclusion, art is the fruit
of association. If sentiment be nourished apart from
men, ambition is kindled among them. If principles
grow in the soil of solitude, actions ripen in the air of
society. He who abides overmuch by himself must carefully keep an open communication between the inner
meditations and plans that occupy his imagination and
the social motives that would fertilize and apply his
energy. Otherwise he is likely to become an idle
dreamer. The currents of his enterprise are in danger
of turning awry and losing the name of action. The

ideas of deeds become the substitutes of deeds: the mental pictures of victories prevent, instead of prepaiing for, actual victories.

Many persons who keep their wits awake in company let them lie dormant when alone, as if sharp study were useless there. But observation and reflection are not more necessary in society than they are in solitude. If seclusion by wisdom is divine, seclusion by ignorance is both vulgar and dangerous. Persons of a rustic and bashful retiredness are at the mercy of all sorts of impositions, lures, false estimates. They are often overawed by the showy and wicked initiates of the world, deferring to them with such fear and wonder as the gorgeous-colored and quiet forest-birds of Brazil might feel if some wild gray ocean-fowl flew screaming from his tempests through their solitudes. While the heart and the spirit stay at home, therefore, let the eye and the mind learn good and evil by travelling much abroad beyond the retreats where virtue and innocence retire to nourish their energies and to protect their delicacy. The less we mix with men the greater our weakness and unwisdom, unless we fill their absence with something better, — grave thoughts and earnest feelings, faith and study. Hermits who watch, aspire, and philosophize, become the truest sages. Then it is not strange —

> That we, in the dark chamber of the heart,
> Sitting alone, *see the world tabled to rs.*
> For the world wonders how recluses know    ˙
> So much, and most of all, how we know them.
> It is they who paint themselves upon our hearts,
> In their own lights and darknesses, not we.

A clear, powerful, assimilating purpose is the sole adequate safeguard against the exposure of those who are alone to idleness, triviality, dawdling reverie. Without this purpose, solitude is a manufactory of vapid visionaries. It is absurd to dilate on the advantages of solitude at the *expense* of the advantages of society. They are supplementary rather than contradictory, and derive their several powers from their mutual contrast.

Goethe has said: "Were there but one man in the

world, he would be a terror to himself." Nay, it may be added, were there but one man in the world, there would be no man in the world. In the absence of humanity he could no more remain man than there could be an island if there were no sea. The single individual is to collective humanity as the little column of mercury in the barometer is to the whole atmosphere. They balance each other although infinitely incommensurate. A quicksilver sea, two and a half feet deep, covering the globe, would weigh five thousand billion tons. That is the heft of the air, — that transparent robe of blue gauze which outsags the Andes and the Alps. Its pressure is unfelt, yet if that pressure were annulled all the water on the earth would immediately fly into vapor. Public opinion is the atmosphere of society, without which the forces of the individual would collapse and all the institutions of society fly into atoms. With every man mankind, or some representative of it, is ideally present in almost every act of consciousness. Often the most important truths are the ones we are least aware of: we act and react on them automatically. Scarcely with more certainty does every movement of a private lung imply the public atmosphere than every act of our souls presupposes the existence of our fellow beings. As Börne said, " Man can do without much, but not without men." The self-esteem of the anchorite is sustained by a subtle conviction that if they knew it men would admire the superiority that enables him to dispense with their society. Saint Antony sought out Saint Paul, the earliest of the hermits of the Thebais, and after much difficulty succeeded in gaining admittance to his cave. The first question of the long-hidden recluse was, " How fares the human race ? "

The ideal dependence of man on his race, even in the extremest instances of withdrawal, has been forcibly expressed by Isaac Taylor in his masterly treatise on Fanaticism : — " Nothing appears too great, sometimes, to be grasped by the conceits of self-importance ; nothing too big for the stomach of vanity ; and yet it is found that the imagination refuses to yield itself, except for a mo ment, or in a very limited degree, to those excitements that are drawn from the solitary bosom of the individual. Man, much as he may boast himself, is by far too poor *at home* to maintain the expense of his own splendid conceptions of personal greatness. Only let some breathless messenger reach the cavern of the hermit, and announce to him that his love of solitude was at length effectively sealed by the utter extinction of the human race, — solitude, from that instant, would not merely lose its fancied charms, but would become terrible and insufferable ; and this man of seclusion, starting like a maniac from his wilderness, would run round the world in search — if haply it might be — of some straggling survivors No conception much more appalling can be entertained than that of a proud demi-god, who, finding within his own bosom an expanse of greatness wherein he could take ample sweep, and incessantly delight himself, should start off from the populous universe, and dwell content in the centre of an eternal solitude."

With the ordinary man the four great units — self, mankind, the material universe or nature, and the intellectual universe or God — are obscurely outlined, confused, with vague and feeble reactions. In the mind of the man of genius these four units are sharply defined from each other, with distinct and intense reactions. For health and peace it is necessary that the relations of these units to each other be truly apprehended and observed. Any discord or insubordination here is sure to breed morbidity and wretchedness.

The jarring of the individual with the whole is so sad and common a disease, the revulsion from the world into a painful and angry solitude is exhibited in the experience of so many superior persons, that the whole subject of its causes, bearings, and cure, deserves careful consideration. The more so as the disease usually passes by the cold and shallow to fasten on those of warm hearts and rich minds. The greatest number of isolated and resentful flingers at the world will be found to be those who first went out to the world with the most impulsive affection and soaring enthusiasm. Disappointment, disgust, and

pride made such a man of Hazlitt, whose quarrelsome
humor and fierce contempt caused most of his acquaintances to regard him as a mere misanthrope, but of whom
Charles Lamb says, — "I wish he would not quarrel with
the world at the rate he does; but, judging him by his
conversation, which I enjoyed so long, and relished so
deeply, or by his books, in those places where no cloud-
ing passion intervenes, I should belie my own conscience
if I said less than that I think him to be, in his natural
and healthy state, one of the wisest and finest spirits
breathing; and I think I shall go to my grave without
finding, or expecting to find, such another companion."

The founder of the Christian religion propounds as
one half of his system of duty the commandment, "Thou
shalt love thy neighbor as thyself." Interpreting these
words either according to their grammatical and logical
force, or according to the laws of a scientific morality,
we must suppose the term *as* to refer to kind and not to degree: thou shalt feel towards him in the same way, though
not to the same extent, as towards thyself. It is impossible for each man to love all men with the same intensity that he loves himself, though he may love them in
the same manner. Besides, were it possible for each
one to love his fellows in exactly the same degree as he
loves himself, it would be fatal. The divine plan of
having each one look out first and chiefly for himself, is
the only plan that could work. Each one has the most
intimate knowledge of his own wants, is best fitted to
supply and defend himself. Give to every one an interest in the welfare of all others as keen and massive as
that he feels in his own, and the exactions it would make
on him would exhaust his powers, and utterly break him
down in futile efforts to respond to their claims. Were
philanthropy universally as strong as self-love, it would
be necessary to legislate against it, invoke public opinion
against it, conventionally enact penalties against disinterestedness, and rewards for selfishness, in order to protect
the whole of society by the preservation of the separate
individuals who compose it. Each primarily for himself,
secondarily for all, is the necessary method of nature.

But there is a strong tendency in man, — a tendency inherited from the conditions of a barbaric past, — for the individual, instead of balancing his love of self with a love for others as much greater in diffusion as it is feebler in degree, to balance it with a general hatred for others, as hostile rivals striving for the goods he wishes to monopolize.  There is in unregenerate man a natural motion to regard his fellows as enemies in the scramble for the fruitions of life ; to turn from them with aversion, and wish them injury, as though their successes were so much taken from him.  It is against this malign tendency that the sublime precept of Jesus finds its application. It is the duty of each to exercise towards his fellow men the same kind of feelings which he directs upon himself. He may not love, forgive, protect them as *much* as he does himself, but he ought to as *sincerely*.  Instead of inverting the self-regarding feelings by antipathy into animosity towards others, he ought by sympathy to reflect those feelings over them, unaltered in kind if thinner and paler in vigor.

It is actually impossible, then, and, if possible, would be ruinous, for us to feel in detail the same degree of interest and affection for the bulk of mankind that we do for ourselves, though we may feel towards them in the same mode ; and there are grand generalities of human beings or human interests for which we can joyously sacrifice our lives.  However instinctive it may be, it is wicked to have one set of feelings for ourselves, and an opposite for our brethren.  This is the fundamental law of morality, and with its fulfilment are bound up both the happiness of the individual and the well-being of the whole.  The violation of this law is more prolific of loneliness and misery than any other cause.  It is written in one of the ancient sagas of the North, " The tree which stands within the village, deprived of its sheltering fellows, droops and fades away ; so it is with the man whom no one loves ; why should he live long ? "

There are many misanthropes in the world, made such by different influences ; and they are all, so far forth, isolated and wretched.  Some are made misanthropes, singuhar as it may seem, by too much tenderness of heart. Their fine, rich, clinging, modest natures want more sympathy than the coarse and careless crowd can give them. Sometimes they crave appreciating respect and kindness so strongly, and, in their neglected state, feel so certain of securing what they desire, if they can only gain an interested attention, that they importunately display to chosen ones their treasures and claims ; as much as to say, "See here, how I love all noble things, my country, my kind, truth, virtue, beauty. Deign to notice the proofs of a mind not poor or vulgar, a heart soft and expansive. Feel towards me as I deserve and want, and there shall be no bounds to my profiting gratitude." Every generous soul will feel more disposed to pity the pain that underlies such an experience, love the sympathy that so needs relief, and admire the ingenuousness that dares thus unveil itself, than to despise it as vanity or hate it as conceit. Yet the latter is the more common treatment it receives. It is not strange that many a grieved spirit, after such cruel misinterpretation, retires into himself apart from the public course, as a fawn who, venturing near a nest of hornets, has been stung, bounds back to the friendly coverts of lake and forest, surprised, terrified, smarting. It is but too natural that he should then feel, as Chateaubriand says, "Why should we open our hearts to the world? It laughs at our weaknesses ; it does not believe in our virtues ; it does not pity our sorrows." Of all the solitudes in society there is no other so deep and fatal as the solitude of those who love too much and desire too much. The occupants of that solitude, unless their tempers have a saintly quality, are apt to be exasperated to acrid returns. Wordsworth gives a forcible description of such an experience in the case of one who owned no common soul and had grown up in lofty hopes.

> He to the world went forth,
> A favored being, knowing no desire
> Which genius did not hallow : 'gainst the taint
> Of dissolute tongues, and jealousy, and hate,
> And scorn, — against all enemies prepared,
> All but neglect. The world, for so it thought,

> Owed him no service ; wherefore he at once
> With indignation turned himself away,
> And with the food of pride sustained his soul
> In solitude.

The one eternal want of man is to feel himself reflected in the souls of his brethren.  If too great a quantity and too high a quality of sensibility, render him too unlike his neighbors, his desires are constantly disappointed, and he is in danger of disliking men, because they hurt him. The following lines are a revelation from the heart of poor Gerald Griffin, the Irish novelist.  The lines are obviously morbidly sentimental ; but in their genuineness they are touching to a gentle spirit, who will respond to them with grieving sympathy, and deem it inhuman to sneer at the experience they express.

> I would I were the lonely breeze
> That mourns among the leafless trees,
> That I might sigh from morn till night
> O'er vanished peace and lost delight.
>
> I would I were the murmuring shower
> That falls in spring on leaf and flower,
> That I might weep the live-long day
> For erring man and hope's decay.

It is impossible that such a spirit should be otherwise than solitary amidst the hard hearts that make up the majority of the world.  He says, describing the life he led, " I used not to see a face that I knew, and after writing all day, when I walked the streets in the evening, it actually seemed to me as if I were of a different species from the people about me."

Men of this temperament are easily induced to withdraw from social scenes in grief and despair, and shut themselves up in such protecting seclusions as they can find or make.  An extract from the diary of John Foster — who certainly was no weakling — betrays such an experience.  " I can never become deeply important to any one ; and the unsuccessful effort to become so costs too much in the painful sentiment which the affections feel when they return mortified from the fervent attempt to give themselves to some heart which would welcome them with a pathetic warmth. My heart then shuts itself up and feels a painful chill. I am glad to be gone to indulge alone my musings of regret and insulation.'' Even the noble Schiller sank for a time into a state of disliking sourness towards his kind. "I had clasped the world," he says, "with the most glowing emotions, and found a lump of ice in my embrace." He then proposed to translate Shakespeare's Timon, affirming that no other piece spoke so eloquently to his heart or taught him so much of the science of life. Bulwer a little exaggeratedly remarks, in allusion to this unhappy passage in the life of Schiller, "It is a state common to all good men in proportion to their original affection for their species. No man ever was, in reality, a misanthrope, but from too high an opinion of mankind, and too keen a perception of ideal virtue."

The touchstone of actual intercourse, too often and rudely applied to natures under the sway of their own ideal creations and demands, is cruel in its effects. Those whose imaginative sensitiveness disqualifies them for comfort in the cold contact of reality, may sometimes wholesomely retreat from an intercourse unsuited for them ; but they must take especial care not to be embittered so as to regard their fellow beings with contempt or rancor. That would only aggravate the evil they suffer. Vain and foolish is it likewise to utter lamentations or blames to the world. Motenebbi, the great Arabic poet, says, " Complain not of thy woes to the public ; they will no more pity thee than birds of prey pity the wounded deer." There is but one medicinal refuge for him who flees from his fellow men because they wound his too tender heart ; and that is a solitude in which he can brood over them in imaginative sympathy. If his imagination be filled with scorn and hate, melancholy and dark indeed is his doom. Goethe, who seems to have experienced almost everything, has depicted this loneliness in his Harz Journey in Winter. " But who is this, apart ? His path vanishes in the bushes, the twigs close behind him, the grass rises again, the desert swallows him. Ah, who can heal the pains of him to whom balsam has become poison? who out of the fulness of love drinks misanthropy? First despised, now a despiser, he secretly devours his own worth in discontented self-seeking."

In addition to those who have become misanthropic by revulsion from a world wherein their affection fails to grasp what it reaches after, there is a class of a darker and intenser type composed of men who have been deceived and abused by those in whom they trusted, or persecuted and wronged by those whose favor they sought. Their personal experience of the tricks, plots, and cabals of mean men, of the slanders and hate of envious inferiors and malignant rivals, of the petty stings of critics, has disenchanted them of their early illusions and led them to form an estimate of humanity as much too base as their original estimate was too exalted. Timon is the representative of these; the noble Timon, who "the middle of humanity never knew, but both ends," whose final relenting before the good Flavius, showed that his nature, "sick of man's unkindness, was yet hungry." Sir Samuel Egerton Brydges, defeated in a suit which he brought for a noble title, — a claim which all but himself thought purely fanciful, — was so soured and envenomed by a perpetual brooding over his great wrong, that he retreated into a haughty exile and solitude, and became almost a monomaniac as well as misanthrope. This is to make the mistake of confounding the whole species with the worst specimens that belong to it, instead of carefully discriminating, so as to keep our reverence for the good and high unharmed in the presence of the low and contemptible. Sometimes one who begins with hating men because he has himself been disappointed and ill-treated by them, goes on to hate them because he believes them intrinsically bad. He forms this perverse belief by putting on all men the stamp of the bad men he has known. He should reverse the procedure. The ideal type of the race, as divinely fair and good, affixed even to individuals who injure us, makes it possible for us to love them. But an ideal type of individuals, as foul and bad, affixed by transference to

the race, breeds a universal disgust and rancor. The
data for forming the typical idea of man are the cardinal
elements of our nature, the great principles of morality,
the choice qualities of our parents and friends, the chai-
acters of the illustrious exemplars in history, the noblest
representations in literature, and the best experiences of
our own hearts. No single tyrannous experience or
person should be allowed to mould it; and all malignant
infiltrations from foes or from our own depravity should
be kept out of it. The typical idea unconsciously affixed
to individuals or the race is what principally determines
the blessedness or the misery of our relations with our
fellow men. Be that idea exalted to its best, as it was in
Channing, and all history becomes a honey-pasturage for
our thought. Be it debased to its worst, as it was in
Schopenhauer, and the whole world becomes the forage
of our spleen.

But, however natural it may be to do so, there is no
justification for those who, when wronged, turn against
mankind with retaliating animosity. They are guilty of
both folly and sin in retorting hate on the level of those
who have injured them, instead of rising above them in
serene devotion to the great ends of existence, — truth,
virtue, faith, God. How much better to look down from
this height on the ignoble, with a divine tenderness of
pity, than to fall in rancorous revengefulness to their
range. Every truly wise and good man, though he may
sometimes fail, will always aim at this, — a mood of bland
and magnanimous benignity towards the most unculti-
vated, the most degraded, the most unjust and unkind.
The effort will often try him, but he is bound to perse-
vere till he conquers every impulse of arrogance. Sir
Walter Scott, in his pecuniary troubles, faced the ignorant and cruel rabble at the court house of Jedburgh, held
his great loving heart down, and calmly braved their
hootings. Agonized by this outrageous injustice heaped
on his misfortunes, with a fore-feeling of his doom of
mental decay and speedy death, he addressed to them,
as he sadly turned away, the words the fallen gladiator
in the Coliseum was accustomed to address to Cæsar, —

*Moriturus vos saluto!* Ah, brave Sir Walter, — noble, tender and true, even in response to the heartless injus- tice of thine inferiors !

It is beautiful to study the examples set by many great men, of magnanimous forbearance and sweetness in the face of aggravated wrongs. Anaxagoras, the friend of Euripides and Pericles, by his bold philosophical teach- ings, offended the superstitious crowd, and more than once came near being condemned to death. At length he was exiled to Lampsacus. When some one there pitied him that he was deprived of Athens, he proudly replied, " Rather Athens is deprived of me." The citi- zens asking him on his death-bed what honors they should pay to his memory, he said, " Let the anniver- sary of my death be kept as a holiday by the school children." And it was so. Phocion bore himself with imposing dignity and serenity before the frenzied mob, who refused to grant him a hearing, and clamored for his death. Asked what message he would like carried to his son, he answered, " That he bear no grudge against the Athenians." No one can forget how Aris- tides wrote his own name on the shell, that the vulgar wretch who was tired of hearing him called the just, might vote for his banishment. When Harvey's book on the Circulation of the Blood came out, " he fell mightily in his practice. It was believed by the vulgar that he was crack-brained ; and all the physicians were against him, and envied him." After describing how much abuse he had suffered, Harvey adds : — " But I think it a thing unworthy of a philosopher and a searcher of the truth to return bad words for bad words ; and I think I shall do better and more advised, if, with the light of true and evident observations, I shall wipe away those symptoms of incivility."

Another class of misanthropes is made up of persons naturally of a savage temper, who bitterly envy others their advantages, but have neither sympathy with their feelings nor interest in their pursuits ; meagre and icy hearts, born under the Wormwood Star. Their own mean and fierce characters, seen in themselves, are prevented by self-love from appearing detestable ; but forming out of their own feelings and conduct their idea of other men, they recognize its detestableness, as seen in them, and are filled with hatred of it.   So the ugly mastiff, seeing himself in a mirror, takes the reflection for another dog, and flies at him in a rage.   Such men instinctively hold themselves aloof from the hearths and bosoms of their kind, cynically sneering at their sentiments, their aims, their doings, their joys and their miseries.   This is the surliest and most repulsive, as it is the smallest, class of men-haters.   Its type is energetically set forth by Shakespeare in Apemantus, the fierce low cur who was

> Set so only to himself,
> That nothing but himself, which looked like man,
> Was friendly with him.

The doctrine of total depravity, as held by Calvin, legitimately nourishes a terrible misanthropy.   Any one who holds this theory in vital consistency, — that all men are naturally utter haters of good, and lovers of evil, detesting God and detested by God, — must become a virtual misanthrope, and desire to escape from the scene of such a demoniac race.   Isaac Taylor has shown, in his profound and terrible analysis of fanaticism, that a rancorous contempt or hatred for the mass of mankind is the appropriate sentiment of him who regards them as religiously cursed and abominable.   " There is a combination of the religious sentiments with the passionate workings of self-love, pride, jealousy, and the sense of personal and corporate welfare, which brings with it the most peculiar and virulent species of misanthropy known to the human bosom ; and an arrogance that far transcends other kinds of aristocratic pride.   With an anathematizing Deity, an anathematized world, and himself safe in the heart of the only Church, the zealot wants nothing that can render him malign and insolent."

But the largest class of misanthropes consists of the lofty men who are repelled and angered by the frivolity of average society.   Turning from their own schemes of advancement and usefulness, their own enthusiasm and

industry, to the pestering jealousies, littlenesses, sloths of
the multitude, the spectacle fills them with pity and scorn.
Accordingly a marked characteristic of most of the great
men who have left an impression of themselves in literature, or whose spiritual portraits have been truly drawn
for us, is the superb feeling they have had of their own
superiority to the crowd.   Dante said,

> To their babblings leave
> The crowd : be as a tower that, firmly set,
> Shakes not its top for any blast that blows.

Michael Angelo said, "Ill hath he chosen his part who
seeks to please the worthless world."   Milton calls the
early history of Britain "a mere battle of kites and crows."
Carlyle entitles the population of Great Britain "twenty-
seven millions, mostly fools"; characterizes the Americans,
from the few specimens who visited him, as "eighteen
million bores"; and generally speaks of all great masses
of men in a tone of supreme contempt.   Walter Savage
Landor uses a host of similar expressions, and adds the
original remark, — which well shows how deeply the soreness had penetrated his noble mind, — that "there is now
no great man in existence."   Even the kindly Emerson
illustrates the temptation of the great to scorn the commonalty, when he speaks of "enormous populations, like
moving cheese, — the more, the worse"; "the guano-
races of mankind"; "the worst of charity is, that the
lives you are asked to preserve are not worth preserv-
ing"; "masses ! the calamity is the masses ; I do not
wish any shovel-handed, narrow-brained, gin-drinking mass
at all."   The influence of such phrases is unhappy.   They
betray in their use an absence of that sympathy which
goes downward for the purpose of lifting upward.   They
betray that proud aspiration which uses the thought of inferiors as a footboard wherefrom to bound into an aggravation of its own superiority.   How much better the
precept of Paul, "Honor all men !"   How much diviner
the sentiment of Channing, "I recognize God even in the
lowest man !"   An American lecturer is reported to have
said, "Every Irishman who lands on our coast is a

wheelbarrow-load of guano for a Western prairie." The absence of sympathy in such an utterance is shocking. The
body of the daintiest philosopher is certainly fashioned
of the same substance, and destined to the same end, as
that of his least advanced brother. The cultivated thinker,
from his height of scholarship and refinement, has no
right to express any sentiment towards his inferior except
sentiments of compassion and magnanimous desire to elevate him. To fling down a bolt of scorn to beat him yet
lower is to show not the spirit of a good man. Contempt
for the plebeian majority, whom Thiers calls "the vile
multitude," is too easy and common in aristocratic minds,
and too pernicious, to need to be expressed in the writings of men of genius.

When men of genius have been poor and unappreciated in their time, suspected and hunted by their contemporaries, this feeling has often easily kindled into
flaming dislike or curdled into acrid disgust. Then the
pride which underlies the genius of such persons leads
them to seek by every means to emphasize their unlikeness from other people, — to intensify their abhorrence
of being confounded with the throng they consider so
despicable. Whatever exhibits their unlikeness they
prize as also showing their superiority. They affect to
despise what others admire, and to admire what others
despise. Their misanthropy is at bottom the resentment of wounded pride joined with injured affection.
If it were pride alone, they would stay indifferently
among men ; but the hurt affection makes them flee
into solitude to hide their anguish and thicken their
armor. Vanity is the vice of the social ; but pride is
the vice of the recluse, and is by much the less amiable
of the two. In the man of vanity the idea of self ex-
pands or contracts according to its fancied dimensions
in the opinion of others. In the man of pride the idea
of self grows from its own centre, and maintains itself
independently of the opinion of others. Vanity acts
piecemeal, like fancy ; pride, in the mass, like imagination. That is acute and fickle ; this, chronic and weighty.
The wounds of mortified vanity are easily healed ; you

have only to reflect its own estimate of itself, and it is soothed and pleased.   But the wounds of offended pride are almost incurable.   The reflection of it in any less glory than it is accustomed to envelop itself in, it resents with lasting anger as an insulting and deadly wrong.   Against the chilling and killing effects of such a demeaning estimate it seeks to protect itself by all possible arts.   Unhappily the most easy and the most effective of these resources is to aggrandize itself in a palace of pride reared on contempt for others.   The man who despises all his race as selfish and sordid must be excessively depraved or excessively proud; for he either reflects himself over them or contrasts himself with them.   It is really curious how instinctively the lonely seek to solace themselves for their unlikeness to the crowd, and for their sorrowful isolation, with considerations that minister to their pride.   Pope adjures some god quickly to bear him to solitude, —

> Where contemplation prunes her ruffled wings,
> And the freed soul *looks down to pity kings!*

They are fond of dwelling on such aphorisms as "the best is always in a minority of one," — each latently or patently feeling himself to be that one.   Thomas Taylor, the fanatically solitary platonist, who by a great mistake was born in the Christian era, bitterly denounces "the attempt to educate the vulgar," as calculated "to disorganize society by making them discontented with the servile situations for which they are meant."   Even unhurting Coleridge complacently speaks of "the *plebification* of knowledge," and says, "I could write a treatise in praise of the moral elevation of Rabelais that would make the Church stare and the Conventicle groan."

Men of the greatest powers of thought and feeling instinctively insist on reality and justice, and are distressed at every violation of these by hypocrisy and pretence.   They must have men and things judged by the intrinsic standards of truth and right.   But they soon learn that in ordinary society men and things are not esteemed for what they are, but for what they seem, or are reputed.  Appearance and interest are constantly put above desert.  The current test is not, How powerfully can a man think ? how purely feel ? how nobly act ? not, What is he in himself ? but, What is thought of him ? what is his position ? how much space does he occupy in the public eye ?  They are pained at discovering the worthlessness of the thin and cold regard which is all that most men at the best will give them.  They are grieved to find how mistaken they were in attributing to the crowd the same reverential and glowing affection which they themselves feel towards the illustrious benefactors of the world ; for the vulgar take delight in seeing that the feet of great men are as low as their own, rather than in seeing that their heads are higher.  They are shocked to perceive the common insincerity of professions of attachment in social circles, the private hate and indifference often really existing between those who in public appear to be friends.  They are disgusted with the obsequiousness and servility of the world before success : —

> This common body,
> Like a vagabond flag upon the stream,
> Goes to and back, lackeying the varying tide.

No wonder the chivalrous Chateaubriand was sickened by the fulsome time-servingness of the successive French gazettes, when Bonaparte had sailed from Elba : "The monster has escaped ! "  " The army has declared for Napoleon ! "  " The Emperor is within three hours of Paris ! "  The indignant revulsion of a noble nature from this cringing and fawning is seen in the anecdote of the high-souled Persian poet, Saadi.  When a friend had been raised to office, and his acquaintances flocked to felicitate him, Saadi stayed away, saying, " I shall go to see him when his office expires ; sure then to go alone."

The delicacy of sensibility and taste belonging to genius, the range through which its powers are able to reach and draw inspiring motives, the tenacity of its sympathies and purposes, and the penetrating earnestness of its demand for conduct of steady sincerity in accordance with the facts, — all four of these qualities set its pos-

H

sessors in distinction, not to say opposition, to the crowd,
who are, in comparison, insensible in their nerves, unclean in their habits, low in their admirations and desires,
frivolous and fickle in their attachments, conventional in
their judgments, and servile or insolent in their manners

When we understand the nobleness of Tiberius Grac-
chus it makes our hearts bleed to see him left to be mur-
dered by the selfish mob in the Senate-house, a martyr
for the oppressed and poor among his countrymen. Ju-
venal, in one of his fearful satires, inveighs against the
Roman populace for the shallow ferocity with which they
turned against the favorite Sejanus the moment his imperial master darkened on him : —

> This is e'er the base practice pursued by the throng ;
> Men who ask nothing, heed nothing, seek right nor wrong,
> But impulsively trample the man who is down ;
> 'T is a fool who would value their smile or their frown.

Well might Petrarch sigh over the sign made by his
favorite hero, Scipio Africanus, when, in exile at Liternum,
he had his tomb built with the words *Ingrata Patria* inscribed on its front.   The illustrious patriots of Holland,
the noble brothers John and Cornelius de Witt, after all
their services, dragged out and massacred in the streets,
and their corpses, in a state of horrible mutilation, suspended to the gallows, are a cruel example of the blind
brutality of the populace.   A thousand powerful expressions of revulsion from the reeking vulgarities of the mob,
from their cruel injustice, from their unprincipled fawning, are familiar to us from the mouths or the pens of
lofty and lonely men, who have now sorrowfully, now
bitterly, denounced the multitude as a  "many-headed
monster," "sweaty citizens," "base and greasy crowd,"
whose opinion and love are "no surer than is the coal
of fire upon the ice, or hailstone in the sun."

> Who deserves greatness
> Deserves your hate ; and your affections are
> A sick man's appetite, who desires most that
> Which would increase his evil.   He that depends
> Upon your favors swims with fins of lead,
> And hews down oaks with rushes.   Hang ye ! **trust ye ?**
> With every minute you do change a mind,
> And call him noble that was now your hate,
> Him vile that was your garland.

Vulgar natures almost invariably affect to look down on their superiors. They are conscious of the difference, and interpret the difference as their own advantage. The mechanical holders by tradition consider the prophetic masters by original insight, heretics and inferiors. Dante was accused of impiety, because he broke the basin in the Florentine Baptistery to save a child who had fallen into it and was drowning; the idolaters of the letter thought Lessing an infidel, and the mummers of a formula called Spinoza an atheist.

The foolish can amuse and indirectly instruct the wise: the wise are irksome to the foolish. For these have only their own faculties and attainments wherewith to measure the faculties and attainments of those; that which transcends or baffles their comprehension rebukes and irritates them, hurts their self-love, and they take vengeance on it by regarding it with hate and affected contempt. It is a fearful wrong, a sort of blasphemy of the Holy Ghost; but it is natural, almost inevitable. The ducks believe the swan that chances among them, an uglier duck. Genius, ridiculed and despised by unfeeling mediocrity, if too modest, sinks in shame and agony; if strong enough, supports itself by a rallying indignation. Only the rarest saintliness can enable great men to see themselves thus outrageously misvalued and scorned by little men, and yet preserve a sweet serenity, rising divinely superior to any heed of the injury.

We have seen, then, that while the crowd are unrefined, superficial, ignoble, and unstable, the man of sensitive genius is pure, profound, magnanimous, and wishes to become ever more divinely rooted and constant in character. As he mixes with the crowd he is exposed to a tendency to become like them, — to be frittered and dragged down to their likeness and level. He knows that no one ever attained to supreme excellence in any art without, in the phrase of Pope, "an inveterate resolution against the stream of mankind." Therefore he has an instinct to shrink from them, to guard himself from all deteriorating sympathies with them. Thus to guard himself from their degrading influence is his duty; but in doing it, it is both his highest duty and interest not to hate or despise them. Their nature, as exemplified in its best specimens, is mysteriously beautiful and great, a divine manifestation; and his own destiny is ideally bound up with theirs beyond the possibility of a real separation. Therefore, in order to honor their Parent, God, in order to bless and help them, and in order to secure his own peace and health of mind, the exceptional man of genius should cherish the utmost respect and kindness for the plebeian mass of his race, a patient tolerance born of magnanimity, and a placid love born of pity, but no egotistic hate, no rancorous scorn, no pharisaic seclusion.

> To fly from, need not be to hate, mankind :
> All are not fit with them to stir and toil ;
> Nor is it discontent to keep the mind
> Deep in its fountain, lest it overboil
> In the hot throng, where we become the spoil
> Of our infection.

Nothing can be more unchristian, to a thoughtful mind more inexcusable, than the swollen haughtiness of Coriolanus towards the crowd of his fellow citizens. When they banish him he exclaims,

> You common cry of curs ! whose breath I hate
> As reek o' the rotten fens, *whose loves I prize*
> *As the dead carcasses of unburied men*
> *That do corrupt my air, I banish you.*

The disposition of sympathy indicated in these words is awful. The greatest men, who have sweet and gentle spirits, will shrink with horror from such a strain of turgid insolence. There is a true and beautiful tone in the sentiment of George MacDonald : "Despise a man, and you become of the kind you would make him : love him, and you lift him into yours." Yet the chasm that yawns between the extremes of human character and attainment, or even between the best specimens and average specimens, cannot be denied. Mrs. Hemans once said, "Life has few companions for the delicate-minded." Old Elwes, on hearing an unfortunate man ask for sympathy in his calamities, turned on him gruffly, — "What do you want sympathy for? I never want anybody to sympathize with *me!*" The distance between the poetess and the miser is as great as the difference between a bird of paradise and a grizzly bear.

A frequent motive for retreating from miscellaneous society is to escape from the pain of conflict or partnership with the mob of backbiters and quarrellers. Sir Walter Scott says, " It requires no depth of philosophic reflection to perceive that the petty warfare of Pope with the dunces of his period could not have been carried on without his suffering the most acute torture, such as a man must endure from mosquitos, by whose stings he suffers agony although he can crush them in his grasp by myriads." It is the sorest trial of the man of a great and loving spirit to be forced into contention with envious detractors. It ought not to surprise us that he sometimes impatiently exclaims, as the magnanimous and unworldly Shelley exclaimed, " How can I run the gauntlet further through this hellish society of men "; that he sometimes cries, as the tender Tennyson haughtily cries,

Be mine a philosopher's life in the quiet woodland ways,
Where if I cannot be gay let a passionless peace be my lot,
Far off from the clamor of liars belied in the hubbub of lies ;
From the long-necked geese of the world that are ever hissing dis-
    praise
Because their natures are little, and, whether he heed it or not,
Where each man walks with his head in a cloud of poisonous flies.

What a man of large mind and liberal temper, who is impartial in thought and sentiment towards all, has to suffer from the pestilential littleness of partisans and bigots is forcibly described in the following extract from a letter by Robertson of Brighton, in which he refers to a criticism on himself.

" I could not help smiling good-humoredly at the writer's utter misconception of my aims, views, and position. If he think that what he calls a philosophic height above contending parties is a position which any man can select for his own comfort and retirement, he miscalculates greatly. If he suppose that the desire to discern the ' soul of goodness in things evil,' to recognize the truth which lies at the root of error, and to assimilate the good in all sects and all men rather than magnify the evil, is a plan which will conciliate the regard of all, secure a man's own peace, 'and of course bring with it great popularity with the multitude,' I can earnestly assure the writer that, whenever he will try the experiment, he will find out his mistake. He will, perhaps, then see a new light reflected upon the expression, 'when I speak of peace, they make them ready for the battle.' He will find himself, to his painful surprise, charged on the one side, for his earnestness, with heresy, and on the other, for his charity, with latitudinarianism. His desire to exalt the spirit will be construed into irreverence for the letter, his setting light by maxims into a want of zeal for principles, his distinction between rules and spirit into lawlessness. He will find his attempt to love men, and his yearnings for their sympathy, met by suspicions of his motives and malignant slanders upon his life ; his passionate desire to reach ideas instead of words, and get to the root of what men mean, he will find treated, even by those who think that they are candid, as the gratification of a literary taste and the affectation of a philosophic height above the strife of human existence. I would not recommend him to try that 'philosophic height,' which he thinks so self-indulgent, unless he has the hardihood to face the keenest winds that blow over all lonely places, whether lonely heights or lonely flats. If he can steel his heart against distrust and suspicion, if he can dare to be pronounced dangerous by the ignorant, hinted at by his brethren in public and warned against in private ; if he can resolve to be struck on every side and not strike again, giving all quarter and asking none ; if he can struggle in the dark, with the prayer for light of Ajax on his lips, in silence and alone, — then let him adopt the line which seems so easy, and be fair and generous and chivalrous to all But if he expects from it, 'of course considerable self-applause and great popularity with the multitude,' I can tell him they are not the rewards of *that* path. Rather let him be content to remain a partisan, and call himself by some name. Then he will be abused by many, but his party will defend him."

It is the mark of a generous soul ever to appreciate at their highest value the merits of his contemporaries, and not confine his admiration to those departed worthies with reference to whom rivalry and envy are impossible. The shadows of the old patriarchs, prophets, and law-givers — Abraham, Moses, Zoroaster, Menu — lie so vast and long across the generations, not because they were so much greater than we ; but because they lived when the sun of history was low in the horizon. And with regard to them we have every motive to employ the aggran-dizing offices of the imagination. The saintly lover of his fellow men will delight to dwell admiringly on the gifts and graces of the highest spirits he knows, because he is free from the littlenesses of vanity and jealousy. In the peace of his spirit, remote from the madding crowd of aspirants who rush through society contending for its notice and prizes, he feels like one who, sheltered in some deep forest, listens from his retreat to the tre-mendous murmur that swells and rolls along the tops of the trees. He avoids every thought calculated to inflame vulgar ambition, cherishes every thought adapted to soften, deepen, and strengthen the heart. Burns pausing over the daisy he had ploughed up, is, for him, a finer picture than Alexander conquering Porus. When he enters into communion with the mighty spirits of the past, he returns to meet his neighbors, not with flattered pride, but in-spired with charity ; he does not transfer their greatness to himself, but to his race. He would reject as a false note the sentiment expressed by Count Oxenstiern in his pleasant Essay on Solitude : — "Occupied with the great minds of antiquity, we are no longer annoyed by contem-poraneous fools."

There has scarcely appeared in history a great genius crowned with brilliant triumphs, who has not been pur-sued by enemies writhing with envy and hate, envenomed at being thrown into obscurity by his superiority. Virgil had his Cornificius, Fannius, Bavius and Mævius, the two latter of whom directed their spleen also against Horace. The wretched cabals and quarrels of literary men are notorious and innumerable. Bach, Handel, Mo-

zart, Beethoven, Weber, Mendelssohn, — all were inexpressibly disgusted and annoyed by the hatreds, plots,
and persecutions of rivals and inferiors.   The same has
been the case with the most celebrated actors : poorer
actors, fancying themselves robbed by these of the applause and profit they deserved, have sought to detract
from their talents and blast their laurels.   So, too, with
the great statesmen, — they have been forced to tread
their proud course amidst the sneers and slanders of
opponents.   In every age envy has dogged the noblest
forms and calumny sat on the sacredest graves.   Nothing can be more foreign to the nature of great genius
than such conduct, — nothing more painful than the experience of it from others.   Nothing can be more sure
to awaken in a rich, sensitive breast a melancholy feeling
of loneliness in the crowd and estrangement from the
world.   True genius, ever incapable of this base bearing
towards its brothers, delights to pay them its homage, —
finds its choicest luxury in giving them encouragement
and love.

When Colbert died the pension of Corneille was
stopped by Louis the Fourteenth.   Boileau hastened to
the king, represented that Corneille was old, poor, sick,
and dying, and offered to resign his own pension in his
favor.   The petty Salieri indeed hated Mozart, but the
lofty Haydn adored him.   Haydn says : — " The history
of great genius is melancholy, and offers posterity but
slight encouragement to exertion ; which is the reason,
alas ! that many promising spirits are disheartened.   I
feel indignant that this peerless Mozart is not yet engaged at some imperial court.   Forgive me if I stray
from the subject, but I love the man too much."

If, however, most men of great genius of whom we know
have been unhappy, it has not been the intrinsic penalty
of their genius, but a consequence of the exasperating
meanness of competitors and the indifference of the unappreciative multitude.   The panacea for their wretchedness is to seek fulfilment and excellence instead of
fame and applause.   It is not aspiration but ambition
that is the mother of misery in man.   Aspiration is a

pure upward desire for excellence, without side-referen-
ces; ambition is an inflamed desire to surpass others.
Great intellect, imagination, and heart, are conditions of
noble joy and content when free from that extravagant
desire for public approbation which so often accompanies
them. The spectacle of poor, starved, heart-broken Chatterton, dead in the London garret, — his legs hanging
over the side of the bed, bits of arsenic in his teeth, —
is not the proper tragedy of genius, but of a morbid hungering and thirsting for social recognition and honor.
Let genius raise itself above the wild chase for human
praise, content itself with the fruition of its powers and
with the serving of men by the fruits of its powers, and
it will be as much happier than mediocrity as it is more
gifted. In one of his creative moods, Carl Maria von
Weber said : — " I cannot understand my happiness. I
seem to wander in a dream where everything is flooded
by a rosy light; and I must touch myself to be assured
that it is true." "I seem to myself to enjoy more with
my eyes in one glance than others do with all their
limbs in all their lives." That is a true trait of healthy
genius. But how terribly it is often perverted into sor
row and agony by an excessive regard for admiration and
fame !

That keen-sighted woman of the world, Miss Mary
Berry, once wrote to a friend : " I am much more disgusted in society by the little impression made by real
merit than by the so often lamented tolerance of vice."
To appreciate general superiority of intellect and excellence of character requires some nobleness of endowments and of aims in the observers, and these are rare
amidst the self-indulging fickleness and frivolity of fashionable circles. Therefore high-minded and original
characters, who cannot stoop to use dishonorable arts for
self-advancement, are often neglected in favor of those
pushing mediocrities who make their way by being always
in the way, so that it is " less trouble to notice them than
to avoid them." Conformity, obsequiousness, especially
inoffensiveness, are more likely than power and desert
to get conventional honors.

6

The grandest writer of late ages,
Who wrapt up Rome in golden pages,
Whom scarcely Livius equalled, Gibbon,
Died without star or cross or ribbon.

Jonathan Edwards never had the degree of doctor of divinity or of doctor of laws conferred on him, while they were showered on scores of his commonplace contemporaries. A clear perception of these facts should comfort in their disappointment the deserving who are wrongfully deprived of the outer prizes of their deserts, and should make them content to forego what the unprincipled win from the conventional. Shall a man really of supreme mark and worth fret at being kept from the titular recognitions which are usually given to factitious claims of name and position, and usually withheld from men of the greatest merit if they be unpopular or obscure?

The character and experience of men depend on the inmost modes of thought and feeling they cherish, their favorite objects and kinds of contemplation, rather than on the sociality or solitariness of their outward habits. Man is a meditating atom, whose happiness or misery lies in his meditations. The cynic, in his isolation of contemptuous hate, was cold, bitter, repulsive, and wretched. The stoic was capable of enthusiasm; could withdraw into a glowing inner life. The man who separates himself from mankind to nourish dislike or contempt for them, has in him a morbid element which must make woe. True content, a life of divine delight, cannot be attained through a sense of superiority secured by thrusting others down; but only through one secured by lifting ourselves up, by communing with the great principles of morality, contemplating the conditions of universal good, laying hold of the will of God. Whoso would climb over a staircase of subjected men into a lonely happiness, will find it misery when he arrives. To be really happy one must love and wish to elevate men, not despise and wish to rule them. There is nothing in which the blindness and deceit of self-love is more deeply revealed than in the supposition with which misanthropic recluses frequently flatter themselves, of their complete detachment

from other men, their lofty freedom.  Spatial separation
is not spiritual independence.  Of all men the man-hater
is the one who is fastened to his fellow-men by the closest
and the most degrading bond.  Misanthropy, as a domi
nant characteristic, if thoroughly tracked and analyzed,
will be found almost always to be the revenge we take
on mankind for fancied wrongs it has inflicted on us,
especially for its failure to appreciate us and admire us
according to our fancied deserts.  The powerful and
savagely alienated Arthur Schopenhauer, who said that,
in order to despise men as they deserved, it was necessary not to hate them, was embittered, almost infuriated,
by disappointment in not obtaining the notice he thought
he merited.  He came daily from his sullen retreat to
dine at a great public table where he could display his
extraordinary conversational powers.  He eagerly gathered every scrap of praise that fell from the press, and
fed on it with desperate hunger.  He sat in his hotel at
Frankfort, in this age of newspapers and telegraphs, a
sublimer Diogenes, the whole earth his tub.  An apathetic carelessness for men shows that we really despise
them, but an angry and restless resentment towards them
betrays how great a place they occupy in our hearts.
Diogenes and Alcibiades were equally dependent on public attention ; the one to feel the enjoyment of his pride
and scorn intensified by the reaction of hate and admiration he called forth ; the other to feel the similar fruition
of his vanity and sympathy.  Stylites made his column a
theatre ; Aurelius made his throne a hermitage.  The
greatest egotists are the most fond both of retirement and
publicity.  There they lave their wounds with the ano-
dyne of self-love ; here they display their claims to admiration.  The truly great and healthy man is not dependent on either, but draws blessings out of both, — resolve,
inspiration, consecration, sanity.  In both he pleases himself by improving every possibility of indulging in sentiments of respect and affection towards his race.
  The great danger of the courters of solitude is the
vice of pampering a conviction and feeling of their own
worth by dwelling on the ignobleness of other men

They are tempted to make the meanness and wretched-ness of the world foils to set off their own exceptional magnanimity. They need especially to guard themselves against this fallacy by laying bare to their own eyes the occult operations of pride and vanity. An efficacious antidote for their disease is a clear perception of the humbling truth of the case, of the ignoble cause of the disease. For it is unquestionably true that the man who despises the world, and loathes mankind, is usually one who cannot enjoy the boons of the world, or has been disappointed of obtaining from his fellows the love and honor he coveted. He then strives to console himself for the prizes he cannot pluck, by industriously cultivat-ing the idea of their contemptibleness. Rousseau de-manded more from men than they could give him. His brain and heart were pitched too high; with the fine in-tensity of their tones the cold and coarse souls of com-mon men made painful discords. Instead of wisely seeing the truth, and nobly renouncing his excessive exactions, he turned against the world and labored with misan-thropic materials to build up his overweening self-love. Of course he was not conscious of this himself. It was a disease, and, fleeing from all antidotes, it fed in solitude; whence he looked abroad and fancied that he saw his contemporaries leagued in a great plot against him. Zim-mermann and Byron, two irascible and lonely spirits most fond of retirement, noticed the danger of a chill shrivel-ling of sympathy in a too isolated life. The former says, "Solitude must render the heart callous." The latter says : —

In solitude
Small power the nipped affections have to grow.

Sir Thomas Browne well says : "He who discommendeth others obliquely commendeth himself." What is the in-evitable inference as to that man's opinion of himself who withdraws from other men because they are unfit for him to live with? Even the gentle Shelley says to his friend Hunt, "What motives have I to write? I *had* motives, and I thank the God of my heart they were totally different from those of the other apes of humanity who make mouths in the glass of time." The pride of Byron burned with a darker fire. His self-exaltation mounts in new strength, Antæus-like, with every reaction from his scorn of others. He describes himself as —

> Not to desperation driven
> Because not altogether of such clay
> As rots into the souls of those whom I survey.

In describing the ruined castles above the banks of the Rhine, he says : —

> And there they stand as stands a lofty mind,
> Worn, but unstooping to the baser crowd.

Again, after speaking of his passionate love of natuie, he adds : —

> Should I not stem
> A tide of suffering rather than forego
> Such feelings for the hard and worldly phlegm
> Of those whose eyes are only turned below,
> Gazing upon the ground, with thoughts which dare not glow ?

Rebelling against all rule or influence from others, insisting that they should passively accept his influence and rule, and by reflecting confirm his estimate of himself, when thwarted in these chronic and deep-sunk desires he grew desolately proud and "forlornly brave," determined to keep himself at his vantage-height. He felt

> Himself the most unfit
> Of men to herd with man, with whom he held
> Little in common.

He once refused to have one of his plays put on the stage, on the ground that its success would give him no pleasure, while its failure would give him great pain. In other words, he would not own that the approval of the public could flatter him, and the tacit superiority assumed by critics who should condemn, was insufferable to him. In such instances, clearly, solitude is not courted for the purpose of noble culture, growth in true wisdom and virtue ; it is sought for self-protection from stings and burdens, for self-fondling, and self-aggrandizement. And these ends are pursued by the recluse at the expense of his species; the lower he can sink them in his esteem the higher he rises in it himself. How much better is Young's maxim: "No man can think too lowly of himself, or too highly of his nature." How much nobler is Jowett's sentiment: "Better not have been than to live in doubt and alienation from mankind." Thousands have been impelled to solitude by resentment, — as the hermit confessed to Imlac *he* was, — where one has been led to it by devotion. The true improvement of our lonely hours is not to cherish feelings of superiority to our neighbors, but to make us really superior by a greater advancement in the knowledge of truth, the practice of virtue, communion with the grandeurs of nature, and absorption in the mysteries of God. He who is continually exercising scorn towards the pleasures of society and the prizes of the world, is one who has failed in the experiment of life and been soured by his failure. The truly successful man appreciates these goods at their genuine value, — sees that in their place they have sweetness and worth, but knows that there are other prizes of infinitely higher rank, and is so content with his possession and pursuit of these latter as to have no inclination to complain of the deceitfulness and vileness of the former. To dwell alone is an evil when we use our solitude to cherish an odious idea of our race, and a disgust for the natural attractions of life. It should be improved, not negatively for dislike and alienation, but positively to cultivate a more earnest love for higher mental pursuits, choicer spiritual fruitions, than the average community about us are wonted to. Scorn for man, disgust for the world, is no sign of strength, loftiness, or victory, but rather a sign of weakness, defeat, and misery. "The great error of Napoleon was a continued obtrusion on mankind of his want of all community of feeling for or with them." He deceived himself in fancying his ruling feelings unlike in kind to those of the bulk of men; they were the same in sort, only superior in scale and tenacity, and in the greater stage on which they were displayed. He showed what a morbid author unjustly characterizes

as "that just habitual scorn which could contemn men
and their thoughts," because, himself vulgarly selfish and
vain, from his high position he saw the unprincipled
selfishness and vanities of other men unmasked and
writhing in virulent struggles. Let him whose standard
is the most august, who,e ideal highest, who has the least
number to sympathize with him, endeavor most strenuously to develop a genial and serene spirit of good-will
for his kind. Let him use imagination, faith, every divine artifice, to dignify and adorn man, ever looking at
him through fair objects and great truths, and communing
with him by their help ; for thus alone,

> Is founded a sure safeguard and defence
> Against the weight of meanness, selfish cares,
> Coarse manners, vulgar passions, that beat in
> On all sides from the ordinary world
> In which we traffic.

The true protection from the deteriorating tendencies
of intercourse with persons of empty minds, shallow
hearts, and idle lives, is in noble and strenuous occupation, and in friendship with the truly good and great.
Priestley says, "that right bent and firmness of mind
which the world would warp and relax, are to be kept up
by choice company and fellowship." Lamb quotes this
in a letter to Coleridge, and adds : — " I love to write to
you. I take a pride in it. It makes me think less
meanly of myself. It makes me think myself not totally
disconnected from the better portion of mankind. I
know I am too dissatisfied with the beings around me. I
know I am noways better in practice than my neighbors,
but I have a taste for religion, an occasional earnest aspiration after perfection, which they have not. We gain
nothing by being with such as ourselves. We encourage
one ancther in mediocrity. I am always longing to be
with men more excellent than myself."

Zimmermann says : — " Our whole existence is occupied with others ; one half of it we spend in loving them,
the other half in slandering them." There is no other
problem in our life so difficult to solve as the problem of
our relation with our fellow men    How much attention

and feeling shall we devote to them? How much shall
we try to lead an independent life in self and nature?
Such a complicated mass of considerations enter into the
subject that it would be too hard a task to answer the
question in detail. But the best conclusions from a great
deal of earnest pondering on it are contained in the following general precepts. First, let him who seeks to be
noble in character and blessed in experience, raise the
interest, respect, and love he gives his fellows, to the
maximum. Secondly, let him reduce his hate, scorn, indifference towards them, to the minimum. Thirdly, let
him strive to take the utmost possible pleasure in their
virtues and joys, and in their esteem and kindness for
him. Fourthly, let him strive to feel the least possible
annoyance from their neglect of him, injustice towards
him, or insults and persecutions of him. Fifthly, let him
endeavor to do practical good to them, and keep constantly in mind, as the sovereign antidotes to misanthropy, the two great maxims of Platonic ethics : No
man is willingly bad ; Virtue may be taught. And finally,
let him labor above all to possess the greatest possible
resources of dignity and happiness in himself, nature, and
God, unexposed to the favor or frowns of capricious man ;
for, after all, the essence of our life, and by far the greater
part of its separate experiences, are alone, — incommuni-
cably alone. Even the strong, wise, healthy, many-sided,
and fortunately-situated Goethe makes this personal confession : — " We may grow up under the protection of
parents and relatives, — we may lean on brothers and
sisters and friends, — be supported by acquaintances, —
be blessed by beloved persons ; yet, in the end, every
man is always flung back on himself, and it seems as
though even God was unable to respond to our reverence, trust, and love exactly at the moment of our need.
While quite young I often experienced that in the most
critical passages the cry is, ' physician, heal thyself';
and how many times I was forced to sigh in anguish, ' I
tread the wine-press alone !' When I looked around to
make myself independent, I found the surest basis to be
my creative talent. The old fable became living in me

of Prometheus, who, separated from men and gods, peopled a world from his own workshop."

It is an abuse of solitude to carry into it the passions,
cares, frivolities, and hypocrisies of society.   Let it be
pure and still ; a cool grotto where the realities of nature
and God may woo the soul away from the hot fen of
emulation and vice.

But unfortunately it is not always to be kept thus clean
and silent.   Though dedicated to sacred presences alone,
the annoyances and temptations that infest artificial
throngs will intrude.   The influences of degradations and
crimes will come.   Every soul that approaches brings its
own qualities and experiences, as well as its own capacities and aspirations with it.   One especial seduction the
solitary should beware of, the tendency to a luxurious
melancholy, self-fondling sorrows.   "Few reach middle
age without receiving wounds which never heal.   We
hide these wounds when we can, or we forget them in
business, perhaps in dissipation ; but they remain with
us still, and in moments of solitude and depression open
to pain us."   It is one of the subtlest and most destructive habits in which those fond of loneliness are tempted
to indulge, to reopen their old wounds in secret, and feel
again their bitter-sweet pains.   We should, on the contrary, dedicate our seasons of still retreat to the cultivation of health, strength, and trusting joy, by a fresh communion there with those principles of truth, those objects
of beauty, those sources of affection, which most elevate
and calm the soul.

Furthermore, a predominant solitariness of mood and
habit has evil exposures in a degree peculiar to itself.
The man of overmuch retirement and self-communion is
especially beset — as we have already partially seen —
by egotism, superstition, morbid views, disregard of the
real interests of this life, and sometimes by an asserting
rebound of the sensual nature in abnormal power.   Wise
and saintly Madame Swetchine, when her dear friend,
the eloquent Lacordaire, had announced his intention
of forsaking the world and burying himself for a long
period in the seclusion of a convent, adjured him not to

do it, warning him that the perfection of true self-detach-ment would be more impossible of achievement there than anywhere else. " Solitude," she writes, " may be good for you, useful, perhaps necessary,— solitude with its *cortége* of calmness, liberty, self-possession ; but not that *isolation* which in removing all barriers would also remove all supports ; would force you to lose that habit of contact with men which is so precious for those destined to live with them and for them ; and would deprive your imagi-nation both of the admonitions of reason and those of sympathy. In all conditions, in all regions, the divine word, ' It is not good for man to be alone,' finds its application." This testimony is the weightier from the profound familiarity of its writer with the opposite side of the truth. For she has said elsewhere, " There are times in life when we have a true thirst for solitude. While I was yet very young my instinct wrote, Solitude is like gold ; the more of it one has, the more one desires."

The lonely man, if full, is quite likely to be full of himself, and to look on others with scorn, or scornfully overlook them. Wrapt in his own idiosyncrasies, out of connec tion with the ordinary characters, views, and plans of men, the aberrations of his individuality uncorrected by their averages, by the common sense of the public, he is exposed to manifold conceits and delusions, of which he often becomes the helpless victim. In frequent inter-course with others our foibles are kept in check by theirs, the tendency of self-esteem to a crotchety exaggeration of insignificant details is neutralized. But the recluse — granting him life and spirit enough — is apt to indulge in hyperbolical estimates of trifles, deeming them intrinsi-cally great because of their factitious importance in rela-tion to himself.

> And now behold his lofty soul,
> That whilom flew from pole to pole,
> Settle on some elaborate flower,
> And, like a bee, the sweets devour ;
> Now, in a lily's cup enshrined,
> Forego the commerce of mankind.

Deprived of really great concerns, this is nature's resource for making him happy. If he sees a humming bird, an animated flame of colors, dart on a vine, tear open the belly of a grape, quench his thirst, and fly away, his bill stained with the blood, — it is the chief event of the week. If he wanders in the forest, far from all civic racket, he pampers his self-importance by feeling that he is " Cæsar of his leafy Rome." To unduly magnify and enjoy the common little things near at hand is the felicitous illusion of superior minds. The much more frequent habit of unduly magnifying and pining for distant and extraordinary things is the wretched illusion of inferior minds. The greatest and wisest minds of all, free from both illusions, see everything as it is, value it at its true worth, and stand firmly poised and self-sufficing in their relations with the whole.

Peace of mind is the great prize of solitude; but it may be lost there as well as gained. It is not necessary for us to be with other persons in order to have inordinate desires, hatreds, envies, and a rebellious will, nurtured and inflamed. Wild crotchets, obstinacy, besotted errors and prejudices, cruelty, revolting sensuality, have too often been the attributes of men sheerly separated from the bosoms and ways of their fellow-creatures. There are anchorites who in sourness and savageness of spirit may match any specimen from the market or the stew.

The evil influences of conventual life have been discussed many times by extravagant partisans on both sides. Perhaps Zimmermann, who handles the subject largely, has held the balance as fairly as any one. The abandonment of the world for the recluse life of the various orders of monks and nuns has so much diminished in our day, — it is likely to be practised so much less still in the future, — that the portrayal of its evils is not nearly so important as formerly anywhere; in a Protestant country hardly necessary at all, and of all countries in the world perhaps the least needed in America. All that is requisite to be said in order to do justice to the subject which we are treating, may be stated within small compass.

First, it is a beautiful thing that there are such places as the Catholic convents and monasteries, where persons, unfitted to struggle with the world, may retreat from its cruel storms, and spend their lives in peace and devotion. There are those of exquisite sentiments, of a tremulous sensibility, whose feelings have been torn and outraged, whose worldly affections have been laid waste by some tragic experience. To face and buffet the cold throngs, to try to sustain themselves amidst interests and passions so alien to their desires, is a lacerating conflict. It is a blessed thing when such are enabled to turn away from a world to which their hearts are dead, and, retiring within the hallowed shelter of the cloister, there pass the residue of their days in pursuits wholly congenial to their cruci-fied affections and their immortal hopes. Altogether be-nign and beautiful to the broken-hearted martyrs of life, who long for it, is the solacing employment of religious seclusion in that divine haven, the monastic retreat in its best forms. When we think of the heartlessness of most worldlings, the fearful excitements, the jading exhaus-tions, the bitter pangs of deception and failure, known by the ambitious and sensitive worshippers of wealth, power, fashion, and pleasure, — we cannot wonder that the utter exemption from all such trials, and the noiseless repose offered in a religious retreat, should often exert a delicious spell on the worn and weary wanderers who approach it. It is easy to enter into the feelings of the poet, who left these words written on the wall in one of the cells of La Trappe : " Happy solitude, sole beatitude, how sweet is thy charm ! Of all the pleasures in the world, the pro-foundest and the only one that endures, I have found here ! "

A strange instance of abandonment of the world for a solitary life is given in the history of Henry Welby, the Hermit of Grub Street, who died in 1636, at the age of eighty-four. This example affords an eccentric illustration of one of those phases of human nature out of which the anchoretic life has sprung. When forty years old Welby was assailed in a moment of anger by a younger brother with a loaded pistol. It flashed in the pan. "Thinking of the danger he had escaped, he fell into many deep consid-erations, on the which he grounded an irrevocable resolu·

tion to live alone." He had wealth and position, and was of a social temper; but the shock he had undergone made him distrustful and meditative, not malignant nor wretched, and engendered in him a purpose of surprising tenacity. He had three chambers, one within another, prepared for his solitude : the first for his diet, the second for his lodging, the third for his study. While his food was set on the table by one of his servants, he retired into his sleeping-room ; and, while his bed was making, into his study ; and so on, until all was clear. "There he set up his rest, and, in forty-four years, never upon any occasion issued out of those chambers till he was borne thence upon men's shoulders. Neither, in all that time, did any human being — save, on some rare necessity, his ancient maid-servant — look upon his face." Supplied with the best new books in various languages, he devoted himself unto prayers and reading. He inquired out objects of charity and sent them relief. He would spy from his chamber, by a private prospect into the street, any sick, lame, or weak passing by, and send comforts and money to them. "His hair, by reason no barber came near him for the space of so many years, was so much overgrown at the time of his death, that he appeared rather like an eremite of the wilderness than an inhabitant of a city." The hermit crab clings not more obstinately to his rock in the sunless corner of some ocean-dell than this crabbed hermit clung to his seclusion in the roaring centre of London. Yet he hardly deserves to be called crabbed ; since, timidly distrusting men, not hating them, he seems to have kept his goodness alive by holy thoughts and kind acts.

Secondly ; there is another class, besides those finding in it a refuge from worldly agony and despair, to whom life in the conventual solitude may be wondrously sweet ; namely, those who have genius enough, exalted passion and ideality enough, to make the doctrines of their creed vivid realities, — the fellowship of saints and angels, the vision of heaven, the contact of God, ever present to them, — their rites open communications with the supersensual sphere. The few rich and ardent souls capable of this, find paradise in the routine of their ritual and the silence of their cells. The experience of such as these, recorded with pens of fire in the pages of a Theodoret, a Palladius, a Basil, a Francis, a Bernard, — compose an eloquence which may well bewitch and electrify tender and soaring souls. To such, — the world all abjured, the fellowship of mankind quite renounced, the tempest of sin and woe roaring faintly afar, — the still and lonely cloister, with its perpetual train of celestial recurrences, may minister health, wisdom, content, rapture. Happy themselves, the only evil is that the use of their powers is lost to the world.

Thirdly ; but when instead of the unhappy or the un-worldly, voluntarily seeking refuge in this heavenly harbor, we have the young and hopeful, overpersuaded or forced into this dismal banishment, with all their ungratified passions throbbing, — when, instead of the imaginative, ardently coveting an unbroken communion with the trans-cendent objects of their beliefs, we have the dry and tor-pid, reduced to a mechanical repetition of forms they are incapable of animating, — then the nun is a victim, the monk a slave, the monastery a prison, its solitude breed-ing diseases and miseries. It is hard to imagine any con- dition more unfavorable to true dignity and happiness than that of those hopelessly separated from the world by their vows and their jail, yet burningly attached to it by the passions that glow and gnaw beneath the placid sur-face of their ceremonial sanctity. In such examples the unemployed and dissatisfied forces of the soul turn in to prey on themselves ; and engender, in some, gross physi-cal vices ; in others, intense spiritual vices ; in all, either an irritable unhappiness or a deathly stagnation. The acrimonious gossip, jealous spite, fathomless pride and contempt, which may be fostered under such circum-stances, are fearfully illustrated by Robert Browning in his " Soliloquy of the Spanish Cloister," and by Isaac Taylor in his chapter on the " Fanaticism of the Sym-bol " ; both of whom, however, fasten on rare exceptions, and with a dark exaggeration.

But it is common with zealous recluses to substitute

foi religion superstition, with its idle solicitudes, morbid scruples, and despotic formalities. Ever since the system was founded some proportion of the recluses have unquestionably been unhappy, the incurable distress of their minds raining in a dreary murmur of sighs through the confessional grate. Yet the number of the unhappy inmates of convents is much less than Protestant writers would have us believe. The most become content at least with the second life of habit. Even those Carmelites who rise at four, sleep in their coffins upon straw, every morning dig a shovel of earth for their own interment, go to their devotions on their knees, never speak to those they see, nor are suffered to be seen by those to whom they speak, and taste food but twice a day, usually get so attached to their mode of life as not to be weaned from it. In the French Revolution, when the Convents were flung open, most of the nuns begged to be allowed to stay and die there. Judge Story, in his poem entitled "The Power of Solitude," has described the scene.

> Hark, from yon cloisters, wrapt in gloom profound,
> The solemn organ peals its midnight sound ;
> With holy reverence round their glimmering shrine
> Press the meek Nuns, and raise the prayer divine ;
> While, pure in thought, as sweet responses rise,
> Each grief subsides, each wild emotion dies.

Beckford saw, in a Carthusian convent in Portugal, a noble and interesting young man who had just taken the vows, and who seemed very sad. " I could not help observing, as the evening light fell on the arcades of the quadrangle, how many setting suns he was likely to behold wasting their gleams upon those walls, and what a wearisome succession of years he had in all probability devoted himself to consume within their precincts. The chill gust that blew from an arched hall where the fathers are interred, and whose pavement returned a hollow sound as we walked over it, struck my companion with horror." The same writer, however, was deeply affected with a different sentiment, when, at the close of a visit to the monks in the Grande Chartreuse, the good fathers accompanied him a hundred paces from their building, and, amidst the frightful scenery of the place, giving him their benedictions, laid their hands on their breasts, and assured him that if ever he became disgusted with the world, here was an asylum.

Those Hindu fakirs, who, withdrawing from society, sit in one posture, year after year, in silence, until they are paralyzed into immovable stocks, illustrate a fanatic abuse of solitude. This is the essential error and evil of the monastic system, especially in its logical result, as seen in such orders as that of La Trappe. The Trappist, on entering the convent, leaves his name behind him, with every other earthly clew, digs his own grave, speaks no more, except to the brethren he meets, the dismal words, " Remember death ! " — admits no news from the rejected world outside. The Jesuit, representative of one pole of the monastic spirit, — no true hermit, but the votary of a grim and dread ambition, — flees from the world only to study it in distant detachment, in order to return into it as a conqueror and ruler of it. The Trappist, representative of the other pole, a preternatural solitary, detesting beauty, fearing pleasure, makes an ideal of despair, prefers ignorance to knowledge, indifference to conquest, and buries himself in an anticipated tomb. He puts death in the place of life. This is the ascetic spirit ultimated. But surely God has placed us here for the purpose of living, not of dying. Dying is not the whole of our earthly destiny, but its act of completion and transition. The monastic regimen was born in death, flourishes in death, culminates in death, tests the value of everything by the standard of death. Whereas the true standard, while we live, by which to test the value of living interests, is the standard of life. True religion is the vivification of the soul, as extended into the unknown ; the religion of La Trappe is the mortification of the soul as spread over the known.

The origin of the Carthusian order was, according to the record in the " Lives of the Saints," on this miraculous wise. A dead man, who had enjoyed the highest esteem among those who knew him in life, while borne to burial, lifted his ghastly face from the bier, and dis-

tinctly articulated, " I am summoned to trial ! " After
a fearful pause, the same voice said, " I stand before the
tribunal ! " A few moments of horror ensued, when this
dreadful sentence issued from between the livid lips : " I
am condemned by the just judgment of God." " Alas,"
cried Bruno, who was one of the train of mourners, " of
what avail is the good opinion of the world? To whom
but to Thee, O Lord, shall I flee ? " And the saint departed, and founded that Order which seeks immortal
life by cherishing the spirit of death enshrined in a form
of the grave. This is not the wisdom and health, but the
fanaticism and disease, of solitude. Better handle roses
than skulls, contemplate the blue freedom of the sky
than the dark narrowness of a cell, think and feel mod-
erately, according to the healthy averages of nature and
life, than with absorbing extravagance on one or two exciting points.

To sum up in a single paragraph this estimate of the
influences of monastic retirement. If the religious vo
tary who exchanges the world for a cell, despises the
world with a too intense predominance, he inflames an
abnormal pride ; and a life fed with scorn must be unwholesome. If he seeks to subdue the world to his
caste-interests, like a Dunstan or a Torquemada, he feeds
an ambition less human and worse than the ambition of
martial heroes, — the Cæsars or Napoleons ; the usual
love of power is replaced by one more unnatural, exasperated, and pitiless, which flatters itself with a heavenly
elevation while drawing its nutriment from infernal roots.
If he sinks into a mere mechanical formalist, it is a very
low and poor type of life, no better than if in the routine
of society. If he becomes the victim of suspicious hates,
wretched spiritual frictions, or of brutal appetites of car-
nality, he will exemplify the most degraded and aggravated, because the most unrelieved, forms of these vices.
But if his earthly affections have been so disappointed
or so ravaged as to make a lonesome and contemplative
state truly soothing and medicinal, or if he has a pronounced genius of religious enthusiasm, the cloistered
solitude may be his truest home.

When one avoids his fellow beings for the purpose of escaping the pervasive opposition and rebuke their presence administers to his egotistic feelings, every step of removal is an injury. Lowell has wisely remarked : — "One is far enough withdrawn from his fellows if he keeps himself clear of their weaknesses ; he is not so truly withdrawn as exiled, if he refuse to share their strength." Then solitude becomes the hot-house of vice. Vices which on the highway were shrubs here become trees, and even exotics are curiously pampered. The peculiarities of any caste of men existing in marked isolation from neighboring humanity, are apt to be a haughty selfishness and conceit. They acquire the habit of Pharisaic exclusiveness, and hold the rights of mankind in abeyance to the interests of their clique. Thus a priest may think less of God and truth than of the ecclesiastical establishment with which he identifies himself. The result, of course, is wholly different when the lonely man spends his thoughts and passions on disinterested principles and plans, themes connected with universal truth and good. The history of monastic ages and of mystical sects is certainly as full of warnings as of examples. Their prevailing influence is to withdraw attention from the general and fasten it on the particular, to absorb the individual in himself, and make him oblivious of the public. They forget that the laws which bind the molecules into wholes have a sovereign importance immensely beyond the molecules.

But *our* chief dangers lie in the opposite quarter, — too much living in a throng, frittering publicity, garrulous disclosures and uneasy comparisons. Our pining is not after loneliness, but after the rush and glitter of crowds. If left to ourselves, we sigh in our desertion, and think it were far happier to be in the throng. But why do we not see that happiness resides in the mind, and is no gift of place? Diocletian and Amurath voluntarily abdicated their thrones, and withdrew into private life, sick of the revolting discoveries they had made, overwearied by the pompous miseries they had found. Charles the Fifth exchanged his kingdom for a cell, and deemed himself the gainer. Philip the Third on his death-bed was heard to sigh, "O that I had never reigned; that I had rather been the poorest man!" Oliver Cromwell declared in one of his speeches, "I can say in the presence of God, in comparison of whom we are but like poor creeping ants upon the earth, I would have been glad to have lived under my wood-side, to have kept a flock of sheep, rather than to have undertook such a government as this." Colbert, the great minister of Louis the Fourteenth, who was proud and fond of his master, but prouder and fonder of his country, was broken-hearted by the alienation of the ungrateful egotist he had served too well, and at the sight of the distress of the people whom he had toiled so hard to shield and bless. On his death-bed he refused to hear the letter his penitent master sent. "I wish to hear no more of the king. It is to the King of kings that I have now to answer. Had I done for God what I have done for this man, I should be saved ten times over; and now I know not what will become of me." Worst injustice of all, the people were as ungrateful as the king. They looked on Colbert as the author of their hardships, instead of recognizing in him their chief friend and benefactor. The great minister was buried secretly by night, for fear the rabble would tear his body from the bier!

The belief that the men of the greatest celebrity are the happiest men, is the inveterate fallacy of shallow minds. The reverse rather is the truth. The fate of Cæsar is a symbol of the fortune of genius; the crown on the brow implies the dagger in the heart. To be persecuted with dislike makes the man of deep sympathetic soul unhappy. And certainly this is the common fate of the great man. Envy scowls at him, and hatred reeks around him. His illustrious genius rebukes littleness, his conspicuous place stirs the venom of obscure ambition, his incorruptible honesty enrages unprincipled selfishness, — and they seek revenge. Papinian, the peerless builder of the Roman Law, who, according to Cujacius, "was the first of all lawyers who have been or are to be, — whom no one ever surpassed in legal knowledge, and no one ever will equal," became unpopular, and was beheaded by Caracalla for his ability and his integrity. Looking over the tragic history of the world thus far, it is obvious that greatness and happiness have rarely been united. " Inquire," says Lavater, " after the sufferings of great men, and you will learn why they are great." In his Dialogue between Nature and a Soul, Leopardi makes the soul refuse the offer of the highest gifts of genius, on account of the inevitable suffering connected with them. Yet it will ever be the characteristic of choice souls to prefer the mournful nobility of the prerogatives of genius, with all their accompanying trials, to jollity and mirth on a more vulgar level. In their view pleasure may be a rose, but wisdom is a ruby. With a thrill of divine valor they affirm that the duty to be noble takes precedence of the right to be happy.

Unquestionably the moral regimen of the hermitage is more appropriate for our case than that of the drawing-room. The frittering multitude of interests and influences is so great, that an economizing seclusion and defence of the soul is one of our greatest needs. The truth that the world is too much with us, is fitter to furnish exhortations to us than the other truth, that it is not good for man to be alone. The word *trivial*, in its etymological origin, is loaded with a forcible lesson. It is derived from the word *trivium*, which denotes the meeting-place of three roads ; a point where idlers spent their time, loitering to see what passed, and to discuss the worthless items and gossip of the day. How much weightier are the suggestions of the word solitude !

### The Uses of Solitude.

Two men, most emphatically, are alone ; the worst man and the best. Judas, hugging the thirty pieces of silver, or throwing them down and retiring to hang himself, is alone ; and Jesus, sitting by the wayside on Jacob's well, with meat to eat that the world knows not of, or going apart into a mountain to pray, is alone. Whenever we feel deserted it is well to trace the cause of the feeling; learn whether the experience be the result of a fault, of a merit, or of some neutral quality unfitting us for such fellowship as would otherwise be ours. The first improvement of our loneliness is to analyze its cause and meaning, — see what *kind* of solitude we are in, and what mode of treatment will be best adapted to the case. These preliminary steps taken, the next duty is to devise the fittest remedies for whatever is painful or wrong in our condition, and endeavor to win the richest compensations from it. A very different regimen should be prescribed for one suffering in the solitude of guilt from that applied to one suffering in the solitude of grief. The former needs the processes of penitence, atonement, reformation : the latter, the ministrations of faith, love, cheerful communion, useful activity. Much of the bitterest loneliness in the world arises from an exorbitant and morbid self-regard, the importunate presence of self in attention. Hawthorne's story of the Bosom Serpent is a terrible illustration. There is a whole class of solitaries simply from shyness, — bashful men like Gray and Cowper, the poets, and Cavendish, the great chemist. A much larger class affect seclusion in consequence of pride. The misfortune of both these classes of sensitive shrinkers is the same, an inability to escape the consciousness of their own personalities as related to the opinions of other people. It is not mere self-consciousness that troubles the trembling sensitive ; it is that self-consciousness imaginatively transferred to another, and exposed to all the variations of the supposititious opinions there. The endless multiplicity of competition in modern society, at every point a prize, at every point a glass, — tends to force us inordinately on our own notice. If we could but gaze at the prize alone, and break or blink the glass ! But unfortunately mirrors prove more fascinating than prizes, and most persons are intent on — themselves. No other article of domestic furniture has been so disproportionately multiplied in modern upholstery as the looking-glass. Parlors, dining-rooms, entries, dormitories, even ladies' fans and gentlemen's hats, are lined with looking glasses. And, not content with this, a recent American newspaper contained the announcement, in a description of several sumptuous banquets, that the host in each instance furnished a photographic likeness of himself, a gift placed in the plate of every guest. The reporter thinks it a most delicate attention, and hopes that the generous givers of dinner-parties will follow the beautiful example and make it a custom! Thackeray, with probing truth, in the vignette to his Vanity Fair, depicts the representative character, stretched at full length, neglecting alike the petty and the sublime objects about him, — puppet, crucifix, church, and sky, — with a melancholy air studying his own lugubrious face in a mirror which he holds in his hand. Yes, this is the malady of the age, — an age of Narcissuses. The curative desideratum is devotion to a divine end, disinterested enthusiasm. Give the victim that, and he will fling off his incubus, his morbid consciousness of self will disappear in a wholesome consciousness of objects.

Sometimes the unhappy subject of this malady, attempting to cure himself by retreating from the crowd where his self-consciousness is disagreeably stimulated, only aggravates the cause in solitude. For he still continues to deal chiefly with his own personality and its private affairs ; and he who does this will find that egotism and its penalties may be more exasperated in the hermitage than in the hall. He must put self in the background, refuse to think of it, escape the haunting torment, by an absorbing occupation with redemptive objects and truths. Petrarch — one of the most eloquent missionaries of solitude — has described this untoward experience in his famous sonnet beginning, "O cameretta che già fosti un porto." He writes : —

> But e'en than solitude and rest, I flee
> More from myself and melancholy thought,
> In whose vain quest my soul has heavenward flown.
> The crowd, long hostile, hateful unto me,
> Strange though it sound, for refuge have I sought, —
> Such fear have I to find myself alone.

That kind of moral solitude which constitutes a painful loneliness does not consist in the state of the soul alone, nor in its circumstances alone, but in a want of adjustment and sympathy between the soul and its circumstances. Pride, chafing, rebelling, despairing, makes loneliness ; aspiration, beating, self-sustained, in thin air, wearied, and falling back on itself, makes loneliness. Then the sufferer, to find relief, must understand what the trouble is, and set himself at work to bring the discord into harmony. If his outer condition is right, or unalterable, he will subject his own wishes and energies to it ; if wrong, strive to rectify it. If some of his facul ties are unduly sensitive, others torpid, he will seek to restore their equilibrium. If he has fancied non-existent circumstances, and of existing circumstances magnified some, depreciated others, and been blind to others, he will zealously endeavor to correct these errors. Thus will peace be won, and the blessed fellowships of existence be re-enjoyed.

In addition to these general directions, there are special resources available in special cases. A purpose is always a companion. An earnest purpose is the closest of companions. To fulfil duties is more than to enjoy pleasures : it carries its own reward.

> To have the deep poetic heart
> Is more than all poetic fame.

There is no bitter loneliness for those affectionately devoted to blessing their fellow-creatures. The keeper of the light-house, when night settles around him, and the tempest holds revelry, as he looks out on the ghastly glare of the breakers, and hears the shrieking of the storm-fiend, finds good company in the thought that the friendly light he trims will warn endangered crews of their peril, and perhaps save them from death. Gifted souls find solace and companionship in their works. It has been eloquently said of Michael Angelo, that he was "a lofty, lonely, lordly spirit, but gentle, sensitive, and overflowing with sympathy, who moved with a benignant complacency among the great forms he called into existence for his own satisfaction." There was a painful, almost tragic solitude in the life Charlotte Brontë led in the bare and sombre old parsonage of Haworth; the long slopes of monotonous hill behind; and before, the dark, dilapidated church, and the rude burial-ground with its mute hillocks and hollows, its tumbling headstones, its piteous grass and desolate paths. The great event of the day to her was the arrival of the post; and she did not dare to let her thoughts dwell on this, for fear it would make her discontented with the duties of the other hours. Yet she must have found a massive comfort, a keen joy, in communing with her own genius, and in composing those powerful books which she flung out upon the world in rapid succession. It should, further more, always be remembered that *sympathy is not the end of our life, but only an accompaniment and help. The true end of life is the fruition of the faculties of our individual being.*

When loneliness of life is caused by superiority of soul, its compensations furnish the proper antidotes for its pangs; and we should be content and happy, not weakly submit to complain. No doubt conceit often pleads effectively to persuade us that such is our case when it is not. Many a self-deceived weakling

> Has stood aloof from other men
> In impotence of fancied power.

We must beware of the subtle sophistry, and not lay the flattering unction where it does not belong. On the other hand, neither modesty nor blindness should be suffered to hide the truth when it is really favorable to a serious complacency. So few live for truth, virtue, progress; so many live for routine, amusement, conformity,— that it is not astonishing for a man of comprehensive lineal activity to surpass the range of those given up to a careless circular activity. A passion for perfection will make its subject solitary as nothing else can. At every step he leaves a group behind. And when, at last, he reaches the goal, alas! where are his early comrades? Let him thank God that his superiors and his peers are there before him. Let him honor these, fraternize with them, and be blessed.

A lost faith is sometimes the cause of a dismal solitude of soul. A sceptic of fine sensibility, robbed of long-cherished beliefs, and provided with no substitute, missing that wonted ministration, may feel as lonely as a pilgrim overtaken by night on an Alpine ice-ocean, — a dark speck of despair between the shining sea of ice and the colder sea of stars, a conscious interrogation-point of fate. His true course is to face his doubts without flinching, boldly follow every clew, make no unfaithful compromise, but traverse the deserts of negation to their end, keeping a spirit open, silent, and watchful for every light of providential direction and every voice of divine reality. He will then find denial but the precursor of affirmation, and disbelief but a process of growth, an extrusion of dead husks for the appearance of living germs. Dogmatic assent will be superseded by spiritual experience, insight will take the place of tradition, and blessed truths richly compensate for the outgrown formularies which it cost him so much pain to abandon. His trial is in leaving the injurious but endeared companionship of beliefs no longer fitted to the wants of his mind, but which he has always supposed indispensable. His reward will be to gain a new companionship of higher and truer views, better beliefs, more accurately adjusted to his real wants as a conscious sojourner in time and a responsible pilgrim to eternity.

> Who ne'er his bread in sorrow ate,
> Who ne'er the mournful midnight hours,
> Weeping upon his bed hath sate,
> He knows you not, ye heavenly powers.

One of the most valuable uses of solitude is to prepare us for society. He who studies, when alone, to understand himself, and to improve himself, to cure his vices, correct his errors, calm and sweeten his heart, enrich his mind, purify and expand his imagination and sympathy, thus makes seclusion a sanitarium, gymnasium, treasury, and church, and takes the surest means to commend himself to his fellow-men. He employs the best method both for giving and securing pleasure when he shall return from his retirement to mingle with others again. Ma-

dame Swetchine says, " I hold it a good thing to forsake
the world from time to time, so as not to lose the relish
for it ; as Rousseau liked to leave those whom he loved,
in order to have the pleasure of writing to them." Those
who nurture any of the malign or unsocial sentiments
when by themselves, become unfit for company; all frequented haunts grow distasteful to them ; they become
jealous, irritable, and wretched everywhere. Advantage
should rather be taken of solitude to assuage every rank-
ling remembrance, all selfish suspicions, to lay emollients
on the wounds of vanity, to foster generous views of our
fellow-beings, and strengthen the benignant feelings.
Then we shall leave our retreat and re-enter the throng,
refreshed and placid, prepared doubly to enjoy the privi
leges of human intercourse. Thus every recurrence of
loneliness, instead of tending to make us misanthropic,
will send us back to our common avocations and pleasures with a wiser interest, a keener relish for all that
concerns our kind. Charles Lamb charmingly describes
such an experience in his little poem on the Sabbath
Bells. A solitary thinker is roaming the hills far from the
walks of men, wrapt in abstruse contemplations, debat
ing with himself hard and elusive questions.

> Thought-sick and tired
> Of controversy where no end appears,
> No clew to his research, the lonely man
> Half wishes for society again.
> Him, thus engaged, the sabbath bells salute
> Sudden. His heart awakes, his ears drink in
> The cheering music : his relenting soul
> Yearns after all the joys of social life,
> And softens with the love of human kind.

Such an improvement of seclusion is the way to make
us attractive to others as well as contented in ourselves.
To be ignorant of yourself, uneasy and exacting, is to be
repulsive no less than miserable. Who would enjoy the
world, must move through it detached from it, coming
into it from a superior position. He must not be weakly
dependent on his fellows, but say to himself, Cannot
God, the Universe and I, make my life a rich, self-sufficing thing here in time ? To command love we must not

be dependent on it; a tragical truth for those who have most need of love. The way to self-sufficingness is the way to public conquest. Happy in the closet is winsome in the crowd. The king of solitude is also the king of society. The reverse, however, is not so true. Many an applauded domineerer of the forum, many a brilliant enchantress of the assembly, when alone, is gnawed by insatiable passions, groans restlessly under the recoil of disappointment. William von Humboldt wrote to his friend Charlotte, "There are few who understand the value of solitude, and how many advantages it offers, especially to women, who are more apt than men to wreck themselves on petty disquietudes." Self-inspection, self-purification, self-subdual to the conditions of noble being and experience, — these form the fitting occupation of our solitary hours. Yet, self must not be the conspicuous object of our contemplation, but great truths and sentiments, moral and religious principles, nature, humanity, and God, the perennial fountains of fresh and pure life. He who follows this course is best qualified to read and interpret the secrets of other souls. He is likewise best fitted to master the world, in the only sense in which a good man will wish to master it. There is no more efficacious mode of observing mankind, than as they are seen from the loop-holes of retreat, and mirrored in our own consciousness. In relation to what is deep and holy, as compared with each other, society is a concealer, solitude a revealer: much, hidden from us in that, is shown to us in this. Amidst a festival the moonlight streams on the wall; but it is unnoticed while the lamps blaze, and the guests crowd and chatter. But when the gossipers go, and the lights are put out, then, unveiled of the glare and noise, that silvery illumination from heaven grows visible, and the lonely master of the mansion becomes conscious of the visionary companionship of another world. Solitude is God's closet. It is the sacred auditorium of the secrets of the spiritual world. In this whispering-gallery without walls, tender and reverential spirits are fond of hearkening for those occult tones, divine soliloquies, too deep within or too faintly far ever anywhere else to

suffer their shy meanings to be caught.  Given a sufficiently sensitive intelligence to apprehend the revelations, and every moment of time is surcharged with ex-
pressiveness, every spot of space babbles ineffable truths.
Silence itself is the conversation of God.   We know that
in the deepest apparent stillness sounds will betray themselves to those who have finer sense and pay keener attention than ordinary.   On the Alps, when everything
seems so deathly quiet in the darkness, place your ear at
the surface of the ice, and you may catch the tinkle of
rivulets running all through the night in the veins and
hollows of the frozen hills.   Has not the soul too its
buried streams of feeling whose movements only the most
absorbed listening, in the most hushed moments, can
distinguish?

What is it to subject a thing, save to extricate yourself
from it, rise apart, and command it from a higher position?   To overcome the world it is indispensable first to
overlook the world from some private vantage-ground
quietly aloof.   Would you lift the soul above the petty
passions that pester and ravage it, and survey the prizes,
the ills, and the frets of ordinary life in their proper per-
spective of littleness?   Accustom yourself to go forth at
night, alone, and study the landscape of immensity ; gaze
up where eternity unveils her starry face and looks down
forever without a word.   These exercises, their lessons
truly learned, so far from making us hate the society of
our fellow-creatures, or foolishly suffer from its annoyances, will fit us wisely to enjoy its blessings ; be masters
of its honors, not victims of its penalties.   If to be alone
breeds in us a sullen taciturnity, it is proof that we are
already bad characters.   The more a misanthrope is dissociated from men, the more he loathes them ; the longer
a pure and loving soul is kept from them, the intenser is
his longing to be united with them.   None are so bitter
and merciless, so abounding in sneers and sarcasms about
society and its occupants, as those most thoroughly familiarized and hardened in its routine.

How certainly
The innocent white milk in us is turned

THE USES OF SOLITUDE.

By much persistent shining of the sun !
Shake up the sweetest in us long enough
With men, it drops to foolish curd, too sour
To feed the most untender of Christ's lambs.

The solitary, if rich in heart, are often the fondest of talking when they find a listener. It eases their fulness. The wand of approaching sympathy melts a channel for the pent-up flood of consciousness to flow off in, as when the ice is broken which had formed over a spring and forced the stream to accumulate far back in its secret runnels. The strength of our desire for contact with society is proportioned, properly, to the keenness of our experience, the wealth of our discoveries, in solitude. No sooner has a truth or principle, unknown to us before, come into our possession, than we are eager to go out armed with this talisman, and test again the natures of men, the phenomena of life, the prizes of the world. It is a natural and wholesome reaction which takes the student back from the monologue of metaphysics to the dialogue of science. Knowledge of self is but half a knowledge of the universe.

A great compensation for any sadness one may feel in being alone is the power his calm environment may be made to yield him. Solitude is the foster-mother of sublime resolves. It is the earth of Antæus, every fresh touch of which emits a thrill of fortifying renewal. Revolutions, sciences, religions, have crystalized in the fervid silence of lonely minds. Had Joan of Arc, instead of cherishing lonely visions in her solitary life, frequented balls, whist-parties, sewing-circles, and the like, would they have given the poor maiden, who could not write her name, power to make that name immortal? The wisdom of the Jesuits is perhaps nowhere better shown than in the tremendous regimen of solitude and contemplation of death under which they put their novitiates. The results of this discipline have been incomparably great, revolutionizing in its subjects the normal forces of human nature. Strength grows in repose succeeding action ; and Pascal says, " We are ridiculous when we seek repose in the society of our fellows." Enthusiasm is no more a growth of the arena than peace is. There is infinitely more apathy in crowded than in deserted places. Overfull hearts turn from the ball to the bower. In the purlieus of fashion indifference often passes for repose, and coldness for power, but warmth for crude-ness, and diffidence for incapacity.

Solitude stimulates and feeds, rather than generates, purposes. They are to be acquired elsewhere, confirmed and fortified here. The design of Mahommed was con-ceived during his journeys among the idolatrous tribes, as the factor of Kadijah; but placed in a focus, and kindled into a contagious flame, among the lone hills of the desert where he so often retired to muse. Purposes and vows are consolidated and intensified by reflection and repetition; and these, driven from the glare and buzz of society, are found in the retreat. Echo is the friend of the lonely. Every great artist, every serious genius, who wishes to create on a rare height, feels the need of a refuge from the trifles and vulgarities of ordi-nary people. Felix Mendelssohn died just as he was in-tending to withdraw for several years into solitude at Rome, to work out in an oratorio his idea of Christ. We cannot enough lament the loss of the feelings that rich, pure, and devout nature would have poured around this sublime theme.

Every marked quality or power of genius is a mental polarity. Nothing is ordinarily so fatal to this as a frit-tering multiplicity of interests, a bewilderment of activi-ties. A throng of objects flitting before the eyes is irri-tating and exhaustive. It is so with the mind too. A rapid alternation of attractions and repulsions dries up sympathy and crumbles thought. To live either in a dis-tracting medley of private motives and efforts, or in the constant frictions and frivolities of society, is to be sub-jected to an influence of the most deteriorating character. Let that influence be continuous and exclusive, and it will soon depolarize the associative points of intellectual power. It is remembrance, admiration, longing, wonder, musing, love, — whose congenial haunts are the still library, the lonely shore, the hill, the glen, the sea, and

.he sky, — that feed and inspire the poet. He would be steadily unpoetized if confined to the barren gossip of saloons, the corrosive emulations of the crowd. The legacy of Wordsworth to our times, in the unique originality of his genius, and its literary expression, perhaps has no grander feature than his wise withdrawal from the miscellaneous contacts and wear of cities to the lakes, — his consecration to the stately care of himself in the bosom of brooding nature, in imaginative sight of mankind, in the transcendent embrace of God. His words of exhortation to others are a noble description of his own life. He says : " It is an animating sight to see a man of genius, regardless of temporary gains, whether of money or praise, fixing his attention solely upon what is intrinsically interesting and permanent, and finding his happiness in an entire devotion of himself to such pursuits as shall most ennoble human nature. We have not seen enough of this in modern times ; and never was there a period in society when such examples were likely to do more good than at present."

The influence of society is distracting or diffusive ; that of seclusion is concentrative. Ruskin says, " An artist should be fit for the best society, and should keep out of it. Society always has a destructive influence on an artist ; first, by its sympathy with his meanest powers ; secondly, by its chilling want of understanding of his greatest ; thirdly, by its vain occupation of his time and thoughts. Of course a painter of men must be among men : but it ought to be as a watcher, not as a companion."

Society, full of multiplicity and change, is every way finite, wasting its force in incessant throbs ; solitude, an unaltering unity, is allied to the infinite. Its repose collects and redistributes the force expended by the struggles that rage and subside unheeded in its measureless realm. Our times want the brooding spirit. When we read, that, one fine morning, Mithridates, king of Pontus, disappeared from his palace, and remained missing for months, so that he was given up for lost ; and, when he returned, it was found that he had wandered, unknown, through the whole of anterior Asia, reconnoitering all the countries and peoples, — we are impressed with a sense of inscrutable solitary force and mystery akin to that imparted by the ocean, as it goes forth, "dread, fathomless, alone." Such souls, huge reservoirs of purpose and power, are rare in any age, but especially rare in ours. Bishop Berkeley said, "In the present age thinking is more talked of, but less practised, than in ancient times." I believe this to be deeply true; reading and other substitutes for thinking have vastly increased in modern times. In power and resources Man undoubtedly has gained, but men appear, relatively, to have lost. There is a tremendous influence in a democratic people to draw up and pull down exceptions to the general level, - - an influence commonly fatal to the production of rich and strong individualities. There is ground for Byron's satirical stroke : —

> Society is now one polished horde,
> Formed of two mighty tribes, the Bores and Bored.

Our *familiar* is rather a nimble and tricksy spirit, like Puck, than that awing genius of brooding silence which kept Socrates, spell-bound, standing fast in one spot all night. We have superabundant impulse, but little patience. It is all come-and-go, and no stay. The dischargers of power are multiplied out of all proportion to the generators of power. The swift succession of events, the thousand vibrating relationships of the age, the incessant teeming of business enterprises, political squabbles, reviews and books, scatter our attention and exhaust our energy. We lie, spread abroad and open, at the mercy of a disintegrating swarm of influences, instead of being gathered into one available mass of purpose steadily directed to its aim. Force enough is wasted in the sterile chatter of conceited criticism to produce much of permanent worth, if it were converted into creative meditation. And so far from seeking to neutralize and correct the evil, strenuously confronting it, resisting it, and curing it by a rectifying regimen of meditation and prayerful seclusion, a withdrawn and studious culture, carefully adapted in each instance to the peculiarities of the person and his circumstances, we make it worse by indulging its instincts, seeking relief in sympathy or escape in confusion, turning our already harassed sensibilities to a distracting medley of meetings, parties, newspapers, novels, theatres.    Crabbe says, with his usual sharpsightedness, —

> Men feel their weakness and to numbers run,
> Themselves to strengthen or themselves to shun.

Exactly the opposite of what they should do.    The like-minded may confirm our disease by flatteringly reflecting it ; the unlike-minded only take our attention from it while, perhaps, it grows more inveterate.    Surely it is better to extirpate than to forget an evil.    Killing time is a poor substitute for improving time.    Dissipation may stem or drown grief, but cannot heal it.    The hospital of loneliness is better fitted for the treatment of wounds than the stadium of competitors and spectators.    Society drains, solitude supplies.    That is the place for expenditure, this for recovery.    This is the place of preparation, that of performance.    The work of a man who leads a life of dissipating publicity flattens into a marsh ; the work of a man who guards his aim by a solitary determination looms into a mountain.    Wordsworth puts into the mouth of his Oswald a true and striking image : —

> Join twenty tapers of unequal height,
> And light them joined, and you will see the less
> How 't will burn down the taller, and they all
> Shall prey upon the tallest.

It is fortunate for us, in this gossiping, headlong age, that, in each recurrence of slumber the soul takes an invigorating bath of loneliness ; that, in the mysterious alternation of our existence, every night the babbling streams of society empty into the oceanic solitude of sleep and dreams.    The great soul, apparently dwarfed to the stature of common men, as soon as alone, dilates again to its native majesty, and begins to hold converse with themes of its own altitude.

7 *

If the chosen one could never be alone,
In deep mid-silence, open-doored to God,
No greatness ever had been dreamed or done :
Among dull hearts a prophet never grew ;
The nurse of full-grown souls is Solitude.

✔ Solitude is the tent of the Almighty, which no thoughtful
man can enter without awe, or need leave without shrift
and an access of strength.

For fine spirits, hurt and weary in the conflicts of much
company, harassed with its fret and worn with its care, to
be alone for a season is a luxurious refreshment, an indescribable solace.   Chateaubriand surrounded himself
in Paris with a few choice friends, profoundly close and
dear, and sensitively avoided the crowd, which he called
" The vast desert of men."   What a eulogy of solitude is
contained in such a phrase as this of Wordsworth : —
" The fruitful calm of greatly silent hearts ! "   Suffering
from a fevered breast and a distempered mind, who would
not find it a benign change to leave the painted harlot,
Fashion, for the sad wood-nymph, Solitude ?   With her
society the sage flees into his thought, the saint into his
God, — and the horrid discord that tore and stunned
them rolls away and becomes a murmur on the horizon.
The nerves, rasped and drained in the collisions of the
crowd, are lubricated and refilled in the repose of solitude.   The fainting consciousness is thus restored, vital
color put into its fading states, the wearing effect of confused voices and calls stopped.   Of course, if the *ideas* of
the things that worried us follow into our retreat, and continue to operate there, these good results cannot be expected ; and the organism must sink under the constant
iteration of demands.   To profit by retirement we must
not allow the ideal equivalents of the goads and loads
that stung and galled us in the thoroughfare to pursue
and irritate us still, but win a true respite from them by
ceasing to think of them.   And we *can* cease to think of
them by occupying our attention with something better,
something soothing and elevating, something grand and
lovely.   The unhappy heart overwhelmed by misfortunes,
when deserted by old associates in whom it had fondly
confided, is apt to add to the crushing weight of the

misfortunes the bitterer weight of a haunting recollection of the desertion, a recollection full of enervating melan choly, or perhaps full of poisonous hate. How much better to forget the desertion and think of remedying the misfortunes! The traveller swept by an avalanche into an Alpine chasm, beyond sight and sound, and left there by his fellows and guides, to die, alone in the frozen gulf, must not fasten in his brain the grievous and despairing thought of his companions gone on their way; he will best do his duty by resolutely turning his mind to a con templation of God and immortality, or to some manly plan for self-rescue. Bonnivard, in a dungeon in the castle of Chillon, was fastened to a ring in a column by a chain four feet in length. He could walk only three steps. He resolved to keep himself alive until some prov idential deliverance should enable him again to serve the cause of liberty. A channel worn three or four inches into the surface of the rocky floor, still visible, marks the pathetic limits of his daily and determined exercise ; and within this circle may be seen three yet deeper cavities, made by the three footfalls which his chain allowed him. After six years he was rescued. It was his indomitable purpose alone which kept him from decay and death.

"Solitude is the home of the strong, silence their prayer." Such is the just and impressive aphorism of Ravignan. Its counterpart would be, Society is the refuge of the weak, speech their confession of defect. Mind and heart grow stronger by drawing strength from their environment. They cannot create, they must im bibe, power. Self-contemplation alone is a jejune and barren process. Nothing is so sterilizing as retirement, when, instead of bringing us into, it cuts us off from, communication with the aboriginal sources of our life. The very measure of genius is its capacity of sympathy with its race, with the universal and disinterested as dis tinguished from the individual and selfish. Every rumi nant must take new food, or the cud itself will fail. The individual does not manufacture force, but derives it from the fountain-head of God through the drenched universe. It will not be imparted to him and accumulated in him

without his co-operation.    To expect the spirit to be
strengthened by foreign appliances, omitting its own action, is absurd.    Would you support a bird with props?
Mental force is communicated in mental food, whence it is
extracted by the digestion of the discriminative faculties,
and secreted by the glands of the soul, and reservoired,
subject to the summons of the will.    This process is none
the less real for being without our consciousness.    The
increase of our power, therefore, if in its primary aspect a gift to us, is in its secondary aspect an appropriation by us.    And in truth these two are one.    Where
both are not, neither is.    The giving from the whole and
the taking by the part are one indivisible act.    Their
mystic consent makes the identity of life, as perception
is the middle term in which object and subject coalesce
and are lost to reappear in their ideal equivalent.    Contact with a mass of humanity or social machinery too vast
for definite reaction, tends to make a rich soul morbid,
starts activities it cannot satisfy.    Contact, on the contrary, with overwhelming natural forces or material
scenery, is salutary, concentrates the soul in a brooding
and assimilative mood which collects its powers and
brings them into equilibrium.    It is said that those who
stand on the floor of Saint Paul's Cathedral hear a strange
sound rising and sinking in the dome, — the aerial reverberation of the combined noises of London.    That multitudinous murmur, the endless roar of earth's hugest city,
must be painfully oppressive to one who thinks of the
guilt, the struggle, the glory and misery represented by
it; the effort to disentangle and track home its moral
suggestions must be wearisome and disheartening.    But
the roaring of the ocean on a forsaken coast, typical of
the everlasting freshness, strength, and grandeur of na
ture, is peace-giving and wholesome, soothes while it
excites.

When these conditions are observed solitude exercises
a nutritious and tonic influence on the higher parts of our
nature, not only refreshing the weary and fortifying the
weak, but renewing loyalty when it is tempted and con-
firming innocence when it falters.    Then loneliness and

listening are the richest nourishment of the soul. Conscience, judgment, better purpose, withdrawing from the storm of seduction, have an opportunity to recover their poise. Its effect is as useful in giving moral direction to our energies as it is in invigorating their source. In the midst of scrambling antagonists, wild with the excitement of the game, man is tempted to forget the moral law, and fancy that course best which points the speediest path to the prize ; but a little reflection, in sober retirement, disperses the perilous falsehoods of the arena ; the dust of busy desire settles ; the clouds and veils of delusion and occasion blow aside ; the steady lights of morality shine out ; and from the stars of solitude eternity sheds sublimer counsels on the soul than are ever discerned by aid of the flaring torches of time.

To one who bravely accepts his public responsibilities, but keeps conscientious watch over the springs of his conduct, society is the open sea where virtue meets its foes, solitude the harbor where it repairs its damages. Careers of loud pretence and manifold showiness have frequently turned out to be hollow, the speedy prey of contempt, while " obscurest lives have the starriest souls disclosed." The wild rose beside the mountain brook has a freshness of beauty, takes from its mighty environments of unadulterated nature a charm, which nothing shown by its haughty sister in the imperial conservatory can atone for the lack of. There are qualities of character, sacred beauties of virgin souls, so infinitely shy in their subtle modesty and refinement, that public exposure profanes and destroys them.

> A charm most spiritual, faint,
>   And delicate, forsakes the breast,
> Bird-like, when it perceives the taint
>   Of prying breath upon its nest.

Solitude is the nunnery of an innocent mind. It is the asylum of those who aspire too high for the sympathies of their fellows, — whose standards are too exalted for the slow and sluggish perceptions of their comrades. Such an one, constantly feeling himself lowered and in-

jured by the judgments to which he is subjected, at
length seeks protection, and strives to sustain himself
on his true level in the only way available ; he withdraws,
wraps the curtain of his royal thought around him, and
lives in a sublime privacy.   The few noble spirits who
make a master-purpose of studying to attain an ideal
perfection are always affecters of solitude.   Shakespeare's Prospero is an imposing example of them, describing himself as —

> Neglecting worldly ends, all dedicate
> To closeness, and the bettering of my mind.

Those who supremely love truth, beauty, and goodness,
form the most illustrious society in the world, and the
sparsest.   Schopenhauer says : " For the most part we
have only the choice between solitude and vulgarity.
The most social men are the least intellectual.   To say
' He is very unsocial,' is almost equivalent to saying,
' He is a man of great qualities.' "
   An ingenuous and heroic man when left alone is in his
confessional and armory, where he sees the pure standard
of duty, repents of his errors, rectifies what is amiss, and
with sincere vows equips himself afresh for the good
fight.   Solitude is the private palæstra where spiritual
athletes put themselves in training for the public contests
of life.   It is there we learn the first, last, greatest lesson
taught by our destiny, namely, patience.   For, in the
words of Parsons, — the excellent words of a true
poet, —

> Patience is the part
> Of all whom Time records among the great,
> The only gift I know, the only art,
> To strengthen up our frailties to our fate.

   To stay in seclusion awhile is to keep a fast of the
spirit.   Its influence is as wholesome and stimulative on
the mind as that of occasional abstinence is on the body.
" To me," Richter says, " a solitary apartment is a spiritual fountain-hall full of medicinal water."   La Bruyère
says, " The misfortunes of men proceed from their inability to be alone ; from gaming, riot, extravagance, dis

sipation, envy, and forgetting God and themselves." The grasshopper leaps about with his pertinacious click amidst the ruins of the Abbey at Ely, where Canute bade his rowers pause off the shore, that he might listen to the monks singing their evening hymn.  To linger there, surrendering the soul to all the pensive morals of the place, must mellow and deepen the heart.  But to spend much time in a gambling-room, in the depraved current of men who come and go, drinking, smoking, staking, swearing, can hardly fail to thicken the rust and grossness on the mind.

The streams flowing directly from the glaciers are turbid and chalky, thick with triturated stones ; but, drawn apart in glens or wayside pools, they become wonderfully clear, having deposited the sediment they before held in solution.  Human souls, withdrawing from the rush and friction of the wearing world, and pausing in quiet places, surrendering themselves into the natural arms of God, Rest and Silence, grow transparent, precipitating the abrasions of life.  They resume their native purity, and at once reveal the bed of consciousness and reflect the blue of heaven.

Let those who feel the lonesome pinings of the heart, beware of yielding to the temptations which would induce them to sink from their high promptings and conform to the average range and custom for the sake of fellowship.  Many have done this, and soon suffered worse than before, and bitterly regretted the degrading com-promise.  The buzz and clamor of unsympathizing throngs are an aggravation, not a relief, of the aching and pining of an affectionate nature.  His craving, taken up and repeated in all these cold mirrors, is so many times flung back, unsatisfied and heightened, into his consciousness.

> The heart that home rejects to crowds may fly ;
> Gay glides the dance, soft music fills the hall :
> It flees, to find the loneliness through all.

The chirruping of millions of crickets breaks the silent monotony of the twilight fields, merely to make the integral solitude multitudinous ; it is not a felicitous ex-

change.    To be capable of a great aim, and of sustained
efforts to realize it, is to be one man taken out of ten
thousand.    They who have this capacity will not be
found hovering about saloons, half crushed in the rush
to the midnight tables of vulgar fashion.    They will
be found rooted in solitude.    For the sustenance of
their spirits they revert from the frothy speeches of
the platform, from the spawn of the press, to the al-
coved treasures and the aristocratic comradery of the
great minds of all ages, the profound masters in the
science of monopathy.    Who would not rather be alone,
and be capable of appreciating the dialogues of Plato,
than be amidst the crew of a Mississippi flat-boat, dancing with boisterous mirth to a Negro melody ?    Accept your pathetic loneliness without shrinking, and pace
the bleak waste without complaint, despite the yearning and the grief.    Travel on, if need be, through the
stony wilderness of despair, without a star, faithful to
conscience and God ; and an omnipresent voice at length
will interpret the desolation into peace, and fill the transfigured solitude with the sweetness of an infinite companionship.

For loneliness not only affords incomparable opportunities for preparation, not only yields strength and rest,
not only ministers to virtue ; it also furnishes rare and
costly joys in unequalled compensation for the pangs felt
in it.    Seclusion and peace are the guardians of innocent
dreams, the nourishers of poetic feeling and holy faith.
The disturbance of rivalry with another, of contempt or
injustice from another, shakes the liquid glass of soul in
which the blessed visions move and the divine joys sleep,
and ruffles them away ; as when in the material world
Wind rides forth with uplifted sceptre,

> And breaks
> The pageants mirrored in the lakes.

The grandest bestowals approach us not when we are
elbowing in the multitude for a conspicuous place, but
when, reticent and receptive, quiet and prayerful, we wait
on destiny in secret.    The king draws near with a throng

amid flaunting banners and salvos of cannon. God comes shrouded in silence and alone. If

> We needs must hunger, better for man's love
> Than God's truth ! better for companions sweet
> Than great convictions !

Jean Paul says, "Great souls attract sorrows as mountains tempests." They also command the same sublime prospects, and bestow the same inestimable benefits on the plains. It is noticeable how fond men of genius are of studying late at night. In the mysterious silence and seclusion of the time their feelings find an exciting affinity, and their thoughts are busiest. The sounds of outer industry have died on the air. The dusky landscapes stretch off into unbroken obscurity. The shapes or the footfalls of passers no longer vary the monotony. Even pleasure and rivalry, sickness and pain, care and avarice, are lulled to rest, and subside into dreams or lapse to temporary oblivion. Then the great poet, sage, saintly student and lover of humanity, solemn and ardent adorer of God, muses and toils. His imagination spreads its powerful vans, and, alone, awe-struck comrade of infinity, he sails on, high over the sleeping hosts of mankind, and far away, beneath the stars.

Wordsworth, who is one of the soundest and carefullest teachers mankind have had, writes : — " I do not recommend absolute solitude as an advantage to anybody. I think it a great evil ; frequent intercourse with the living world seems necessary to keep the mind in health and vigor." But we are indebted to the same deep and patient master of experience for the following paragraph, — the solemn and burning burden of a prophet : — " It is an awful truth, that there neither is nor can be any genuine enjoyment of poetry among nineteen out of twenty of those persons who live, or wish to live, in the broad light of the world, — among those who either are, or are striving to make themselves, people of consideration in society. This is a truth, and an awful one, because to be incapable of a feeling of poetry, in my sense of the word, is to be without love of human nature and reverence for God."

K

The nature of the man exposed to a constant round
of fashionable living undergoes a crisping and hardening
process, a veneering process, which destroys its susceptibility to the impressions of natural beauty, mystery, and
piety.   He outgrows — or rather dies out from — the
power of wonder, love, and enthusiasm.   This encase-
ment of blunting usage makes him incapable of the
purest range of emotions, cuts him off from admiration,
pity, every deep and fresh impulse of unselfish experience.   Robbed thus of sentiment, dry, conceited, sur-
feited, he never thinks to ask in adoring wonder and
delight, WHO it is that polishes the eye of the antelope,
pins the rainbow to the cloud, and, as often as Night
returns, sets on her sable brow the silent crown of stars.
On the contrary, he who seeks every opportunity for sequestered reflection, — who pauses, apart, when the mur
muring tides of toil are as still as if there were no life,
to commune with the ultimate facts of experience, is
likely to be open to the poetic and religious lures of mystery, quite sure to learn the value of simplicity and independence, and to be happy in himself, not resting for his
content on complicated and precarious sets of conditions.

The happiness supposed to belong in those hallowed
refuges from the storms and cares of the world, the abbeys and convents of the Middle Age, is charmingly indicated, as Montalembert has remarked, in the names the
monks gave their religious houses.   Good Place, Beautiful Place, Dear Island, Sweet Vale, Good Rest, Blessed
Valley, Haven of Delights, Valley of Peace, Bird's Nest,
Valley of Salvation, Way of Heaven, Sweet Fountain,
Brightness of God, Happy Meadow, Blessed Wood, Consolation, Dear Place, Joy, Crown of Heaven.   The pro-
nunciation of these delicious words is like the dripping
of successive drops of honey on the tongue ; and, with a
half melancholy, half luxurious heart-ache, one almost longs
to loose himself from all other ties, and go make his everlasting home in one of these still retreats.

Many a visitor lingering beyond the intended limits of
his stay in mountain vales, the secluded homes of innocence, truth, frankness, and health, has bid them farewell

with an aching heart and with tears. Does any one leave a crowded ball-room or supper-party so? The picture drawn by Robert Burns, with his artless but powerful touch of sweet simplicity, must wake a response in every unperverted bosom.

> How blest the Solitary's lot!
> Who, all-forgetting, all-forgot,
> Within his humble cell,
> The cavern wild, with tangling roots,
> Sits o'er his newly gathered fruits,
> Beside his crystal well!
> Or, haply, to his evening thought,
> By unfrequented stream,
> The ways of men are distant brought,
> A faint collected dream.

The intensely poetic feeling for the secret haunts of nature, the semi-religious love of the lonely and sublime scenes of glens and mountains, the solacing and restorative charm of nature for hurt and over-sensitive minds, — this marked phase of modern experience is quite a recent growth. The deserted and impressive scenery of the world used to be awful to men. Henri Martin, speaking of the feeling towards nature expressed in the French literature of the seventeenth century, says, " The smallest solitary valley was a *horrible solitude*, the smallest rock, a *frightful chaos.*" And he adds that, "it was the excess of their sociality, the absolute necessity of conversation, that gave rise to this abhorrence of the desert." But a deeper and darker influence than this Parisian gloss would imply, lay under the experience. It was the general cruelty and terror and superstition that were abroad. It was a lurking belief in diabolic spirits and agencies, — that *demoniacal possession of nature* in which the pagan mythology died out in mediæval christendom. The heathen deities, fauns, dryads, oreads, that once tricksily danced over the classic landscapes, under the influence of the popular Christianity had changed into devils haunting every dark and remote place, making solitude fearful. Rousseau, Saint Pierre, Chateaubriand, free from this superstition, wounded by men, bearing their passionate souls into the retreats of nature, finding balm, comfort, health, bliss in communion with her, by the creative genius so eloquently breathed in their writings spread abroad the new and delightful sentiment afterwards deepened by Wordsworth, Byron, Shelley, and scores of other gifted authors, and now become so common.    Bernardin Saint Pierre, whose famous "Studies of Nature" have been so influential, celebrates in this work his own grateful experience.    At the time his "Paul and Virginia" had melted all hearts, from the palace to the cottage, he felt his own heart breaking within him as he wandered, poor, sick, homeless, alone, and despairing.    "The ingratitude," he says, " of those from whom I had deserved kindness, unexpected family misfortunes, the loss of my small patrimony, the blasting of my hopes, had made dreadful inroads upon my health and reason.    I found it impossible to continue in a room where there was company.    I could not even cross an alley in a public garden if several persons had collected in it.    When alone, my malady subsided.    At the sight of any one walking to the place where I was, I felt my whole frame agitated, and was obliged to retire.    I often said to myself, My sole study has been to merit well of mankind; why do I fear them?" He was restored in body and mind by following the advice of his friend Rousseau.    " Renouncing my books, I threw my eyes upon the works of nature, which spoke to all my senses a language that neither time nor nations have it in their power to alter.    Thenceforth my histories and my journals were the herbage of the fields and meadows.    My thoughts did not go forth painfully after them, as in the case of human systems ; but their thoughts, under a thousand engaging forms, quietly sought me.    In these I studied without effort the laws of the Universal Wisdom."    The irritability was stolen from his temper, the soreness from his mind, the wounds from his affections, and he grew well and happy,—healed by the accords, sanctities, and repose of Nature, who smiled on her votary,

> Mild druid of her woodpaths dim,
> And laid her great heart bare to him.

It is strange how deeply linked the love of nature is

with the souls of her frank intimates, — how fondly, even
in their dying moments, they yearn over her familiar spots
with memories unwilling to separate. It is touching to
read of Robin Hood shooting his last arrow to the place
in the forest glade where he wanted his grave to be.
Wilson, the ornithologist, wished that he might be buried
in the woods, where the birds would sing above his grave.
The inmost fibres of the heart mystically respond to the
romantic peace of the immemorial antiquity of nature.
And there is always a double charm in the protected repose and loveliness of lonely nature when it is experienced
as a contrast with the turmoil of civilization, the wrongs
of unkindly men, the sorrows of life. Thomson cries, in
his " Hymn on Solitude," —

> O, let me pierce thy secret cell,
> And in thy deep recesses dwell !
> Perhaps from Norwood's oak-clad hill,
> When meditation has her fill,
> I just may cast my careless eyes
> Where London's spiry turrets rise,
> Think of its crimes, its cares, its pain,
> Then shield me in the woods again.

When the gentle and holy Silvio Pellico left his prison,
prematurely aged and broken, he felt that the only boon
left for him was serenity. He withdrew from politics,
from the world, devoting himself to his parents and to the
religion of a peaceful inner life. He did this in no spirit
of hostility, but in a spirit of resignation. He said, " I
read, I think, I love my friends, I hate no one, I respect
the opinions of others and preserve my own." As quoted
by Tuckerman, in his beautiful sketch of the Italian Martyr, he wrote to his friend Foresti : " I have learned that
but little is needed to beautify existence save the society
of the loved and honorable." Yet so misunderstood and
persecuted was he by personal critics, that he was forced
to say, " I left Spielberg to suffer another martyrdom in
my own country, — calumny, desertion, and scorn, which
have stripped all earthly illusion from life." The religious consolations of loneliness, a resignation to retirement in forgiveness and faith, are the choicest resource
left for such a man in such circumstances.

There can be no equable, sufficing happiness except in a self-ruled and withdrawn spirit, — a spirit that, in the idea of God, busy with impersonal and eternal objects, in dear retirement, reposes on great bases of truth, nobleness, and peace. I hardly know of a more touching proof of this than is afforded by the experience of the illustrious Madame Récamier. After forty years of unchallenged queenship in French society, constantly enveloped in an intoxicating incense of admiration and love won not less by her goodness and purity than by her beauty and grace, she writes from Dieppe to her niece : "I am here in the centre of fêtes, princesses, illuminations, spectacles. Two of my windows face the ballroom, the other two front the theatre. Amidst this clatter I am in a perfect solitude. I sit and muse on the shore of the ocean. I go over all the sad and joyous circumstances of my life. I hope you will be more happy than I have been."

It is certainly a potent neutralization for the gnawings of envy and depression to know that the pluckers of the great prizes and the occupiers of the great seats in the eyes of the world are not the happiest men, — in truth are generally the least happy men. The common multitude, who cannot conquer in the arena of social rivalry, should therefore contentedly stand aside from the struggle, and either admire or pity the victors, never hate or envy them. Pope Adrian the Fourth, in his Philosophical Trifles, says : "I know no person more unhappy than the Sovereign Pontiff. Labor alone, were that his only evil, would destroy him in a short time. His seat is full of thorns, his robe stuck with points, and overwhelmingly heavy. His crown and tiara shine, but it is with a fire that will consume him. I have risen by degrees from the lowest to the highest dignity in this world, and have never found that any of these elevations made the least addition to my happiness. On the contrary, I feel it impossible to bear the load with which I am charged." The close of the life of Aristotle, in a very different sphere, enforces the same moral. This prince of all true thinkers, loaded with immortal glory, was compelled to flee suddenly and by stealth to Chalcis, in order to save his life, and spare, as he said, the Athenians a new crime against philosophy. There, it is believed, this great man, in his old age, wearied with persecution, poisoned himself. The venerable Hildebrand, the greatest of all the Popes, after the herculean labors of his self-devoted and mighty career, crushed by an accumulation of hardships, said : " I have loved justice and hated iniquity ; therefore I die in exile." It is impossible to ascribe a slight importance to the moral lesson taught by the consenting sighs of so many of the masters of the world.

The greatest men — creators of philosophies, founders of religions, conquerors of nations — have ever been fond of a certain remoteness and privacy, — cultivators of solitude. Thus they have walled about their mysterious personality, clothed themselves with an alluring prestige which the curiosity of the crowd constantly sought, but was never permitted, to break into. Such men neutralize the sorrowfulness of their isolation by feeling that if they are alone, it is because they are so high ; also that it is their own choice, since they could easily stoop if it so pleased them. But they will not cast off their wondrous prerogatives. Who would not gladly have the celestial fire in his breast, even if it does sometimes "pain him with its burning"? How much feeling there is in these words of Michael Angelo ! —

> That which the good and great
> So often from the insensate world may meet,
> That evil guerdon did our Dante find.
> But gladly would I, to be such as he,
> For his hard exile and calamity
> Forego the happiest fortunes of mankind.

And the spirit of the gifted man often commands royal society when to the spectator's eye he is most deserted. Musing on the tracks and signals of his great predecessors he loves them wonderfully well. As he goes on his way alone, his pent-up walk

> Widens beyond the circles of the stars,
> And all the sceptred spirits of the past
> Come thronging in to greet him as their peer.

The selectest privilege of solitude, its most delicious charm, is liberty. Schopenhauer says, "Who does not love solitude, loves not freedom; for constraint is the inseparable consort of society." It was their morbid dislike to obey the becks and whims of others, conform to that average sympathy which is the rule of conventional breeding, which made Petrarch and Rousseau such lovers of solitude. The spirit of moral liberty dwells in solitude. As her votaries approach her altar there, all external constraint is gone. We are free. We shed our stiff awkwardness, every fatiguing posture, reserve, and grimace, draw a long sigh of delightful relief, and feel through all our powers the luxurious flexibility and naturalness of a perfect ease. To be alone is to be free to act unconstrainedly; and in a world of artificialities this is a happiness rarely to be enjoyed elsewhere. Escaping from the little interests, little passions, vexatious restraints of society into solitude, we seem to recover a lost good once native to us. The innumerable impressions of aboriginal freedom, of untamed nature, accumulated in the ancestral organisms whose experiences dimly vibrate in our own, constitute a basis for many weird reactions. In each of us the wild man sleeps at the bottom of every drop of blood; and in many an emotion, strangely vague and strong, he rises into consciousness. To the overtasked citizen of a feverish and suspicious society, weary, uneasy, ambitious, what an irresistible sense of relief, refreshment, and enchanted liberty, there is in the thought of exchanging hot drawing-rooms and noisy thoroughfares for such a scene as the traveller describes in the Bay of Seven Islands, in the virgin remoteness of cool Labrador! "The magnificent sandy beach of the east side of the bay, with its fringe of beautiful white and balsam spruce, forming the boundary of the forest which covers the flat country in the rear, is a fine camp-ground, ample enough for ten thousand Indian lodges. On a summer day, with a gentle breeze blowing, it becomes a delightful but very lonely lounge: and with the sea in front, the calm bay at your feet, the silent forest just behind, backed by the everlasting hills, which, inconceivably desolate and wild, stretch for a thousand miles towards the west, it is a fit spot for old memories to renew themselves, old sorrows to break out afresh."

Happy is he who, free from the iron visages that hurt him as they pass in the street, free from the vapid smiles and sneers of frivolous people, draws his sufficingness from inexhaustible sources always at his command when he is alone. Blest is he who, when disappointed, can turn from the affectations of an empty world and find solace in the generous sincerities of a full heart. To roam apart beside the tinkling rill, to crouch in the grass where the crocus grows, to lie amid the clover where the honey-bee hums, gaze off into the still deeps of summer blue, and feel that your harmless life is gliding over the field of time as noiselessly as the shadow of a cloud; or, snuggled in furs, to trudge through the drifts amidst the unspotted scenery of winter, when Storm unfurls his dark banner in the sky, and Snow has camped on the hills and clad every stone and twig with his ermine, is pleasure surpassing any to be won in shallowly consorting with mobs of men. If you are so favored as to have a friend, worthy the name, whose eye brightens and whose heart replenishes yours, in whose nature you find the complement and touch the equilibrium of your own, that is a very different affair. Exception is to be made in such a case. Every man with a healthy heart will endorse the charming thought of La Bruyère, thus versified by Cowper in his excellent poem, " Retirement."

> I praise the Frenchman, his remark was shrewd,—
> How sweet, how passing sweet is solitude !
> But grant me still a friend in my retreat,
> Whom I may whisper, Solitude is sweet.

Whoever is fond of receiving *great* impressions, expansive exaltations of consciousness, cannot fail to be irked and galled by the littlenesses and the festering jealousies of the crowd. Like Daniel Boone, he will gasp for breath within the conventionalities of society, and with a sigh of boundless relief rush to the wildernesses of nature and lonely thought, throwing his soul open to the fresh fellowship of field, forest, mountain, stream and star. The praised, aspiring Maurice de Guérin writes in his journal :

ʃ "The longer I live and the clearer I discern between
true and false in society, the more does the inclination to
live, not as a savage or a misanthrope, but as a solitary
man on the frontiers of society, on the outskirts of the
world, gain strength and grow in me.  The birds come
and go, and make nests around our habitations, they are
fellow-citizens of our farms and hamlets with us; but
they take their flight in a heaven which is boundless, but
the hand of God alone measures to them their daily food,
but they build their nests in the heart of the thick bushes,
or hang them in the height of the trees.  So would I, too,
live, hovering round society, and having always at my
back a field of liberty vast as the sky."

A precious prerogative of retirement and stillness is the
rejuvenation of the soul, the sentiments, the ideal faculties,
when years have heaped their scars and burdens on us,
when cares and sorrows have depressed our energies and
undermined our hearts with gnawing distrusts.  In the
company of others we are reminded of our rebuffs, our
disappointments, our age.  But in solitude the shames of
memory are flung off, no fear of ridicule represses imagination and affection.  We spread our wings there, and soar
with the old joy we knew when we were young and credulous.  This profitable delight is strikingly described in the
following passage by Leopardi, one of the loneliest of men,
who knew full well that whereof he wrote when he penned
it : "The habit of soliloquy is so confirmed from day to
day that when restored to intercourse with men its subject
feels himself less occupied in their society than in solitude.
And I do not think that this companionship with self is
confined to men used to meditation, but that it comes
more or less to all.  And more to be separated from men,
and, so to speak, from life itself, is useful.  For man, —
even when wise, enlightened, and disenchanted by experience of all human things, — accustoms himself to admire
them again from a distance, whence they appear much
more beautiful and worthy than close at hand ; forgets
their vanity and misery ; begins to form himself anew, and
almost to recreate the world to his liking ; to appreciate,
love, and desire life ; and with these hopes, if he is not

entirely deprived of the expectation of restoring himself to the society of men, nourishes and delights himself as he was wont to do in his early years. In this way solitude almost fulfils the office of youth. It makes the soul young again, restores the power and activity of imagination, and renews in the man of experience the blessings of his first inexperience."

One of the uses of solitude is preparation for death. Schopenhauer said, "My solitary life has prepared me better than most men for the lonely business of dying." It is a terrible pain to imagine ourselves absolutely cut off from sympathy with our fellow-men. Under the premonition of so fatal a loss the soul feels as if it were fainting away into infinite vacancy. Chaucer describes Arcite going away from his heart's queen, Emily, to be

> In his cold grave
> Alone, withouten any company.

To our natural instincts this is, perhaps, the deepest meaning of death, — to be thus darkly and utterly sundered from our kind. To rehearse the act in idea robs it of its terror, and of some of its dismalness, by familiarizing us with it. It is a bracing moral regimen, an exercise helping us toward a free personal detachment, often to fling ourselves forward in imagination to the time when thousands of men will be merry with their wives, children, and friends, laughing over their nuts and wine, while we shall be "walking alone along the lampless and frozen ways of death." Separated from men by a secluded life, it is easier for us to wean ourselves from the thought and love of all that society which death will end. Lacordaire wrote from his retreat at Sorèze to his friend, Madame Swetchine, " Every day they announce to me, on the part of those I have formerly known, rejections of opinion and changes of front, which give me the vertigo. O, how happy I am in being far from the spectacle! God in giving me this solitude has fivefold rewarded the labors of my life, and I ask Him for only one thing more, death!"

Common natures in their social relations suffer most

from want of objects to reciprocate their manifestations
of affection, and thus pacify the hunger of their hearts.
Baffled in this natural quest they pine.  The remedy for
such wounds is magnanimity.  There is no escape in
recoil ; there is cure in self-forgetfulness.  Be willing to
love and serve without noticing whether there is appreciative return or not.  This is a magical balsam for the
bruises of self-love.  It can even carry the peace of solitude into the heat and laceration of society.  The folly
of asceticism makes self-denial, the wisdom of morality
makes self-subjection, the law of man.  To live in the life
and feel the good of all, we need not to renounce, but only
to rule, our own.

The divinest souls, yearning abroad among their kind,
chiefly feel the need, not of a return for what they give,
but of objects to lavish their exuberant tenderness on
So copiously furnished, they ask not a supply but a vent.
This is seen in the examples of the highest characters of
history.  They accept their lot serenely, contented to impart alone so long as not thwarted in that.  The truly
great man remembers that ordinary men come to him as
they go to a fountain, "not to admire its stream, though
clear as crystal, but to fill their pitchers."  When kept
from bestowing, however, there must be profound anguish,
as is shown in the experience of Jesus, in that cry of
transcendent pathos, "O Jerusalem, Jerusalem, thou that
killest the prophets and stonest them which are sent unto
thee, how often would I have gathered thy children together, even as a hen gathereth her chickens under her
wings, and ye would not ! "  Then there is one resource,
and no more.  Its secret breathes in the sublime declaration, " And yet I am not alone, for the Father is with me."
Simply to relieve the heart we continue to sigh long after
we have abandoned all hope of a response, as the gray
bird perched on the tree-top sings his lonely plaint although the silence of the woods brings him no answer.
But profoundly beneath all conscious recognition we feel
that there is an invisible auditor, God, who marks every
pang, understands our love, sympathizes with our struggles and tears.

With what do we wish to live most neighborly? Not with beings merely fashioned in our form. No ; but with the forces that feed our noblest life. Primarily, with the intimate Divinity who inspires, commands, and loves us, and secondarily, with the persons who, serving his purposes as we ought to, may help us by their example and fellowship to do the same. And surely it is often true that *these* are nearer to us and more communicative when we are alone than when we are in company.

> O lost to virtue, lost to manly thought,
> Lost to the noble sallies of the soul,
> Who think it solitude to be alone !
> Communion sweet, communion large and high,
> Our reason, guardian angel, and our God !
> Then nearest these when others most remote,
> And all, ere long, shall be remote but these.

Earnestness feels the contact of indifference as a profanation. The deeper and richer that earnestness is, the more grateful and sufficing solitude becomes. Brooding over and pursuing its own purposes it is an inexhaustible source of true delights. But without a dedicated spiritual life within, loneliness is a famine breeding death. What moral profit did the solitude of their cave administer to the Seven Sleepers of Ephesus? Yet the natural affinity of aspiration and wisdom for retirement is clear to every observer. If you saw two persons intently reading, and, looking over their shoulders, found that one was absorbed in Lola Montez on the " Arts of the Toilet," the other in Saint Bruno on the " Delights of Solitude," you would infer a great difference in their respective characters, — a difference of mental dignity decidedly to the advantage of the latter. Most men live blindly to repeat a routine of drudgery and indulgence, without any deliberately chosen and maintained aims. Many live to outstrip their rivals, pursue their enemies, gratify their lusts, and make a display. Few live distinctly to develop the value of their being, know the truth, love their fellows, enjoy the beauty of the world, and aspire to God. Why are not more persuaded to join this select class? The first condition of desiring it is the removal of vice, shallowness, distraction, and indifference : and for this the regimen of solitude, in some form, is indispensable.

Spirituality did ever choose loneliness. For there the far, the departed, the loved, the unseen, the divine, throng freely in, and there is no let or hindrance to the desires of our souls. Memory, the glass in which we gaze on the absent, is called into requisition least where the present are thickest. Solitude is our trysting-place with the dead. God be thanked no earthly power can close that retreat or bar us from the sinless fellowship it holds. There, whenever I turn to the past, comes to meet me the mother — too delicate for this harsh world — who died so young, or ever I knew to give her the love she needed. There the bright and beautiful little brother, who followed her so soon into the darkness, with his masses of golden curls, his deep eyes, his winsome ways. There the transfigured boys and girls, friendly playmates of my unpolluted years, whom the Angel bore away and embalmed forever, before time could dim the glory on their brows or mix one adulterating ingredient with their ingenuous affections. There, too, the beloved youth, my eldest born, so good and true, just treading in the Castalian dew and flowers when wrapt in the sable fold of eternity. And close by him, snatched away in the same week, the darling daughter, sweet prattler, the youngest born, whose life was a brief frolic of beauty, innocence, and joy ; the little angel, who but flew from God to God across my path. Ah, how it makes the heart ache to remember that such things were and are not! But that ache is welcome, as a signal of the deeper life prophetic of its own future ease. And never will I refuse the still invitation to the tryst of the dead.

There are some lines in the sombre and jagged but powerful poem of an author whom it is now the fashion to underrate, which may serve to conclude these reflections on the uses of solitude. Young says of the dark solitude of night, that it

> Is the kind hand of Providence stretcht out
> Twixt man and vanity : 't is reason's reign,
> And virtue's too ; its tutelary shades
> Are man's asylum from the tainted throng.

The world's infectious ; few bring back at eve,
Immaculate, the manners of the morn.
Something we thought, is blotted ; we resolved,
Is shaken ; we renounced, returns again.
Nor is it strange : light, motion, concourse, noise,
All scatter us abroad ; thought, outward bound,
Neglectful of our home affairs, flies off,
And leaves the breast unguarded to the foe.
Ambition fires ambition ; love of gain
Strikes like a pestilence from breast to breast ;
Riot, pride, perfidy, blue vapors breathe ;
And inhumanity is caught from man,
From smiling man.    A slight, a single glance,
And shot at random, often has brought home
A sudden fever to the throbbing heart,
Of envy, rancor, or impure desire.
We see, we hear, with peril ; safety dwells
Remote from multitude ; the world 's a school
Of wrong, and what proficients swarm around !
We must or imitate or disapprove ;
Must list as their accomplices or foes ;
That stains our innocence ; this wounds our peace.
From nature's birth, hence, wisdom has been smit
With sweet recess, and languished for the shade.
This sacred shade and solitude, what is it ?
'T is the felt presence of the Deity.
Few are the faults we flatter when alone.

It is worth much time and study to understand all the
varieties of loneliness, from that of the dying savage
writhing on the ground under the sky, to that of the rahat,
or Buddhist saint, hovering on the verge of Nirwána.
There are two fearful specifications of the historic loneliness in human life which will load with pain every
sympathetic heart that ponders them.   The social loneliness of the Galleys is one.   Who can appreciate the
awful mass of unshared agony, shrinking apart, and deadly
languishment caused by that barbaric penalty which combined in one overwhelming woe the collected horrors of
exile, foul degradation, physical hardship, and hopeless-
ness?   The individual loneliness of imprisonment is the
other.   In connection with every royal house, every ruling
priesthood, in all civilized lands, there have been political
and ecclesiastical prisons.   And these prisons have never
long been empty.   Their dumb dungeons have been
crowded with noble or criminal aspirants — with bards,

patriots, thinkers, divine champions — whose thoughts
went out to free and strengthen the world, whose souls
exhaled into eternity through their dark bars.   The gen
erous reformers and heretics, kings, high adventurers,
artists, inventors, — from Joseph in Egypt to Kossuth in
Austria, — from Paul in Rome to Raleigh in London, —
from Tasso to Lovelace, — from Hebrew Daniel to Bohe-
mian Huss, — would form a list of names of extraordinary
interest.   The list would be one of portentous length.
And who but would sigh to think that all these have lain
under chains, in dismal cells, utterly cut off from friendly
contact with mankind, feeling the bitter weight and loneliness of their fate until every one of them breathed the
Psalmist's petition, " Lord, lead my soul out of prison ! "

It is no contemptible feat to appreciate the causes and
characteristics of all the kinds of isolation out of the average community of human life, higher and lower, — from
the stealthy solitude of smugglers unloading their boat
in a retired cove, while sea and gale roar together, and
fitful clouds drift over, and the stars shine dimly through
the rifts, — to the holy solitude of vestals who celebrate
their devotions, an ocean of crime and wretchedness, shut
out, surging around their walls, while starlight and torch-
light kiss in the storied windows, and the organ peals its
triumphant notes from the choir, and the plaintive strains
of the penitential hymn die along aisle and arch.

But it is especially grand and strengthening to know
the great lonely personalities of the past ; to be sympathetically caught up to the height where they stand, outlined against infinitude and eternity ; to be acquainted
with them as they were in their glory and their gloom,
their grandeur and their grief.   For, vast as were their
gifts and their strange joys, many times through the most
authentic utterance of their experience

> There sobs I know not what ground tone
> Of human agony.

This is true of them all, whether it be Moses, composing
his mournful psalm in the desert ; Plato, twice imprisoned,
once sold as a slave, forced to teach his deepest and dear·

est thoughts under an esoteric veil ; Empedocles, musing over his doubts upon Etna ; Tacitus, in the midst of decaying Rome, wrapt in the hermit-robe of his individuality, painting those sombre and Titanic pictures of crime and ruin ; Dante, slowly crystallizing the singular force and tenderness of his genius in a fabric of immortal verse with the stores of his learning and all the pangs his afflicted spirit knew ; Bacon, bequeathing his fame to his countrymen "when some time be past"; Pascal, sighing from his cell at Port Royal, " Man is so unhappy that he is weary without any cause for weariness "; Shakespeare, overflowing in the soliloquy of his sonnets, the dense obscurity that shrouds his life so indicative of its loneliness ; Kant, pacing the limits of his iron logic-castle, or writing in his memorandum-book a little before his death, with reference to that month being the shortest in the year, "O happy February ! in which man has least to bear, — least pain, least sorrow, least self-reproach "; Schleiermacher, passionately breathing his Monologues ; Mazzini, venting the fiery sadness of a prophet-heart in his exile-addresses ; or Rothe, immured in the colossal system of ethics his architectural faith and reason have built for his unshared residence. Such men as these, and their sparse peers, form no social groups ; but, far-scattered in time and space, each one is "as alone as Lyra in the sky." They are, in a sense, mediators between God and the crowd. They lift the gaze of their species. Above their separated epochs and lands, each on his own throne, Vyasa, Zoroaster, Gotama, Confucius, Pythagoras, Cæsar, Newton, Goethe, Swedenborg, and the rest, — poets, heroes, saints, philosophers, — radiant victors over sloth and sin and error and all the blind tyrannous brood of misery, — they afford to lower men at once example and inspiration, goal and guidance. He who has elevated himself into real fellowship with these solitary heads of great men, rising at wide intervals above the herd of common names, eternal conquerors of the oblivion which has made the rest its prey, need ask no other testimony to his achievements, no other reward of his toils and sorrows.

At the top of his mind the devout scholar has a holy of holies, a little pantheon set around with altars and the images of the greatest men. Every day, putting on a priestly robe, he retires into this temple and passes before its shrines and shapes. Here, he feels a thrill of awe; there, he lays a burning aspiration; further on, he swings a censer of reverence. To one, he lifts a look of love; at the feet of another, he drops a grateful tear; and before another still, a flush of pride and joy suffuses him. They smile on him: sometimes they speak and wave their solemn hands. Always they look up to the Highest. Purified and hallowed, he gathers his soul together, and comes away from the worshipful intercourse, serious, serene, glad, and strong.

### Conclusion.

SINCE, in the particular tendencies of the present time, our weakness lies in the direction of a gregarious miscel-lany and loquacity of life, instead of an undue seclusion, it behooves us to court, rather than to repulse, solitude. Innocence is better than eloquence, self-sufficingness a costlier prize than social conquest, self-denial and oblivion an aim of diviner sweetness and height than self-assertion and display. A holy man may well rejoice to be delivered by obscurity from the blame and praise of the world, glid-ing unnoticed to his end like a still rivulet under the grass in some sequestered vale. But few in our days sincerely wish this. And yet it were wisdom and religion to covet it, and slay our deadly foes, lust and vanity, with the bright weapons of the saint, renunciation and faith.

> Opinion is the rate of things,
> By which our peace doth flow;
> We have a better fate than kings,
> If we but think it so.

Make we then frequent withdrawals into meditative lone-liness and silence. And with reference to this let us re-member what Marcus Antoninus so well says: " Nowhere either with more quiet or more freedom from trouble does a man retire than into his own soul, particularly when he has within him such thoughts that by looking into them he is immediately in perfect tranquillity. Constantly then, give to thyself this retreat, and renew thyself; and let thy principles be brief and fundamental, which, as soon as thou shalt recur to them, will be sufficient to cleanse the soul completely, and to send thee back *free from all discontent with the things to which thou returnest.*"

There is, therefore, for the tonic discipline or the repose of solitude, no need of going to any remote hermitage. The game of *solitaire* may be played as effectually in the drawing-room or on the sidewalk as in cell or desert.

> There needs no guards in front and rear to keep the crowd away;
> Superior height of life and soul will hold them all at bay.

One thought, one recollection, one emotion, one sigh, — and you may be as far from the comrades who are talking and laughing around you, as though mountain-ranges intervened or oceans rolled between. In any company let me but think, Now, soft and faint, the starlight is falling along the shattered colonnades of Karnac, — and I am alone. Let me glance upward where the ghostly moon is swimming through the noontide air, — and I am alone. Let fancy go forth to the snow-laden flanks of the Alps, where the sombre pine-ranks are waving like hearse-plumes over the corse of nature, — and I am alone. Let a dream of heaven rise in imagination, a glimpse of the faces of the unforgotten dead pass before the mental eye, a sense of the presence of God rise into consciousness, — and, it matters not where the place is or how many noisy claimants press around, I am instantly and unutterably alone.

Even if we wish it not we must sometimes be alone. It is our duty to see to it that we are prepared to be alone profitably and cheerfully, without weariness and without fear. The difference in human solitudes is immense. The solitude was pleasing which Archimedes knew, — so absorbingly occupied with his mathematical problems as to be unconscious of the capture of his Syracuse, and to be slain by the Roman soldiers sooner than forsake his fascinating work. But how painful, to a man of his sensitive warmth, was the solitude of the brave and generous Canning in his premiership at the close of his career, — too proudly refined for democratic intimacies, cast off by his Tory associates on account of his liberal statesmanship, idolized by thousands who could not personally approach him, pitilessly persecuted by his surrounding enemies, stung at every pore, at once slowly bleeding and freezing to death on the height of his power. The loneliness felt by the subject of morbid superstition and terror resembles the landscape in the gloomy gorges of the Grand Chartreuse ; while the loneliness felt by the subject of healthy faith and awe before the unknown realities of being, resembles the scene on the white roof of the Milan Cathedral, when some visitor, climbing thither by moonlight and gazing on the forest of statues, feels as though a flight of angels had alighted there and been struck to marble. We cannot always live in public. There are secrets and moments we can never share. We should be familiar with the necessity, and make it grateful. We should cultivate in thought its serene, contentful aspects, and guard against its oppressive, fearful aspects. The gifted man, isolated in proportion to his superiority, if needing sympathy, feels, Schopenhauer strikingly says, as though men had forsaken him ; if self-sufficing, as though he had succeeded in running away from them. What a contrast of misery and blessedness in the states of these two ! Who is it that sits on the world as lightly as a gull on the ocean, except he who has learned by solitary thought to detach his affections from the prides and vanities of society, and often to lose himself in the fruition of a transcendent faith ? To be separated by ascetic superstition is to know the loneliness of Arsenius, who, after being tutor to the Emperor Arcadius, went into the desert, and for fifty years made his life one long solitary prayer. To be separated by the remorseful memory of crime, is to know the loneliness of Milo, when caught by the fingers in the rebounding oak he would split, and left as a prey to the wild beasts. To be separated by absorption in some sweet care, is to know the loneliness of Izaak Walton trouting in a secluded glen. To the guilty and debased soul there may come a loneliness like the solitude of a volcanic peak, full of boiling lava and smoke. To the virtuous and trustful soul there may come a loneliness like the solitude of a spring in the desert, where, all night long, the wild children of nature successively slake their thirst; the fawn and the panther, the lion and the elephant, — and the moon comes there, sees her fair face, and departs smiling.

Die away, then, vain murmur of tongues! Retreat, hollow hum of toils and cares! Fade out, cold procession of alien faces! Begone, fair seductions, that excite, then deceive and desert your victim! Cease to vex any more this poor brain and heart, ye restless solicitations to things that can never suffice, and that perish so soon! Disappear all, and leave me awhile alone, with my soul and nature, my destiny and my God!

# SKETCHES OF LONELY CHARACTERS

## PERSONAL ILLUSTRATIONS OF THE GOOD
## AND EVIL OF SOLITUDE.

———◆———

## BUDDHA.

About six centuries before the beginning of the Chris-
tion era, in Kapilavastu, a royal city of India, the Prince
Gotama, or Sakya Muni, was born in the palace of King
Suddhodarna, lord paramount of the Aryan race. Heir
to the regal glories of the house of the sun, brought up in
a powerful and splendid court, amidst the utmost richness
and refinement of poetic and metaphysical culture, he
was profusely supplied with all that could gratify the
senses or develop the mind. He was endowed with
surpassing personal beauty and with the noblest traits of
character. At an early age he showed such extreme
thoughtfulness and sympathy that his teachers foretold
his destiny to become a recluse. The king took every
precaution to prevent this catastrophe. But in vain.
Fate would be fulfilled.

One day, while riding out, he saw a decrepit old man,
half bent to the earth, tottering along with great difficulty.
Learning, on inquiry, that all men who lived to a great
age were subject to these infirmities, the prince sorrowfully meditated for a long time. After an interval he met
a beggar suffering from a loathsome disease. This spectacle of sores and pains brought his former reflections
back in double intensity. And when, a few months later,
he happened to behold a dead body in the last stages of
decay, and was told that this was the unavoidable end of
all men, he was horrified at the evils of existence. The
keenest sense of the vanity of worldly pleasure and magnificence took possession of him. Melancholy contemplations on the ghastly circle of birth, growth, decline,

disease, death, corruption, constantly occupied his soul. But at length, in one of his excursions beyond the palace, he saw a hermit, becomingly clad in a simple robe, walking cheerfully along the road with a staff and almsdish, in perfect health, with a serene and smiling face. The prince asked who this was. And being told that it was a recluse, who by religious withdrawal and meditation had freed himself from the cares and miseries of ordinary mortals, he at once matured his previous ruminations in an indomitable resolution to imitate the example thus set, detach himself from bondage to the disgusts of human existence which the gorgeous masks of his station had vainly hidden, and try to discover the means of eternally delivering himself and all men.

With transcendent strength of self-denial he fulfilled his purpose on the very day of the birth of his first child, when the whole palace and city were ringing with festivities. Pausing on the threshold of the room where the sleeping mother and babe lay in their loveliness, he gazed on them a moment, then turned away, forsook without a sigh the most seductive prizes of the world, and, accompanied by a single servant, whom he soon dismissed, started for a desolate forest, far from all the attractions and distractions of his former experience. What a picture it is of spiritual prowess, a peerless personality, a divine consecration, and an inexpressible mental loneliness, — the musing Gotama, only twenty-nine years old, in the fragrant bloom of his royalty, — after the three sad sights of helpless age, repulsive sickness, and putrescent death, and the pleasant sight of the happy ascetic, — stealing away from queen and child and palace, exchanging the diadem and golden robe for the alms-bowl and clout. and sitting down under the bo-tree in the deep woods to think out the doctrine of salvation !

For six long years Gotama persevered, in strict seclusion from the world, in the practice of all known austerities for reducing the flesh to its lowest influence, and arts for raising the mind to its grandest power, constantly striving to vanquish every selfish desire, every earthly attachment, and achieve a knowledge of truth. Astounding descriptions are given of the penances he underwent, the agonies he endured, the temptations he withstood, the repeated failures he experienced before his final victory. His figure, the authority he acquired, and the part he played in subsequent history became so prodigious, and the imaginative fertility and credulity of the Asiatic races were so teeming and unchecked, that it would have been unnatural if he had not been surrounded with a glittering cloud of supernatural attributes and feats, if he had not been obscured under masses of fictitious marvels. The deifying wonder that has wrought on the biography of Jesus, the collective miracles ascribed to him, are the merest trifle in comparison with the overwhelming powers attributed to Buddha. One consequence is a common doubt whether there really ever was such a person. But the doubt is not valid. The soundest historical criticism must admit that the prince and sage, Gotama Buddha, once lived, and that we possess, enveloped in stupendous perversions and exaggerations, a trustworthy knowledge of his life, character, doctrine, and influence. Stripping off the mythical accretion, we discern, under the distortions of the miraculous, the unmistakable indications of the natural and the true. Behind the grotesque exaggerations of the legendary monstrosity, we trace the affecting features of a genius and hero of the most exalted order.

Casting away the sceptre and laurel, retreating into the solitude of the wilderness, year after year Gotama maintained his pursuit of a perfect insight and emancipation, determined never to falter till he had solved the problem of existence. He had grown up in a country and age where innumerable rival sects, both in philosophy and religion, lived side by side, with universal tolerance, but engaged in keen debate. Hindu faith and metaphysics, represented by masters whose comprehensiveness and subtility of thinking have scarcely been surpassed, included every form of speculative opinion, both sceptical and dogmatic, dualistic and monistic, — polytheism, monotheism, atheism, pantheism, sensualism, subjective ideal· ism, objective idealism, absolute idealism, nihilism. No fineness, no stretch, no complexity of dialectics was un-

known to the Brahmanic sages.   Gotama went over these
varieties of thought with consummate vigor and patience.
He analyzed the nature of the soul and its constituent
faculties with exhaustive profundity and acumen, fearless-
ly scrutinizing the powers and experiences of human
nature, following every clew of logic and of intuition to
its furthest reach.   He canvassed the dogmatic beliefs
and religious rites of the Brahmans with startling audaci-
ty.   Long baffled, at variance in his own thought, dissatisfied and unsettled, worn to a skeleton with incessant
thinking and privation, at last the hour of triumph broke,
the end of his tremendous toils was accomplished, he believed he had attained the sum of truth, free from admix-
ture of error.   He was thirty-five years of age, when,
called Buddha, — the *Awakened*, the *Illumined*, — wiser
than the wisest, higher than the highest, he began to
teach his system for the salvation of all living creatures
from the miseries of existence.

  While the greatest teachers and leaders of our race are
most the fathers of the future, they are also most the sons
of the past.   No one, however originative, can be independent of his educational inheritance.   Freely as Gotama rejected or modified established views, and added
new ones, the foundation and motive of his system were
the same as those of the system in vogue when he arose.
From a combination of causes one predominant style of
thought and feeling had prevailed in India for ages.
The stagnant despotism of the government, the fixed
cruelty of the institution of caste, the oppressive heat
and languor of the climate, the weary monotony of usages,
the tenacious, passionate sensibility of the people, the
rich brooding meditativeness of the Hindu mind, con-
spired to produce an intense feeling at once of the burdens of life and of the profound unreality of sensible
objects.   The habit of thinking all natural phenomena
mere dreams and illusions, all existence an odious penance, nourished for many generations, had taken deep
root and secured vivid development in the whole Hindu
race.   No other nation was ever so priest-ridden, or accepted so besottedly the creed and ritual imposed on

them. They believed that the visible universe, filled with
created beings, from gods to insects, was a congeries of
deceptive appearances, in which all creatures were en-
tangled in a whirl of miseries. As soon as one died who
had not attained emancipation, he was born again in some
other form, to repeat the horrid routine. The supreme
sigh was to be freed from the chain of births and the
wheel of illusion. The means of this deliverance the
Brahmans monopolized in their own caste, with their exclusive possession of the Veda and the Sacraments.
They taught that there was but one real Being ; every
other existence was an illusion, removable when the soul,
by adequate penance, worship, and meditation, came to
pierce the blurring veils of sense, and recover the lost
knowledge of the identity of its own true self with the
sole Being. All who had not this knowledge could only
practise the prescribed ceremonies, and accumulate merit,
until some fortunate link of the chain of transmigrations
should bring them within the priestly class. Nature,
therefore, was considered a torturing round of illusions,
through which all creatures whirled in the circuit of transmigrations, hopeless of escape save through the door of
the Brahmanic caste. This fearful monopoly the priestly
hierarchy had managed for many centuries with monstrous self-complacency and a crushing popular ceremonial, using the key of knowledge for themselves alone,
seeking no converts in other castes, dispensing no redemptive light on other lands.

Gotama started from the same cardinal principles, but
with an abrupt difference of spirit and method. If the
system he constructed was more eclectic than original,
his wonderful moral sympathy and personality stamped it
with a startling freshness of form and novelty of power.
His four fundamental propositions were: *There is sorrow,
Every living creature feels it, Deliverance is desirable, Pure
knowledge is the only possible deliverance.* He first diverged
into sharp contradiction with the Brahmans by flinging
away, as worthless and burdensome, their cumbrous ceremonial law with its superstitious prayers, sacrifices and
austerities, and translating the substance of their abstruse

philosophy into a brief formula of salvation. Secondly, he diverged from them in the purity and expansiveness of his morality. His personal and didactic ethics were as noble as have ever been exemplified. He placed in the foreground of his system all the practical virtues, such as justice, veracity, purity, benevolence, reverence. He taught self-sacrifice in its highest form, and recommended the practice of every virtue on disinterested principles. When he had acquired his own deliverance, his mind burned with the divinest pity for others, with tender and heroic desires to redeem all from their sorrows. His was the first missionary religion that ever appeared on earth. Before him no religionist had ever dreamed of converting a foreign people to his form of worship. Religion was a family or national treasure scrupulously guarded from strangers. Not even the lowest grade of Hindus, the Sudras, would admit a foreigner into its ranks. But this great reformer, with an unequalled boldness of generosity, commanded his disciples to traverse the earth with the free offer of salvation to all. He was inspired by an unprecedented feeling of brotherly sympathy for the whole race. The earliest teacher of whom there is proof that he extended the sense of duty from the household, the village, the tribe, the nation, over all castes and outcasts, to the widest circle of mankind, is Gotama Buddha. It is his imperishable honor to be the first man historically known to have distinctly propounded the idea of humanity. Six centuries afterward Jesus conceived that idea with still deeper inspiration, and preached it with still greater effect. But it is wonderful that Buddha should have clearly declared it so long before, and the world will always owe him a debt of revering gratitude for the fruits it has borne in the followers of his faith.

While Gotama agreed with the Brahmans that the world was a prison and lazar-house, life an evil, deliverance a good, and pure knowledge the means of deliverance, his theory of what that pure knowledge was stood in extreme opposition to theirs. They taught that by penance, prayer, sacrifice, and reflection, man might attain the perception of the one divine Reality, and through that perception extricate himself from the time-medley of change and illusion, break the bond of metempsychosis, and, absorbed in the Godhead, be born no more. Gotama taught that by the practice of disinterested virtue and indomitable thought man might detach himself from all desire, and so neutralize the attractions that hold him in this wretched sphere as to fly away into a state of unconditional exemption. He believed strictly in no God, no absorption, no transmigration, no real self. But he had equivalents for all these : he recognized the phenomena which the Brahmans had generalized under these terms, only he sought by a sharper analysis and a wider intuition to give a sounder explanation of them. Like Hume, Spinoza, and other subtle masters of thinking, Buddha fancied he saw the delusiveness of all selfhood, saw that the soul is no substantive unit, but merely a current of states, its sole identity consisting of the accumulated mass of associations in experience, the organic conditions of memory. Accordingly, when the organism goes to pieces in death the soul is extinct, as a harmonious consensus ceases with the extermination of the related parts. The attainment of this knowledge, that the soul is a *process*, closing with death, and not a substance, capable of re-peated births and lives, is the first great step in Gotama's doctrine of salvation. This is the essence of his metaphysics, which he affirms, illustrates, and enforces without end.

But he could not wholly throw off the influence of the habits of thought embedded in the Hindu mind by thousands of years of intensely repeated meditations. The concentrated substance of these habits was intertwined with the doctrine of transmigration. Gotama furnished for the transmigrating soul which his remorseless analysis destroyed, a substitute as plausible to the mental state of his hearers as it is strange and incredible to us. He maintained that when one died who had not achieved a perfected insight and virtue, the desire that remained, the love of finite things, the cleaving to existence, *produced another being* endowed with the exact desert, good or bad, **left** behind by the departed predecessor. Thus though

there is no surviving soul in man, yet the law of retribu-
tion holds over; the fearful vortex of births is preserved
full, the detestable kaleidoscope of illusions is kept twink-
ling.   He attributes a kind of individuality to the *karma*
of every being, the aggregate of his actions during his
existence, the sum of his merit and demerit.   And this
*karma*, or collective moral worth of a man, when he dies,
is transferred intact to his successor.   It is a striking
example of what was the almost invariable error of the
ancient metaphysicians, — regarding an abstraction as an
entity.

Gotama saw that there could be no illusion without
some reality behind it to cause it.   If the soul or self
regarded as an integral entity was an illusion, there must
be some force to sustain the process of life on which that
illusion rested.   Now this force he interprets as a cleaving to existence, a subtle desire to be and to feel.   This
cleaving to existence is itself the result of ignorance.   In
consequence of ignorance, there is an accumulation of
merit and demerit; in consequence of merit and demerit,
consciousness is produced; in consequence of consciousness, the mental faculties and the body are produced; in
consequence of the mind and the body, sensations are
produced; in consequence of sensations, desire is produced; in consequence of desire, attachment is produced;
in consequence of attachment, birth is produced; in consequence of birth, grief, discontent, vexation, decay, and
death are produced.   Thus originates the complete catena-
tion of evils.   Whenever one of these constituents ceases
to be, the next in the series ceases to be, and the whole
combination of sorrow ends.

The method Buddha proposed for destroying the cleaving to existence was by removing the ignorance which
caused it.   This ignorance he would remove by destroying the self-love, the personal desires, the enslaving attachments, which blind men to the two truths that all
finite being is essentially evil, a painful turmoil of changes,
and that eternal deliverance from it is the absolute good.
This fatal love of self, this profound clinging to things
he would overcome primarily, by revealing to man ·he

phenomenal nature of the soul, that he is only a brief and complicated process of states, the new individual to whom his karma is to be transferred being an utterly separate person with no remembrance of him whatever ; and secondarily, by the most persevering emphasis and contemplation of all the disgusts and horrors of experience. In this manner he aimed to detach man from false delights, wean him from the folly of selfish affection, lead him to lose himself in an infinite surrender and repose, cause him in disinterested sympathy for others to labor to break the unhappy series of existences, dissolve the dark combination of woes, and unpeople the worlds by peopling Nirwána.

The meaning of this last word, Nirwána, is the key of Buddhism, alike in its own essence and in its distinction from Brahmanism and from Christianity. The attainment of Nirwána is regarded as the fulfilment of the highest possible destiny of man. The highest possible destiny of man, to the mind of the Brahman, is the identification of the self of the seer with the Soul of the Universe. To the Christian it is the immortal blessedness of the personal soul in a beatific world, the translation of the conscious individual to the society of the redeemed and the presence of God in heaven. What is it to the Buddhist? Is it to become identical with empty infinitude? The Brahman would say, with ultimate insight, I am God. This, at bottom, is the creed of every thorough-going idealist, such as Vyasa, Plotinus, or Hegel. The Christian would say, with filial trust, I am an inextinguishable personal spark struck out by God, a favored and indestructible child of the Infinite Spirit. This is the consistent creed of all who regard the soul as a finite immaterial entity. The Buddhist would say, with perfected detachment, I am nothing : and would carry out the legitimate consequences of the thought. Holding that his soul or selfhood has no substantial, but only a phenomenal, being, that it is but the point of convergence of the forces of the organism, yet believing that that phenomenal centre of consciousness is fatally bound to a continued succession of lives, and exposed in every life to innumerable loathsome evils until he so perfectly perceives the delusiveness of its substanti-

9                                                                          M

ality and so completely sheds all the affections begotten
by the illusion as to dissolve the karma and annihilate the
cleaving to existence, — he sets himself at work to secure
this end, to dissipate the spell of ignorance, break the
chains of desire, and achieve an absolute detachment, an
absolute indifference to everything.

Nearly all Christian writers, nearly all Western philosophers, who have studied this system, — so completely opposed to their own modes of feeling, — have been horrified
by it, filled with astonishment at it.   Even so flexible and
wide a scholar and thinker as Max Müller says, it seems
" a religion made for a mad house," and stands amazed
before the almost incredible fact that such religious power
should have been exerted, such moral benefits conferred,
by a teacher whose whole doctrine is summed in the dark
code of atheism and annihilation.   But atheism and annihilation are very different experiences to the Buddhist
and to the Christian, and exert very different influences.
The place occupied in the mind of a theist by the idea
of God, or by the idea of immortality, in the mind of an
atheist is occupied by something else : and this substitute
may fulfil for him the office of the idea whose place it
holds.   However strange it may appear, Nirwána, godless and empty as it is, largely discharges for the disciples
of Gotama the functions discharged for us by the ideas
of God and immortality.   It is the inspirer of their toils
and aspirations, the receptacle of their exhaling worships.
To appreciate the real nature and influence of the doctrine we must not stand on the outside with disdainful
superiority, but enter the interior with charitable humility
and curiosity and sympathy, and try to reproduce its relationships as they live in the bosoms of its advocates.
We must, for the time, divest ourselves of our own speculative and emotional peculiarities, and invest ourselves
with those of the ancient Hindus, and of Gotama himself.

Proceeding in this spirit, we shall first conceive of all
finite existence as made up of unreality, pain, and imper-
manence.   Next we shall conceive of all ignorant beings
as inextricably fastened by their ignorance and their de-

luded desires in this heaving collection of misery. Then we shall conceive that a perfect salvation for them all is possible and unspeakably desirable. Still further, we shall conceive that that salvation is to be won by a certain mode of thought, feeling, and action ; by the patient practice of every social virtue and self-sacrificing discipline. We shall see how that reflective insight, sympathy and self-surrendering aspiration, carried through the constant practice of the five great meditations of kindness, pity, joy, disgust, and indifference, will ripen into a perfect detachment and equipoise, — sure signal of the destruction of the productive cleaving to existence, infallible precursor of the eternal release, absorbing foretaste of Nirwána. And finally we shall, by an adequate contemplation of the illustrations he uses, so familiarize ourselves with the Buddhist's habit of sentiment that it will no longer baffle or be repulsive to us, but we shall enter into it as he does himself, putting it on and off at pleasure. Such an exercise, a mental freedom and force competent to the sympathetic conquest of modes of thought and feeling so wholly foreign from our own, is an achievement of the most honorable kind. Nothing can be more emancipating, expanding, and enriching in its effect.

The moral regimen of Buddhism is self-renunciation, disinterested sympathy, the common virtues of life, and meditative aspiration, carried to their last terms, for the purpose of escaping the intolerable evil of existence and winning the absolute good of Nirwána. Thus conceived, as it is by its votaries, so far from wondering at its effects we must see that no effects are too great to ascribe to it. To trace the proper working of any system of religion we should look at the system as it lies in the minds and hearts of its disciples, not as it is impoverished and degraded in the travesty presented by ignorant and hostile observers.

This difference in the quality of the same mode of thought when regarded by different persons is curiously illustrated in Jean Paul's critique on the moral influence of the subjective idealism of Fichte, and in Fichte's own estimate of it. This is Jean Paul's awful vision :

" Around me is a wide petrified humanity : in the dark unpeopled stillness no love glows, no admiration, no prayer, no hope, no aim.  I, so all alone, nowhere a single throb of life, nothing around me, and besides myself nothing but *nothing*, am only conscious of my lofty Unconsciousness ; within me the dumb blind working Demogorgon is concealed, and I am it.  So I emerge from eternity, so I proceed into eternity.  And who knows me now and hears my sorrow ?  I.  Who knows me and hears it to all eternity ?  I."  Compare with this horror the glowing picture drawn by the master himself :  " In this point of view I become a new creature.  The ties by which my mind was formerly united to this world, and by whose secret guidance I followed all its movements, are forever sundered, and I stand free, calm, and immovable, a universe to myself.  No longer through my affections, but by my eye alone, do I apprehend outward objects, and am connected with them ; and this eye itself is purified by freedom, and looks through error and deformity to the True and Beautiful, as on the unruffled surface of water forms are more purely mirrored in a milder light."

There are four paths leading by prolonged and arduous exertions to the fruition of Nirwána.  Through these paths Gotama sought by his system to guide all beings to the shoreless ocean of exemption, to the wall-less city of rest.  These four paths, Sowán, Sakradágámi, Anágámi, and Arya, are only divisions, at different approximations to the goal, of the one straight and narrow way, namely, the dissolution of the whole linked series of sorrow by the extinction of its two earliest terms, ignorance and desire.  By the "destruction of the hundred and eight modes of evil desire," the Buddhist "rescues himself from birth, as from the jaws of an alligator."  It is impossible to discover any way in which it is "desirable to hold a red hot bar of iron" ; so one who has fully contemplated the evils of existence can see "no form in which existence is to be desired."  It is delicious for one who has been "broiling before a fire" to "escape into the coolness of an open space" ; the evils of existence are the fire, Nirwána is the cool open space.

Is it true, then, that the religion which has had the most numerous following of all the historic religions has made an atheistic annihilation at once its God and its Elysium? Astounding as the proposition may be, so in the form of statement it is to us. But we may be quite sure so it is not in the substance of faith to its votaries. Let us, therefore, instead of turning away in scorn, or shuddering with horror, try to discern the meaning of Nirwána in the theory of life and death held by Gotama Buddha.

In the outset we must grasp the fact that the Oriental Buddhists loathe existence as the sum of evil, the Western Christians cling to it as the one good ; the former yearn towards extinction as the sum of good, the latter shrink from it as the one evil. This direct antagonism of faith and feeling between them and us is a result of historic causes, — race, climate, institutions, and other influences. To appreciate the truth in the case we must not begin by proudly assuming that we are wholly right, they wholly wrong. We must impartially endeavor to discern how far both may be right, how far each may be wrong. On reflection, it is clear, first, that those who follow their natural primitive instincts, dwelling on the known goods of experience, will cleave to life with blind exaggerating greed and tenacity ; their self-consciousness and selfish desires will gather into a ruling object of regard everything pertaining to the fruition of personality. Those, on the contrary, who, under the domination of an ascetic recoil, select for constant contemplation the known evils of experience, aggregating and emphasizing them, will naturally acquire a morbid dislike of life, an habitual weariness and loathing of it, as made up of the evils which exclusively fill their vision. Now, obviously, the truth lies between the two views ; and the wise and healthy style of conduct lies between the two extreme courses which they legitimate. Our present existence, which is by no means to be confounded with the entire range of universal life, is neither pure good nor pure evil, but a mixture of them, good in its essence and intent, evil in some of its accompaniments. It is, therefore, not to be supremely loved, nor supremely hated. With an accurate discrimination of its good and evil it is to be soberly valued, carefully improved, and meekly resigned at last in consonance with that Order of the Whole, which must be incomparably better and more important than any atomic part. The universal and absolute detachment which forms the soul of Gotama's theory of life, is the fanatical exaggeration of a sacred truth into a noble error. The true process and purpose of life is the fruition of function. Renunciation, the highest attribute of a moral being, is a function of free self-consciousness for the sake of co-ordinating, refining and enhancing the other functions. To make it a devouring end in itself, and use it for the suppression of all function, is a supreme perversion. The genuine purpose and destiny of man in this life is the self-ruled and harmonious fruition of the functions of his being, not their self-abnegated extermination. The real office of renunciation is not cessation and destruction, but regulation and fulfilment. It should free him who exercises it from slavery to all lower and worse standards, to the service of higher and better ones. Every self-denial should be the instrumental transition to a greater and purer gratification. Detachment from evil is a means ; the only end is attachment to good. The generous illusion in the Buddhist theory of salvation is that it makes an all-engulfing end of that which is truly but a means. Detachment from the transient and individual arrogates the place of attachment to the eternal and absolute.

And yet, though this error is correctly ascribed to the terms of the theory as set forth by its dogmatic expounders, it is experimentally neutralized in the hearts of those who practise the system, as it plainly was in the experience of him who first propounded it. All the renunciations and detachments of Gotama were prompted by and taken up into one supreme attachment, namely, that marked by the word Nirwána. All his desires were swallowed up in the one desire to be without desires. And *the desire to be desireless, carried to such a pitch of harmony and equilibration as to fancy itself extinguished in its own fulfilment, is*

NIRWÁNA. It is clear on reflection, that however closely, according to the ordinary conceptions of language, the Buddhist idea of Nirwána and the Christian idea of annihilation appear to correspond with each other, metaphysically and morally their fundamental meanings to the Eastern and the Western mind are in world-wide variance. By annihilation we mean a boundless negation, the deprivation of all being ; and we regard it as a blank horror. By Nirwána the Buddhist thinkers mean a boundless affirmation, the resumption of that relationless, changeless state of which every form of existence is the deprivation ; and they regard it as an infinite entrancement. Immortality and annihilation are words we use to mark our ignorance of the destiny of man after death, our ignorance of that limitless abyss of potentiality which is the foil to the visible creation. Imagination appropriates the attractive elements of the known to make the one mask beautiful ; and we love it ; appropriates the repulsive elements of the known to make the other mask hideous ; and we hate it. In like manner Gotama masks his ignorance of the dismal night against which all created things stand in relief, the unknowable, infinite side of our destiny, with the word Nirwána. And if that mask be formless and colorless, and yet he has the energy of faith to look towards it with unconquerable love and longing, the feat is wonderful rather than absurd, and he deserves to be admired. Instead of presuming to look down on this cosmopolitan hero of the mysteries of human life and destiny as a deluded inferior and unbeliever, we should see that there was much in his example both of faith and conduct so far superior to our attainment that we are scarcely competent to emulate it.

In self-sacrificing detachment from the collective seductions of the earth, with disinterested sympathy for all creatures, he forsook the throne of an empire for the tree of an anchorite in the forest, and persevered for years in the search for truth by meditations so profoundly abstracted, that, his biographers say, if a trumpet had been blown close to his ears he would no more have heard it than if he had been dead. Having completed his investigations, and compacted the results in a teachable form, he took up his residence in a monastic school, and began to gather disciples. For nearly fifty years he never ceased to proclaim his doctrine of salvation to all who would listen. He made frequent journeys over the country, preaching his system with an energy of conviction, an earnestness of appeal, a variety of illustration, and an emphasis of example, which combined with other co-operating influences to work a revolution in some respects the greatest ever wrought by one man. For when the dying old Gotama, at the age of eighty, under the sal-tree near Kusinara, saw Nirwána, his monasteries were dotting the hills, the yellow cloaks of his monks fluttering in all the roads, of India. And the system continued to spread rapidly over one nation after another, drawing swarms of converts, until it became, as it is at this moment, after the lapse of twenty-five hundred years, the most numerously followed of all the religions that have ever prevailed on the earth. The ignorant myriads of his followers, unable to understand or be satisfied with the transcendent abstractions of the system, deformed its teachings by the addition of their superstitious notions, and ended, in many cases, by deifying the sage himself and painting a new paradise in the abyss. Still, in all its forms, the religion retains much of the metaphysical speculation, and more of the sublime ethics, of its founder. The man who could do this, — overthrow the exclusive despotism of the Brahmanical hierarchy with his spiritual democracy, revolutionize surrounding countries, make his philosophy the religion of half the world for over a score of centuries, compelling innumerable multitudes of disciples to forego all the world for the self-denying repetition of his example, — this man must have had not only a personality, but also a faith, commensurate with these astounding effects. Gotama Buddha stands out as one man from amidst thousands of millions.

To stigmatize such a man, in the opprobrious sense of the words, as an atheistic eulogizer of nothingness, a godless unbeliever, is manifest injustice. Absolute pure being is nothing definite, is no *thing*. It is All. As Spinoza, with other metaphysical masters before and after him,

has said, every determination of being is a negation ;
every attribute or quality affirmed of it is a limitation.
Now Gotama's doctrine of the extinction of existence
means the removal of limitations, the destruction of all
obstacles to the return into that pure being, whereof, as
indicated by the word Nirwána, he himself says, " We can
affirm nothing, neither that it is nor that it is not, since it
has no qualities." It is that conditionless state, the idea
of which it bewilders the faculties of thought to conceive
and baffles the resources of language to express ; although
the writings of every deep speculative philosopher, from
Heraclitus to Hamilton, deal familiarly with it. The
scientific idea of force is the idea of as pure and mysterious a unity as the One of Parmenides. It is a noumenal
integer phenomenally differentiated into the glittering universe of things. The Christian who asserts that the Unknowable Cause of All is an intelligent and affectionate
Father, a personal counterpart of himself dilated to immensity, would brand as an atheistic nothingarian the
scientist who pauses with the idea of a unit of force and
denies substantive validity to everything else. And yet
to the philosopher who has adequately thought his way
to that conception, with the fit emotion, it is unquestionably a conception of overwhelming religiousness, capable
of yielding an unsurpassed measure of authority and
trust, of awe, sweetness, and peace.

Every representation of God, salvation, or heaven, is a
state of mind. To every man his highest apprehensible
reality stands for God. The purest, serenest, most sufficing state of mind he knows stands to him as the representation of salvation. The perpetuity of that state of mind
is his heaven. Apply this to Gotama Buddha. When as
the result of his exercises he had freed himself from un-
balanced desires, risen above the disturbing sphere of
worldly things, and in the perfect triumph of detachment
and indifference secured *an ecstatic equilibrium of the constituents of consciousness*, experimentally equivalent to the
extinction of consciousness, his mind a waveless sea without shore, fixing the unruffled idea of that state for eternity, he projected it to infinitude and called it Nirwána.
9 *

And thus, while according to our notions he believed in no God and no future life, Nirwána was to him at once God, salvation, and heaven.  The power of this faith inspired him to break the bonds of passion and vanquish the temptations of the world as easily as the arrow of a skilful archer cuts through the shadow of a tree.  It is the most wonderful psychological phenomenon in the history of the human race.  For Gotama Buddha, teeming with repose of strength, wisdom, and bliss, advanced towards Nirwána, in what seems to us the most absolute conception of loneliness, the most awful thought of solitude, that ever dawned on the mind of man, — no Personal Ruler of the Universe through which he was travelling, but an inflexible moral law treating every one exactly after his deserts ; and at the goal, no comrade, no object, no idea, no feeling, — only one unbounded, unbroken, and eternal blank.  But if in substance of thought Nirwána and annihilation are the same, they are wholly different in the form and color under which they are apprehended, and in the mode of feeling and spirit of life which they produce.  The perception of the indivisible unity of real being and the purely phenomenal nature of the self, — in the faith of Buddha this is the matchless diamond whose discovery sets every prepared slave free.

### CONFUCIUS.

AN important place must be granted to Confucius in any list of the illustrious lawgivers and exemplars of mankind. He also deserves mention among the great lonely men in the history of the world.  At the age of three years he was deprived of his father, received public office at twenty, began his course as a teacher and reformer at twenty-two, and lost his mother at twenty-four.  He is represented as weeping bitterly for his mother and paying her every possible tribute.  He was grieved by the cruelty and injustice of the rulers, and by the irreverence and viciousness of the people, and labored hard to teach both classes their duties as well by example as by precept.  His

moral purity, his learning and vigor, while drawing attention and disciples to him, also provoked the envy of rival
teachers, and the distrust of the officials over him. He
was dismissed from office ; and, in disfavor, in private
life continued his studies and labors for fifteen years, —
from his thirty-fifth to his fiftieth year. Then for five
years he was restored to the confidence of his sovereign.
He at length lost his post as minister of the court in Loo,
through the influence of some wantons who induced the
ruler to violate and resent the austere precepts of the
sage and abase him from his honors.

We catch impressive glimpses of his character in the
sayings he has left. The Master said, " The superior
man has dignified ease without pride ; the mean man has
pride without dignified ease." The Master said, " Some
men of worth retire from the world because of disrespect
and contradiction." In his fifty-sixth year the injured
Confucius turned from the seat of his fond hopes and
started upon his.exile. As he went along he looked back
on Loo with a melancholy heart, and gave vent to his
feelings in these verses : —

> O, how is it, azure Heaven,
> From my home I thus am driven ;
> Through the land my way to trace,
> With no certain dwelling-place ?
> Dark, all dark the minds of men !
> Worth comes vainly to their ken.
> Hastens on my term of years ;
> Desolate, old age appears.

For thirteen weary years he wandered from province
to province, using his faculties and his renown to the
utmost, but lamenting the want of Court position and
patronage to give his teachings more effect. Once he
said, " If any of the Princes would employ me, in the
course of twelve months I should have done something
considerable." At another time he said, " Am I a bitter
gourd ? Am I to be hung up out of the way of being
eaten ? " The world did not deal kindly with him ; for
in every province which he visited he met disappointment ; now suffering from poverty, now from deserted-

ness, now from persecution. Once he pined so sorely
for home and friends that he cried aloud, " Let me return,
let me return." Again he is said to have been several
days without anything to eat. While tarrying in Wei he
was so annoyed by applications to solve petty questions
and settle disputes that he exclaimed, " The bird chooses
its tree, the tree does not chase the bird,"—and prepared
to depart.

Just then came his recall to Loo. He was sixty-nine
years old. The remaining five years of his life he spent
in peace ; but not as he would have preferred. Denied
any place of rank and authority, his counsels set at naught,
he reluctantly turned away from his plan of tranquillizing
and perfecting the State through the Sovereign and the
Law, and devoted himself to the slower moral accomplishment of the same end by completing and transmitting his literary works. Perhaps one may understand
something of his disappointment in being obliged to
abandon a legislative and executive mission for a purely
didactic and moral one, from the following tribute paid to
him during his life by Tsze-kung, one of his disciples.
Tsze-kung said, " Were our master in the position of the
Prince of a State, he would plant the people, and forthwith they would be established ; he would lead them on,
and forthwith they would follow him ; he would stimulate
them, and forthwith they would be harmonious ; he would
make them happy, and forthwith, multitudes would resort
to his dominions ; while he lived, he would be glorious ;
when he died, he would be bitterly lamented."

Early one morning, it is said, he rose, and with his
hands behind his back dragging his staff, moved about
by his door, crooning, " The great mountain must crum-
ble, the strong beam must break, and the wise man wither
away like a plant. In all the provinces of the empire
there arises not one intelligent monarch who will make
me his master. My time has come to die." He went to
his couch and never left it again. He expired on the
eleventh day of March, four hundred and seventy-eight
years before the birth of Jesus. Legge, the best of his
English biographers, — from whose great work on the

Chinese Classics the chief data for this sketch have been drawn, — has painted the closing scene well, and moralized on it not unkindly, though, possibly, in a tone a little too professional and conventional.

If the end of the great sage of China, as he sank behind the cloud, was melancholy, it was not unimpressive. He had drank the bitterness of disappointed hopes : the great ones of the empire had failed to accept his instructions. But his mind was magnanimous and his heart was serene. He was a lonely old man, — parents, wife, child, friends, all gone, — but this made the fatal message so much the more welcome. Without any expectation of a future life, uttering no prayer, betraying no fear, he approached the dark valley with the strength and peace of a well-ordered will wisely resigned to Heaven, beyond a doubt treasuring in his heart the assurance of having served his fellow-men in the highest spirit he knew and with the purest light he had.

For twenty-five centuries he has been as unreasonably venerated as he was unjustly neglected in his life. His name is on every lip throughout China, his person in every imagination. The thousands of his descendants are a titled and privileged class by themselves. The diffusion and intensity of the popular admiration and honor for him are wonderful. Countless temples are reared to him, millions of tablets inscribed to him. His authority is supreme. He is worshipped by the pupils of the schools, the magistrates, the Emperor himself in full pomp. Would that a small share of this superfluity had solaced some of the lonesome hours he knew while yet alive !

## DEMOSTHENES.

IN spite of his burning patriotism, great statesmanship, and unequalled oratoric triumphs, Demosthenes impresses us as one of the lonely personalities of history. His exceptional ethical depth and fervor, his pronounced strength of character, the determination he formed in his orphaned youth to secure justice on his guardians for the neglect and wrong he had received from them, his tireless devotion both to the service of his country and to the art of eloquence, the stories of his long retirement in a cave, and of his solitary pacings on the stormy sea-shore, the bitterness with which a host of unscrupulous enemies pursued him through his whole career, — all combine to show that he was a man marked by a manifold isolation from his contemporaries. How he must have felt this, when for political reasons the Areopagus, with such foul injustice, decreed him guilty of pecuniary corruption ! When his haters, leagued with the rabble, had secured his banishment, it is said that he shed tears as he went. And during his exile in Egina, he went every day to sit on a cliff by the sea to gaze towards his beloved country. After the destruction of her liberties, the emissaries of the tyrant tracked him to the temple of Poseidon, where, turning at bay, he swallowed poison, and died at the foot of the altar. At a later day his penitent countrymen, whose eyes too late were opened to his nobleness, built him a tomb with the inscription, O Demosthenes, had thy power been equal to thy wisdom, the Macedonian Mars would never have triumphed in Greece !


## TACITUS.

WHEN we read the ominous lines in which Tacitus has described the corruptions and cruelties of his countrymen, we form to ourselves a picture of the historian as a lofty and sombre soul, turning with angry disgust from the stews and theatres and streets of Rome, from the dissemblers, informers, plotters, poisoners, sycophants, revellers, and murderers around him, — to live in his own thoughts. Regardless of immediate advantages, despising the arts of popularity, he turned in sorrow and scorn from the atrocities and beastly vices of pretors and emperors, and gave himself to the lonely task of transmitting to future times the terrible record of his own. This side of his life is revealed in his history. A softer and fairer phase of his soul, as it pleases us to imagine, appears expressed in the sentiments he puts into the mouth of Maternus, in his dialogue concerning oratory. "Woods and groves and loneliness afford such delight to me that I reckon it among the chief blessings of poetry that it is cultivated far from the noise and bustle of the world, without a client to besiege my doors or a criminal to distress me with his tears and squalor. Let the sweet Muses lead me to their soft retreats, their living fountains, the melodious groves, where I may dwell remote from care, master of myself, under no necessity of doing every day what my heart condemns. Let me no more be seen in the wrangling forum a pale and anxious candidate for precarious fame. Let me live free from solicitude, a stranger to the art of promising legacies in order to buy the friendship of the great ; and when nature shall give the signal to retire, may I possess no more than I may bequeath to whom I will. At my funeral let no token of sorrow be seen, no pompous mockery of woe. Crown me with chaplets ; strew flowers on my grave ; and let my friends erect no vain memorial to tell where my remains repose."

## LUCRETIUS.

THE eloquent and mighty Lucretius, lifted far from the vulgar ignorance and superstition of his time, revolving sublime thoughts and emotions in his powerful mind, leaving to posterity scarcely a trace of himself, save his burning and wonderful *De Natura Rerum*, was as solitary in his time as though he had lived in an aerial car, anchored miles above Olympus. We wonder what frigid and distressful isolation of his warm heart, or what maddening sorrow, led him, at the early age of forty-four, to open into the abyss the forbidden door of suicide. His story and his end furnish another illustration of the truth, that, out of an hundred great men, with ninety and nine the penalty is more than the prize ; the wreath on the head is less felt than the thorn in the bosom. It is to be hoped that he enjoyed a happy friendship with the Memmius to whom he addressed his poem. We recognize the proof of a noble heart in the generous enthusiasm with which he praises Epicurus, Empedocles, Ennius, and others of his illustrious predecessors. We greatly revere the humanity and heroism he showed in his powerful labors to free men from the dreadful curses of superstition so rife in his time. In the celebrated lines which form the opening of his second book we cannot but believe we see a partial picture of himself.

> How sweet to stand, when tempests tear the main,
> On the firm cliff, and mark the seaman's toil!
> Not that another's danger soothes the soul,
> But from such toil how sweet to feel secure!
> How sweet, at distance from the strife, to view,
> Contending hosts, and hear the clash of war!
> But sweeter far, on Wisdom's height serene,
> Upheld by truth, to fix our firm abode,
> To watch the giddy crowd that, deep below,
> Forever wander in pursuit of bliss.
> O blind and wretched mortals! — know ye not
> Of all ye toil for, Nature nothing asks,
> But for the body freedom from disease,
> And sweet unanxious quiet for the mind?

When we think of the immense mind of Lucretius escaping into the Invisible, it affects us as though some lone planet had rolled off the flaming walls of the Universe, and sunk into the night.

### CICERO.

CICERO is not only one of the most shining and attractive personages of Roman history, he is also one of its most original characters. The classic world furnishes not another example of such a splendid combination of talents, personal interest, a dramatic career, a tragic end, and immense fame. The rich complexity of his traits, his sensitive vanity, his ardent patriotism, his genial humanity, his philosophical tastes, the empassioned mobility of his moods, his love of natural scenery, the rapid alternations of his hankering for society and his desire for solitude, give to his spiritual portrait a more modern cast than that of any other of the ancients. He had that dominant self consciousness, that swift vehemence of action and reaction on the contrasted thoughts of self, friends, foes, country, mankind, duty, destiny, which we are accustomed to regard as a morbid peculiarity of the genius of later times.

Although of a burning ambition, and intensely subjective, he was endowed with the noblest susceptibilities for ail greatness, goodness, truth and beauty. He lacked that consolidated pride or rebutting self-sufficingness necessary for stability in his giddy position. He was too fond of notice and display ; loved too well to flatter and be flattered ; and was inconsistent with himself, boastful in prosperity, supplicating in calamity. His very weaknesses, however, were at bottom more closely allied to virtues than to vices. They arose not from selfishness or cruelty, but from an over-strong regard for the love and honor of his fellow-men. They sprang chiefly from the sympathetic vigor of his imagination, which successively presented to him the different aspects of things, persons, parties, policies, opinions, so vividly that each was ideally assumed for the instant. His frequent waverings were largely due to the fact that, unlike Cæsar, Pompey, and the most of his great contemporaries, he was troubled with a sensitive conscience. He could not think of single objects, detached from each other and from associations. The laws of contrast and affinity, powerfully active in his mind, were ever distinguishing and joining all to which he turned his attention. The desire to be remembered, to be loved and admired, was inseparable from his idea of himself. Pictures in his own mind of the appearance he should make in the eyes of mankind had an influence always too strong for his peace, perhaps sometimes too strong for his virtue. But surely this is a fault more gracious and venial than that stolid complacency which will ask nothing from society, or that contempt which scorns to be depressed and elated by the condemnation and approval of others.

The thoughts and feelings of Cicero seized him with such absorbing energy, that, for the moment, he was possessed by them, and impelled to give them an expression of proportionate power. Accordingly, in his orations, in his essays, especially in his letters, he freely pours him-self out. He has no secrets from his friends, holds nothing back from his pen. His brother Quintus, after receiving an epistle from him, writes, " I see you entire in your letter." Few of the characters of antiquity could bear this unreserved exposure as well as he does. If he loses something by it in the respect of the censorious who criticise him, he gains more in the love of the generous who judge him. He transports us into the midst of the scenes he describes, into the midst of his own soul. In estimating others we give the spectator data for estimat-ing us ; and the heart of that man is not to be envied who can, without a glow of loving admiration, read those honeyed and golden pages of Cicero which have sweet-ened the hours and enriched the souls of so many of the greatest men of succeeding ages.

It was as natural that such a man should sometimes recoil from the drudgery, hate, envy and hollowness that accompanied the career of ambition in the capital of the world, as it was that he should irresistibly covet both pri-vate friendship and public place and applause. Weary of the conflicts of the great days of the forum, half sick even of the shouts and laurels of the crowds, he turned his back on the dust and roar of Rome, and, with a joy like that of a modern poet, sought the shelter of some secluded villa. He had villas in the most retired and beautiful spots, at Antium, Arpinum, Formiæ, Tusculum, where he loved often to retreat to soothe his ruffled nerves, and to study and write. On reaching one of his country houses he is delighted with the fresh beauty of everything, and with the deep peace. He flies to his books, ashamed of having left them. His love of solitude is so great that he never finds himself solitary enough. His clients and acquaintances come after him, until he cries in vexation, " This is a public promenade, and not a villa." On the arrival of two bores he writes to his dear Atticus, " At the moment of this writing Sebosus is announced. I have not finished my groan before they tell me that Arrius also has come. Is *this* quitting Rome ? Of what use is it to fly from others and fall into the hands of these ? "

In these country retreats Cicero enjoyed the echoes of his fame, and in genial fellowship with nature, letters, and his absent friends, refreshed himself for further struggles, or labored to lift his reputation higher, and by means of philosophical works transmit it to the latest ages. Here, in his adversity, he vainly strove to forget the world, to care nothing for the detraction of his enemies and the neglect of his fellow-citizens. The effort was vain ; for, wherever he went, he carried Rome in his mind, with all its passions and plots. He could never forget the Roman people, nor be indifferent to their opinion of him. The hate and contempt of his rivals were as torturing to him as the love of his bosom friends was delicious. His exile was as agonizing as his coronation had been ecstatic. He was made for extremes, unfitted for the serene medium where self-content dwells. Too painfully vexed and hurt by the reactions of his excessive self-esteem, romantic imagination, and spiritual wealth, on the ignorance, envy, and coldness of careless and selfish men, he sought refuge in solitude. Yet this solitude was not true solitude. It was but the supersedure of actual companionship by an ideal one in connection with which he fought over the battles of the Senate, relived the triumphs of the past, and imagined greater ones for the future. It is obvious that he often knew, in all its revulsive force, the sharp loneliness of being flung back on himself, inwardly wounded and deserted. Every criticism disturbed, every sneer stung him. He was driven to both extremes, — to contend in the suffocating throng, to meditate and sigh in isolation ; and in both he was happy and unhappy, belonged with the most social and with the loneliest of men. The wittiest and most eloquent man of the Roman world, who was fonder of the festive board of friendship, or shone more conspicuously in the thick of swaying multitudes ? Musing on the sea-shore of Sicily, going brokenhearted into banishment, weeping in the dense woods of Astura, stretching out his neck to the sword of the wretch who pursued him, who more sadly solitary ?

Admit that he sought personal glory too keenly, and weakly shrank from ridicule and neglect ; the fault may be forgiven, the weakness is even not wholly unlovable in him.  Heap up all the accusations to which he is obnoxious, allow to them everything that truth can ask; still the fact remains that he was a miracle of genius and industry, an ardent and illustrious lover of his country, of philosophy, of literature, of humanity, and of virtue, whose works have scattered delight and benefit over many nations, through many centuries.  It is an ignoble and a hateful task to try to tarnish his record and create scorn for him.  His fame clusters with the affections of the greatest and best men for two thousand years.  It is a luxury to add to that tribute the homage of one more throbbing heart.

He has been stigmatized as a coward.  It is unjust. For one with his rich imaginative sensibility it was an act of transcendent courage to turn from the ideal air of philosophy and hurl those fearful Philippics amidst the very daggers of the myrmidons of Antony.  And, when his last effort for republican liberty had proved futile, did he not heroically die on the altar ?

The chief peculiarity of the character of Cicero, in a historico-biographical point of view, is the extent to which he anticipated the modern habit of over-sensitive literary genius, the deliberate portrayal of himself and his feelings in his writings ;  the eagerness with which he strives to show himself worthy of affection and honor, and to secure this prize alike from his contemporaries and posterity.  In this respect he is the prototype of Petrarch, who again is the prototype of Rousseau and of all who trace him in his line.

Those who wish to make a study of this great man, — so sweet and commanding despite his foibles, — will find no lack of helps in the works of his numerous biographers and crities.  Middleton, in his *Life and Letters of Cicero*, treats him with idolatry ; Niebuhr, in his *Vorträge über Römische Geschichte*, with enthusiasm ; Drumann, in his *Geschichte Roms*, with hate ; Abeken, in his *Cicero in Seinen Briefen*, with generous impartiality ; Mommsen, in his *Römische Geschichte*, with insolence ; Merivale, in his *History of the Romans under the Empire*, with a fairness rather severe than merciful ; Forsyth, in his *Life of Cicero*, with loving candor ; Boissier, in his *Ciceron et Ses Amis*, with affectionate justice. This work of Boissier is the most interesting, emotional, and just of the whole. One lays it down with the feeling that Cicero — the brilliant, brave, boastful, shrinking, timid, vain, garrulous, learned, wise, unhappy, tender, pious, immortal Cicero — deserves to be blamed somewhat, pitied a little, excused a great deal, admired more, praised and loved most of all, by the world of his fascinated and grateful readers.

## BOETHIUS.

The author of the " Consolation of Philosophy," Boethius, has a place singularly by himself among men, in the fame of his beautiful work. After holding at the court of Theodoric, king of the Ostrogoths, the offices of Consul and Senator, with brilliant ability, he fell into undeserved disfavor with his sovereign. His unflinching honesty, together with his conspicuous kindness, courage, and watchfulness, brought a pack of informers and other base men against him. A sentence of confiscation and death was passed on him unheard. During his imprison-ment he wrote that precious treatise on the solaces of wisdom, which has strengthened many a kindred sufferer from injustice since. He underwent a horrible death, being first tortured by a cord drawn around his head till the eyes burst from their sockets, and then beaten with clubs till he expired. When we trace the lofty medita-tions with which he comforted himself in his prison, and compare his sweet, generous mind and heroic virtues with the brutal ferocity of the jealous mediocrities around him, a tragic loneliness associates itself with the figure of him presented by the historic imagination.

## DANTE.

Dante Alighieri is the most monarchic figure in lit-erary history. Awe and Love now accompany the shade of the untamable Ghibelline on the journey of his fame, as he pictured Virgil guiding his steps through the other world.  That stern, sad, worn face, made so well known to us by art, looks on the passing generations of men with a woful pity, masking the pain and want which are too proud to beg for sympathy, extorting, chiefly from the most royal souls, a royal tribute of wonder and affection.

Some one has said that Dante was "a born solitary, a grand, impracticable solitary.  He could not live with the Florentines ; he could not live with Gemma Donati ; he could not live with Can Grande della Scala."  The truth in the remark is, perhaps, a little misleading.  It is certainly not strange that an exile should be unable to live at home with the victorious party of his persecutors ; that a man absorbed in an ideal world should ill agree with a prosaic and shrewish wife ; or that the demeaning favors of a patron should gall a generous spirit.  Dante was no separatist, either in theory or in native temper of soul, though he was lonely in experience and fate.  The inward life was to him the only constant end ; the ecstasy of the divine vision the only sufficing good.  Memory, thought, and faith were his three cities of refuge.  His intellect was too piercing, his disposition too earnest, his affections too sensitive and tenacious, his prejudices and resentments too vehement and implacable, for satis- factory intercourse with others to be easy.  "He delight- ed," Boccaccio says, "in being solitary and apart from the world, that his contemplations might not be interrupt- ed.  And when he was in company, if he had taken up any subject of meditation that pleased him, he would make no reply to any question asked, until he had con- firmed or rejected the fancy that haunted him."  Ben- venuto da Imola speaks of his having been seen to stand at a book-stall in Siena, studying a rare work, from mat- ins till noon ; so absorbed in it as to be unconscious of the passing of a bridal procession with music and love- poems, such as he especially delighted in.  Owing to the extraordinary scope, intensity, and pertinacity of his states of consciousness, he was both an exceedingly lov- ing and magnanimous, and an exceedingly irascible and revengeful man. If he was sensitively exacting, he could also be regally self-sufficing. To such a nature fit society would be delicious, but hard to find ; unfit society, easy to find, but insufferable ; solitude, a natural refuge, not less medicinal than welcome.

The different kinds of spiritual loneliness meet in a more striking combination in Dante than in almost any other man. He knew, in a distinguishing degree, the loneliness of individuality ; for he had a most pronounced originality of character, all of whose peculiar features the circumstances of his age and life tended to exaggerate. Altogether, with his towering self-respect, his deep sense of his own prophetic office, his soft, proud, burning reveries, it would be hard to find a more intrinsically isolated personality. He knew the loneliness of genius, his mind being of a scale and altitude far aloof from those about him. Among the peaks of human greatness, the solitary cone of the intellect of Dante shoots highest into the sky, though several others touch a wider horizon and show a richer landscape. He knew the loneliness of love. The wondrous fervency and exaltation of his sacred passion for Beatrice, no one else could enter into : he could speak of it to no ordinary comrade. In his own words, " The first time I heard her voice, I was smitten with such delight that I broke away from the company I was in, like a drunken man, and retired within the solitude of my chamber to meditate upon her." He knew the loneliness of a passionate, idealizing grief. He says, " I was affected by such profound grief, that, rushing away from the crowd, I sought a lonely spot wherein to bathe the earth with my most bitter tears ; and when, after a space, these tears were somewhat abated, betaking myself to my chamber where I could give vent to my passion unheard, I fell asleep, weeping like a beaten child." And again he says, —

> Ashamed, I go apart from men,
> And solitary, weeping, I lament,
> And call on Beatrice, " Art thou dead ? "

He knew the loneliness of an absorbing aim. The production of his immortal poem, in which heaven and earth

were constrained to take a part, and which, he says, kept
him lean many years, implies immense studies and toil.
Such an exhaustive masterpiece is not more a result of
inspiration than of unwearied touches of critical art.   He
knew the loneliness of exile.   Banished by party hate, he
always yearned after his dear Florence; upbraided her
that she "treated worst those who loved her best"; and,
in his very epitaph, called her the "of all, least-loving
mother."   He wandered in foreign lands, from place to
place, almost literally begging his way, "unwillingly showing the wound of fortune," tasting the saltness of the
bread eaten at other men's tables, and at last dying in a
strange city.   He knew the loneliness of schemes and
dreams reaching far beyond his own time, embracing the
unity and liberty of his country; over whose distraction
and enslavement others slept in their sloth or revelled in
their pleasures.   And finally, he knew the loneliness of a
transcendent religious faith, which his imagination converted into a vision ever recalling his inner eye from the
gairish vanities of the world.

Before Dante was driven out by his fellow-citizens,
Beatrice had died; his best friend, Guido Cavalcanti,
had died; and he had lost, by the plague, two boys, aged
eight and twelve years.   Carrying these scars, and another
as dark, inflicted by the disappointment of his patriotic
hopes, he went forth never to return.   Although he awak
ened interest everywhere, his tarryings were comparatively
brief.   He knew his own greatness.   His unbending king-
liness, his serious and persistent sincerity, unfitted him for
intercourse either with vapid triflers in the crowd, or with
haughty mediocrities in high places.   God made him incapable of fawning, or playing a part.   He must appear
as he was, act as he felt, speak as he thought.   It is obvious from his history that he profoundly attracted the
superior men with whom he came in contact.   This is not
inconsistent with the fact that speedy breaches occurred
between him and nearly all of them.   He broke with
some because they betrayed the cause of his country;
with others, on account of personal incompatibilities.
Who possessed fineness and tenaciousness of spiritual

fibre, richness and energy of mental resources, sobriety and loftiness of imaginative contemplation, to act and react in unison with the soul of Dante Alighieri?

He had a warm intimacy with the imposing and brilliant military adventurer, Uguccione della Faggiuola, and offered him the dedication of the "Inferno." There appears to have been a strong attachment between him and Giotto. One cannot look on the recovered portrait of Dante by Giotto, without feeling that it must have been drawn by a hand of love. Benvenuto da Imola relates, that one day, when Giotto was painting a chapel at Padua, — the wondrous frescos which at this day make the traveller linger on them with a sweet pain, unwilling to tear himself away, — Dante came in, and the painter took the poet home with him.

When first banished, he was generously welcomed in Lunigiana by the Marquis Morello Malaspina. Before long, however, he went to enjoy the splendid hospitality of the young lord of Verona, Can Grande della Scala. In a letter to Can Grande, dedicating the first cantos of the "Paradiso" to him, he says, "At first sight I became your most devoted friend." He lays down the proposition, that "unequals, as well as equals, may be bound by the sacred bond of friendship." In support of this, he gives several arguments ; one of which is, that even the infinite inequality of God and man does not prevent friendship between them. The grandees at the court looked down on Dante from their titular elevation : he looked down on them from his intrinsic superiority. One day, Can Grande said to him, concerning a favorite buffoon, "How is it that this silly fellow can make himself loved by all, and that thou, who art said to be so wise, canst not?" Dante replied, "Because all creatures delight in their own resemblance." The offended poet departed. He paid a long visit to Fra Maricone in the convent of Santa Croce di Fonte Avellana, where he wrote much of his matchless poem. Later he found a pleasant refuge with his good friend, Bosone da Gubbio, in the castle of Colmollaro. But his last, kindest, most faithful patron and friend was the noble ruler of Ravenna, the high-souled and cultivated Guido Novello da Polenta. Here he spent the last seven years of his life, furnished with a fitting home, his wants supplied, treated personally with deference and love, employed in honorable offices. When he died, his remains were honored with an imposing funeral. His body, robed as a Franciscan friar, lay in state in the palace of the Polentas ; his hands resting on the open Bible ; a golden lyre, with broken chords, lying at his feet. The erection of a becoming monument was prevented only by the misfortunes and banishment of Guido himself.

In spite, however, of these exceptions, Dante's word is true, " It is rare for exiles to meet with friends." The picture of him in Paris, deserted, destitute, hungry ; sitting on straw in the Latin Quarter listening to the University lecturers ; admitted, after extemporaneously defending propositions on fourteen different subjects, to the highest degree, and obliged to forego the honor for lack of means to pay the fee, yet consoled by the hope of an enduring fame, is pathetic and exciting. How touching, too, are his words in the treatise " De Vulgari Eloquio " ! — " I grieve over all sufferers ; but I have most pity for those, whoever they may be, who, languishing in exile, never see their native land again, except in dreams." Yet, with the force of his invincible soul, he rallies upon divine resources, and enjoys ideal substitutes and equivalents for what he is deprived of in actuality. " Shall I not enjoy," he exclaims, " the light of the sun and the stars ? Shall I not be able to speculate on most delightful truth under whatever sky I may be ? "

There are truly two Dantes, — one, the young Dante of the " Vita Nuova " ; the other, the mature Dante of the " Divina Commedia." The first is represented in the portrait by Giotto, with its meditative depth, feminine softness and sadness ; the second, in the more familiar traditional effigy, with its haggard, recalcitrant features, iron firmness, and burning intensity, its mystic woe and supernal pity. Both of these characters are abundantly revealed by his own pen, since almost everything he wrote has an autobiographic value, both direct and indirect. He often narrates the events of his life, and records his feelings and judgments, in the first person. Furthermore, the contents of his works take the form of experiences passing through his soul, and reproduced by his art in stereoscopic photographs that at once reflect the delicate lineaments of his genius and betray the tremendous power of his passions.

The dominant characteristic, in a moral aspect, of the younger Dante, — of Dante as he was by nature and culture, — is the tenderest and most impassioned ideal love, frankly exposing itself on every side, and seeking sympathy. He speaks, confesses, implores, with an exuberant impulsiveness of self-reference like that of Cicero, whom he studied and loved; and he describes his painful consciousness of loving and thirst for love, with a fulness of self-portrayal like that of Petrarch. This phase in the character and life of Dante has been for the most part overlooked; but no one can read his "Vita Nuova" and his "Canzoniere," with reference to this point, and fail to recognize it. Free from the foibles of Cicero and the extravagances of Petrarch, fully possessed of what was best and most original in them, Dante, in his first literary development, is the true link between the humane philosopher of Rome and the romantic poet of Vaucluse. He had the learned scope and effusive sympathy of the one; and he had the clinging, introspective Christian sentiment and faith of the other. The Romantic Literature, — between which and the Classic Literature Petrarch stands with a hand on either, — that glorious outbreak of the spirit of chivalry and letters and song, under the breath of the Provençal bards, contains little or nothing of value which may not be found clearly pronounced in the youthful poems of Dante. He says that, when his lady passes by, —

> Love casts on villain hearts a blight so strong,
> That all their thoughts are numbed and stricken low;
> And whom he grants to gaze on her must grow
> A thing of noble stature, or must die.

Humboldt has expatiated on his sensibility to the charms of nature, as evinced in the truth and grace of his incidental descriptions. Tradition also proves his love of valleys, forests, high prospects, and wild solitude, by identifying many of his tarrying-places during his exile with the most secluded and romantic spots. The inextinguishable relish of revenge and disdain, the ferocity of hate embodied in such passages as the description of Filippo Argenti, by which Dante is popularly recognized, are not more unapproachable in their way than the numerous passages of an earlier date in which he expresses his love, his unhappiness, his craving for attention and sympathy, are in theirs. Nothing can surpass the confiding softness of his trustful and supplicatory unveiling of the tender sentiments of his heart. He shuts himself "in his chamber, and weeps till he looks like one nigh to death"; his "eyes are surrounded with purple circles from his excessive suffering." — "Sinful is the man who does not feel for me and comfort me." He even takes "the most distasteful path, that of invoking and throwing myself into the arms of pity." — "Seeking an outlet for my grief in verse, I composed the canzone beginning —

> The eyes that mourn in pity of the heart
> Such pain have suffered from their ceaseless tears,
> That they are utterly subdued at last :
> And would I still the ever-gnawing smart
> That down to death is leading all my years,
> Forth in wild sobs must I my misery cast.

"In order that the conflict within me might not remain unknown, save to the wretched man who felt it, I resolved to compose a sonnet which should express my pitiable state." — "My self-pity wounds me as keenly as my grief itself."

> My bitter life wearies and wears me so,
> That every man who sees my deathly hue
> Still seems to say, "I do abandon thee."

Such was the native Dante, exquisitely affectionate, sensitive, confiding, melancholy, lonesome, baring his weaknesses, and yearning for sympathy.

What an incredible exterior change, when we turn from

this romantic portrait, and contemplate the elder Dante; Dante as he became in self-defence against the cruel injustice and hardships he endured! Then he blushed with shame : a look from Beatrice made him faint : he said, " My tears and sighs of anguish so waste my heart when I am alone, that any one who heard me would feel compassion for me." Now, encased in his seven-fold shield of pride, he scorns the shafts of wrong and of ridicule, saying, "I feel me on all sides well-squared to fortune's blows." He never lost his interior tenderness for humanity ; his enthusiasm for the sublime sentiments of poesy and religion ; his vital loyalty to truth, beauty, liberty. But, towards the frowns of his foes and the indifference of the world, he put on an adamantine self-respect which shed all outward blows. He incarnates, as he is commonly seen, an unconquerable pride, lofty as the top of Etna, hard as its petrified lava, hot as its molten core, but interspersed with touches of pity and love as surpris- ingly soft and beautiful as though lilies and violets sud- denly bloomed out of the scoriæ on the edge of its crater. His contemporary, Giovanni Villani, describes him as " a scholar, haughty and disdainful, who knew not how to deal gracefully with the ignorant." He himself, in his great poem, makes his ancestor Cacciaguida foretell, that of all his future calamities, what will try him most is " the vile company amidst which he will be thrown." Disgust and scorn of the plebeian herds of aimless, worthless men, however, never became an end with him, a pleasure in itself, but merely a means by which he protected him- self against the wrongs and lack of appreciation he suf- fered. They served as an ideal foil by which he kept himself on the eminence where God had set him, — saved his nobility and dignity from sinking even with his for- tunes. This is what distinguishes the office of a generous pride from that arrogant and poisonous egotism which feeds itself with misanthropy. The pride which nourished the virtue and undying usefulness of Dante, which helped to keep his genius from decay, and alone kept his will from drooping, has no alliance with the stung and exud- ing conceit of selfish men-haters. This is why the *hau-*

*teur* is grand in him which in a Menecrates is ludicrous, and in a Swift detestable.

In the twenty-fifth canto of the last part of the " Divina Commedia," Dante prophesies that he shall return to ungrateful Florence, and receive the laurel-wreath beside the font where he was baptized. Then, in present default of this fruition, he makes St. Peter crown him in Paradise. What a royal comfort to give himself this ideal meed! What matchless courage to dare to paint the fruition with his own hand, and hold the picture before mankind! He always felt himself in others with wonderful keenness, and passionately coveted love, and its phantom, — fame. But after his disappointments and exile, he would not bend to ask for either. In the free realm of the soul he imperiously appropriated them, and bade posterity ratify the boons.

The progress of his poem mirrors the perfecting of his character. In the " Inferno " he says : —

> Now needs thy best of man ;
> For not on downy plumes nor under shade
> Of canopy reposing, *fame* is won,
> Without which whosoe'er consumes his days
> Leaveth such vestige of himself on earth
> As smoke in air or foam upon the wave.

But at length, in the " Paradiso," weaned from the fretful Babel, calmly pitying the ignoble strife and clamor, he looks down, from the exalted loneliness of his own religious mind, on the fond anxiety, the vain arguments, the poor frenzies of mortal men.

> In statutes one, and one in medicine,
> Was hunting : this, the priesthood followed ; that,
> By force of sophistry, aspired to rule ;
> To rob, another ; and another sought,
> By civil business, wealth ; one, moiling, lay
> Tangled in net of sensual delight ;
> And one to wistless indolence resigned.
> What time from all these empty things escaped,
> With Beatrice, I thus gloriously
> Was raised aloft, and made the guest of heaven.

The beginning is the most easily appreciated by the vulgar ; the end is the least popular, because it is the most original and marvellous. The "Inferno" is sculpture ; the " Purgatorio," painting ; the " Paradiso," music. The scene rises from contending passions, through purifying penance, to perfected love. An excited multitude, gazing, wander with him through the first ; a smaller and quieter throng accompany him over the second ; a select, ever-lessening number follow him up the third ; and at last he is left on the summit, alone, rapt in the beatific vision.

## PETRARCH.

SOME peculiarities, generally in literary history traced to Petrarch, have given him the reputation of more originality as a man and as an author, more novelty and power of character, than he really possessed. Still his influence, both personal and literary, has been remarkable. And his tender philanthropy, ardent patriotism, romantic melancholy, the music of his plaintive though monotonous lyre, combine to lend a deep interest alike to his person and his story.

The love of friends, the chivalric love of woman, the love of fame, the love of books, the love of the great men of the past, the love of nature, the love of solitude, — these were the dominant sentiments in the soul of Petrarch. Of course all these sentiments had been felt and expressed many times before. Chivalry, which in its essence is an imaginative heightening of sympathy, gave them an especial enrichment and refinement, a vividness and an exaltation not known in previous ages. The Troubadours, the immediate predecessors of Petrarch, had sung the chief of them with variety and emphasis, borrowing something from the classic traditions, but adding more through that union of ecclesiastical Christianity and Germanic feeling which formed the peculiar genius of knighthood. In the works of Petrarch the sentiments of classic philosophy and poesy blend with the sentiments of the best Christian Fathers who had written on the monastic life, and with the sentiments of the Provençal

bards.   His originality and importance consist, first, in
the peculiar combination he gave to these pre-existing
ideas and feelings ; secondly, in the new tone and accent
lent to them by his personal character and experience ;
and thirdly, in the fresh impetus imparted for their repro-
duction and circulation in subsequent authors by the popularity of his writings and by the conspicuousness of his
position as the reviver of letters at the close of the Dark
Age.

The strength of Petrarch is his sympathetic wealth of
consciousness.   His learning, eloquence, and love of liberty, his gentleness and magnanimity, his purity, height,
and constancy of feeling are admirable.   He says : —

> And new tears born of old desires declare
> That still I am as I was wont to be,
> And that a thousand changes change not me.

His weaknesses are an exorbitant, all-too-susceptible
vanity, the prominence of a complacency forever alternat-
ing between fruition and mortification, the painful mingling of an effeminate self-fondling with a querulous self-
dissatisfaction.   The Petrarchan strain has been caught
and echoed interminably since his day.   The morbid subjective school, in some sense founded by him, has been
continued by Rousseau, St. Pierre, Chateaubriand, the
young Goethe, Byron, Lenau, and scores of other powerful authors, who have carried it much further than he, and
made it more and more complicated by additionally inter-
weaving their own idiosyncrasies.   Still above the jar of
tones the fundamental chords he sounded are clearly dis-
tinguishable ; a troubled excess of sensibility, exaggerated
aspirations, separation from the crowd, a high-strung love
of nature and seclusion, all grouped around an unhappy
and importunate sense of self.

Petrarch was fitted by his poetic temperament to enter
into the charms of the withdrawn scenes of nature, beautiful and wild landscapes, with an intensity uncommon in
his day ; in ours, partly through his influence, more frequent.   The unaffectedness of his taste for nature is
shown by the exquisite loveliness of the sites he chose for

his residences at Vaucluse, Parma, Garignano and Arqua. For sixteen years he spent much of his time in the picturesque seclusion of Vaucluse. This romantic valley, with its celebrated fountain, sixteen miles from Avignon, will forever be associated with his tender passion and his charming fame. In this profound retreat, amid this rugged scenery, "in a shady garden formed for contemplation and sacred to Apollo," or in a deeper grotto at the source of the swift Sorga, which he was "confident resembled the place where Cicero went to declaim," he roamed and mused, he nursed and sang his love for Laura. He said his disgust of the frivolousness and heartlessness, plottings and vices of the city drove him for the soothing delights of the country to this retired haunt, which had the virtue of giving freedom to his heart and wings to his imagination. After his frequent journeys on literary and state commissions to the courts of princes in famous cities, he always hurried back to his beloved Vaucluse, comparing his condition to "that of a thirsty stag, who, stunned with the noise of the dogs, seeks the cool stream and the silent shade." Here he passed much time alone, among the rocks and defiles, and by the brink of the fountain; also much time with his friend Philip de Cabassole. These two friends often strolled through the valley and over the cliffs, discussing literary and philosophical questions, until their servants, alarmed at their long absence, went forth with torches to seek them.

Petrarch always had a sincere fondness for solitude, a deep familiarity with its true genius. Few have written on the subject so well as he in his treatise on the "Leisure of the Religious," in which with such glow and sweetness he depicts the advantages of the monastic life; and in his elaborate dissertation "Concerning the Solitary Life." The latter work was sketched in his early manhood, but not completed till twenty years afterward. The argument of it is that the true end of life for every man is perfection; and that the distractions, insincerities, corruptions of crowded society are fatal to progress in this; while the calmness, freedom, and devout meditation of solitude are highly favorable to it. Whenever he touches

on this theme the pen of Petrarch seems impregnated
with the softest fire.  Born for solitude, enamored of
leisure, liberty, reverie, and ideal virtue, he fled the noise
and pestilential vice of cities with horror, and sought the
silence and purity of the fields and the woods with a
depth of pleasure which his pages clearly reflect.

> Still have I sought a life of solitude —
> This know the rivers, and each wood, and plain —
> That I might 'scape the blind and sordid train
> Who from the path have flown of peace and good.

After secretly fleeing back to Vaucluse, he writes to a
friend : " I had resolved to return here no more : in justification of my inconstancy I have nothing to allege but
the necessity I feel for solitude."   At another time he
writes : " The love of solitude and repose is natural to
me.  Too much known, too much sought in my own
country, praised and flattered even to nausea, I seek a
corner where I may live unknown and without glory.  My
desert of Vaucluse presents itself with all its charm.  Its
hills, its fountains, and its woods, so favorable to my
studies, possess my soul with a sweet emotion I cannot
describe.  I am no longer astonished that Camillus, that
great man whom Rome exiled, sighed after his country.
Solitude is my country."

The pictures in the imagination of Petrarch — as afterward was the case with Rousseau — were so vivid and so
delightful that his own undisturbed reveries gave him the
most satisfactory employment.  His ideal enjoyments by
himself, with none to contradict, nothing to jar or vex,
were a more than sufficient substitute for the usual intercourse of men.  It was a necessity with him to express
what he thought and felt, to mirror himself in sympathy
either actual or imaginary.  To restrain his emotions in
disguises or in bonds, to accept commands from others,
was ever intolerably irksome to him.  These are the very
qualities to make vulgar society distasteful, solitude delicious.  " Nothing is so fatiguing," he says, " as to converse
with many, or with one whom we do not love and who is
not familiar with the same subjects as ourselves."   " On

the mountains, in the valleys and caves, along the banks of the river, walking accompanied only by my own reflections, meeting with no person to distract my mind, I every day grow more calm. I find Athens, Rome, Florence here, as my imagination desires. Here I enjoy all my friends, the living, and the long dead whom I know only by their works. Here is no tyrant to intimidate, no proud citizen to insult, no wicked tongue to calumniate. Neither quarrels, clamors, lawsuits, nor the din of war reach us here. There are no great lords here to whom court must be paid. Avarice, ambition, and envy left afar, everything breathes joy, freedom, and simplicity." These sentiments were sincere expressions. The apparent inconsistencies with them shown in his life, his frequent intimacies with great personages and brilliant courts, merely prove that there was also another side to his soul; that in spite of his own belief that he was weaned from the public and sick of celebrity, he really had all his life strong desires for congenial society, usefulness, honor and fame.

At the very time that he told the King of Bohemia that his chief desire was " to lead a secluded life at its fountain-head among the woods and mountains, and that when he could not go so far to find it, he sought to enjoy it in the midst of cities," he was engaged in composing a "Treatise on Illustrious Men." He wrote letters to Homer, Varro, Cicero, and other great men, as if they were still alive; and said that he strove to forget surrounding vexations by living mentally with the renowned spirits of the past. He went into society to enjoy his friends, to serve his country and the cause of letters, and to win glory. He went into solitude not from dislike or indifference to men, but as an escape from galling restraints, or from distressing censures and injuries. His sensitiveness to public opinion, even to the most trifling criticism of the most insignificant persons, was excessive in the extreme. His unrivalled celebrity brought his character, his writings, his actions, into all men's mouths. The wretchedness thus caused him was unendurable, and he fled from it to the bosom of nature. He had written " Four Books of Invectives against Physicians," exposing the impositions and absurdities of the profession in his time. This bold and serviceable work brought a swarm of attacks on him. He said, "I shall bury myself in a solitude so profound that care and envy will not be able to find me out. What folly! can I expect to find any place where envy cannot penetrate?" After being crowned Laureate in Rome — the first repetition of that august ceremony for thirteen hundred years — he says, "It only seemed to raise envy and deprive me of the repose I enjoyed. From that time tongues and pens were sharpened against me." He cared too much for the opinion of men, not too little. He yearned to love and admire, to be loved and admired. "I esteem myself happy," he once writes, "in having quitted Venice for Padua. There I should have been suspected ; here I am caressed."

Led by too much of his personal experience of the world to think mankind at large set against virtue and wisdom, and against the votaries of virtue and wisdom, his character, as Ugo Foscolo has said, sometimes wears a tint of misanthropy by no means natural to him. Really he had "more of fear than hatred, more of pity than contempt for men." He was one of those unfortunate men whose self-complacency is so unstable, whose sympathy so keen, that they are afraid of those they love. His kind acts were innumerable. He owned the only known copy of Cicero's treatise "De Gloria," and lending it to his decayed schoolmaster, to be put in pawn for the temporary relief of the poor old man, it was irreparably lost to the world. "As a man," he said, "I cannot but be touched with the miseries of humanity ; as an Italian, I believe no one more keenly feels the calamities of my country." He was not, in any reprehensible sense, an egotist, far less a misanthrope, in his love of isolation. He lacked — and this was his central weakness — what made the strong Goethe so sound, the mature Wordsworth so content, namely, a direct life in the objects of nature, freedom from brooding on morbid sensations : and a direct life in the general truths of humanity, freedom from a wearisome attention to personal details. He himself confesses to the mistake of taking "life in details rather than in the gross."

He read and wrote with hardly the slightest intermission. He reflected and brooded till he lost his health of body and mind, and life became "sicklied o'er with the pale cast of thought." A dreadful *ennui* devoured him. He imagined that "a weariness and disgust of everything naturally inhered in his soul." He said, "I conceived that to cure all my miseries I must study them night and day, renouncing all other desires ; that the only way of forgetting life was to reflect perpetually on death. In kindred strain he sings,

> Ceaseless I think, and in each wasting thought
>   So strong a pity for myself appears
> That often it has brought
>   My harassed heart to new yet natural tears.

Again he says, "I am weary of life. Whatever path I take I find it strewn with flints and thorns. Would that the time were come when I might depart in search of a world far different from this wherein I feel so unhappy." And once more, at a later date, he writes, "I start up in wildness, I speak to myself ; I dissolve in tears ; I have visions which inflict on me the torments of hell." This was near the end. His last composition was a letter to his friend Boccaccio, which closed with the words, "Adieu, my friends! Adieu, my studies!" He was found dead in his library with his arm resting on a book. Distinguished honors were paid to his remains and his memory. At this day in popular fame he stands at the head of all the poets of love, his name wedded to that of the Laura he has immortalized. Scholars make grateful acknowledgment of the signal services he rendered to the cause of learning. Psychologists recognize him as one of the few whose characters have contributed a distinctive historic influence to following times.

The richness of his mind, the burning passions of his heart, lent to the coldness and fickleness of average men a stronger repulsion, and invested him with a double isolation in giving him superior company of his own. "Beholding, on the shores washed by the Tyrrhene Sea, that stately laurel which always warms my imagination, through impatience I fell breathless into the intervening stream. I was alone and in the woods, yet I blushed at my own heed-lessness; for, to the reflecting mind, no outward witness is necessary to excite the emotion of shame." These strik-ing words touch the secret of the haunting unhappiness of Petrarch, namely, his intense sympathy, that presence of his fellow-beings in imagination from which he could not free himself, of which even his apparent misanthropy it-self was but one of the disturbed symptoms. Such an experience as that just quoted was relatively unknown in classical antiquity. Egypt, Judæa, Greece, Rome, had no such characters as Zimmermann, Senancour, Chatterton, Chopin, Heine, David Gray. The self-gnawing wretched-ness of such men is the product of a later civilization, of our Christian epoch. What is the cause of this melan-choly moaning and fading of men of genius, so familiar to us now? What makes the unhappiness of Christian genius in comparison with the clear content and joyful-ness of the best type of ancient Pagan life?

It is a consequence of the enormous enhancement of sympathy. In antiquity, the family was the unit of life; outside of it, the individual had comparatively few re-sponsive tendrils of feeling. Christianity, with the progress of civilization, and the universal intercommunication of the nations of the earth, has generated a powerful feeling of the relation between the individual and the total race. Jesus identified himself with all the afflicted members of humanity: "Inasmuch as ye have done it unto one of the least of these, ye have done it unto me," — a sentence which has been unspeakably influential on the historic sentiment of the last eighteen hundred years. Shake-speare makes Antony say of the murder of Cæsar : —

> O, what a fall was there, my countrymen !
> Then you, and I, and all of us fell down.

This feeling of entire humanity in each person devel-oped an unprecedented, mysterious, objective sympathy which has since often oppressed sensitive minds as " the

burden and the weight of all this unintelligible world."
We read, it is true, in the Sanscrit Mahabharata and
Ramayana, in the Persian poets, and in the Arabian
Nights' Entertainments, expressions of feeling as deep,
fine, and vast as anything in modern Christian literature ;
but it is something very different which they express. It
is either personal affection, as when a lover is represented
fainting away at a frown, falling dead under an unkind
word ; or it is a response to ideas of a transcendental
faith, an ecstatic idealism, a pantheistic theosophy. The
pining and swooning emotions of the finest Orientals are
subjective, resulting either from love of a particular person, or from mystic devotion. But the emotions we are
dealing with are objective, although neither personal nor
metaphysical in their object. They are really unrecognized reactions on the vague general idea of humanity.

Now, by means of literature, newspapers, telegraphs,
interlacing ties of business, travel, kindred, friendship,
innumerable mutual interests, a man of sensitive genius
lives constantly as it were in the ideal presence of all
mankind. Public opinion is a reality as solid to him as
the globe, its phenomena as influential as sunshine and
darkness. Where life used to be direct, it is now reflective. Consciousness, once made up of single lines, now
consists of a mazy web. The immense complication of
actions and reactions, distinctive of modern experience,
produces a mass and multiplicity of feelings not yet harmonized, to be harmonized with difficulty and slowness,
but infallibly productive of painful desires and sorrows
until harmonized.

Furthermore, the healthy objectivity of Greek life be-
tokened the well-balanced adjustment of man's desires
with his earthly state. Ecclesiastical Christianity threw
discredit and darkness on the earthly lot by its overwhelming portrayal of the worthlessness of the evanescent
present in comparison with the everlasting glories of
heaven. The doctrine of immortality engendered a sentiment of correspondent proportions, which, unable to
renounce this world and patiently wait for the other, attempted to dilate the prizes of time to the capacity of its

demands.  A vast, hungry sense of incongruity resulted, prolific in disease and unfathomable misery; a sick and sore introspectiveness, a devouring greed for love and admiration, a frantic effort, in the phrase of Bacon, " to cure mortality with fame."  The increase of sympathy consequent on the ideas of the unity of the human race and the community of human life has made the experience of the modern masses of men happier than that of the ancient masses; but its unharmonized excess has created the unhappiness of that class of exceptional men of genius of whom the unhappy Petrarch stands as the first popular literary representative.  In his eloquent " Trionfi " he nobly depicts the great periods in the experience of the soul.  First, Love triumphs over Man ; secondly, Chastity triumphs over Love ; thirdly, Death triumphs over both ; fourthly, Fame triumphs over Death; fifthly, Time triumphs over Fame ; and finally, Eternity triumphs over Time.  The man of large and fine genius, before he can cease to be unhappy, must, in his own soul, go through all these triumphs into the last one, by self-denial and the firm subordination of his impulsive sensibilities to the unchangeable conditions of destiny.

In personal intercourse Petrarch was one of the most fascinating of beings.  His friends idolized him, " welcomed him with tears of joy as though he had been an angel.  One high duty of writers of genius he fulfilled with signal effect, — that of softening and refining the feelings of the vulgar.  The other duty of great men, to be healthy and happy, that they may inoculate the needy world with sanity and joy, he is, perhaps, more to be pitied than blamed for failing to fulfil.

They say his strains tend to effeminate his countrymen. Well, there are plenty of influences in the other direction, military, political, mercantile, mechanical.  Not without good effect does his soft and softening strain mingle in the harsh roar of toil, trade, ambition, and battle.  In consideration of his great love his offences must be forgiven.  They *are* forgiven and forgotten in the affections of multitudes of readers, who, gratefully cherishing his worth and service, blend his name alike with the thought of loneliness and the memory of Laura.

They keep his dust in Arqua, where he died :
The mountain-village where his latter days
Went down the vale of years ; and 't is their pride —
An honest pride, and let it be their praise —·
To offer to the passing stranger's gaze
His mansion and his sepulchre ; both plain
And venerably simple, such as raise
A feeling more accordant with his strain,
Than if a pyramid formed his monumental fane.

And the soft quiet hamlet where he dwelt
Is one of that complexion which seems made
For those who their mortality have felt,
And sought a refuge from their hopes decayed
In the deep umbrage of a green hill's shade,
Which shows a distant prospect far away
Of busy cities, now in vain displayed,
For they can lure no further ; and the ray
Of a bright sun can make sufficient holiday,

Developing the mountains, leaves, and flowers,
And shining in the brawling brook, whereby,
Clear as its current, glide the sauntering hours
With a calm languor, which, though to the eye
Idlesse it seem, hath its morality.
If from society we learn to live,
'T is solitude should teach us how to die.

## TASSO.

THE noble Torquato Tasso, fearing that his fathei would be displeased if he stole time from his legal studies, hurried into seclusion, and, secretly devoting all his leisure hours to the muse, produced his brilliant poem of Rinaldo before he was eighteen years old. His fervid fancy, fondness for study, exquisite sensibility, and intense desire of popular love and fame, while they made him keenly crave society and friendship, compelled him to know much solitude. His enemies, envious critics, dictatorial patrons, and literary censors, persecuted him with endless vexations of the most exasperating sort, bitterly attacking his style, insisting on the omission of what he felt to be the best passages of his poems, and circulating grossly altered and mutilated editions of them in spite of all his protestations. Escaping from the annoy-

ances and unappreciation he suffered at Ferrara, he wandered for several years from city to city, " the finest genius
of his time, a prey to sorrow and disease, his splendid
fancy darkened by distress, his noble heart devoured at
once by the agony of hopeless love and the ambition of
literary glory." When he returned to the court of Al-
phonso, expecting affectionate welcome and honor, he
was met with rude neglect. Insulted and derided, he
gave vent to his indignation in such terms as caused the
Duke to have him put under guard in an asylum for pau-
pers and madmen. The misery of such a spirit, so tender
and so proud, so surpassingly alive to the breath of human
opinion, when subjected to the foul injustice and severity
with which the haughty heartlessness of his master here
pursued him, can hardly be conceived by ordinary minds.
His own descriptions of it are indescribably touching.

> There is no solitude on earth so deep
> As that where man decrees that man shall weep.

He writes to his dear friend Gonzaga, " The fear of being
perpetually imprisoned here increases my melancholy,
and the squalor of my beard, my hair, and habit, exceedingly annoys me. But, above all, I am afflicted by solitude, which even in my best state was often so tormenting
that I have gone in search of company at the most un-
seasonable hours." Incarcerated for seven long and cruel
years, his loneliness was so great that his disturbed mind
created for itself the belief that a familiar Spirit was in
the constant habit of coming to hold high and kind communion with him.

All the historians of Tasso agree in eulogizing " his
candor, his fidelity to his word, his courtesy, his frankness, his freedom from the least tincture of revenge or
of malignity, his attachment to his friends, his gratitude
to his benefactors, his patience in misfortune, his mildness
and sobriety, his purity of life and manners, his sincere
piety. None of his foes seem to have been able to charge
him justly with a single moral stain." He was extremely
sensitive to slights, exacting of the respect due to him.
This was his single ungracious quality. Few things are

more cruel than that so highly loving and gifted a soul should have had such numerous and rancorous enemies that his life was embittered and burdened by them until he was quite weaned from it. When a guest of Rome, lodged in the Vatican, waiting to be crowned with laurel, — the first poet so honored since Petrarch, — he sighed to flee away and be at rest. Growing very ill, he obtained permission to retire to the Monastery of Saint Onofrio. When the physician informed him that his last hour was near, he embraced him, expressed his gratitude for so sweet an announcement, and then, lifting his eyes, thanked God that after so tempestuous a life he was now brought to a calm haven. The Pope having granted the dying poet a plenary indulgence, he said, " This is the chariot on which I hope to go crowned, not with laurel as a poet into the capitol, but with glory as a saint into heaven."

## BRUNO.

GIORDANO BRUNO, an exceedingly brave, sensitive, loving soul, for these very qualities, which in more favorable conditions of society would have blessed him with dear comrades and popular admiration, was made an outcast and an exile. Intensely desirous of wisdom and nobleness, unflinchingly loyal to reality, detesting falsehood and indifference, a burning worshipper of truth and freedom, in an age of despotism and conformity he was naturally considered dangerous, and was put under ban. Lonely in his loftiness of unterrified thought, hunted from nation to nation, with brief respites, unfriended, save by a few generous exceptions like Fulke Greville and Sir Philip Sidney, the integrity of his own soul was his unquenchable comfort, and the presence of the Infinite Spirit of Eternal Verity was his inseparable companionship.

The tonic of his veracious health and cheer is breathed in the words he speaks : " To have sought, found, and laid open a form of Truth, — be that my commendation, even though none understand. If, with Nature, and under God, I be wise, that surely is more than enough."

Imprisoned, mercilessly tortured, kept for over two years from the sight of all human faces save hostile and mocking ones, with divine resolution refusing to deny a thought or recant a word, he was at last burned at the stake. In some lines of his own, written with prophetic anticipation of this very end, he says, "Open, open the way! Ye dense multitude, spare this sightless, speechless face all harsh obstructions, while the toil-worn, sunken form goes knocking at the gates of less painful but deeper death!"

Genius often brings with it into the world a feeling of melancholy strangeness, if not of estrangement, — a mysterious homesickness of soul. It feels itself a foreigner on the earth. The features of Bruno — in the portrait transmitted to our times — are affectingly expressive of this. He looks like one whose affections had been repulsed by an unworthy world, and whose soul had found strength by divinely rallying itself upon God. As we gaze on his strong, sad, lonely lineaments we are reminded of what he himself says in one of his sonnets : — "You may read the story of my life written in my face."

## VICO.

VICO, the great founder of the science of history, was one of the loneliest minds of his century. A more profound or original thinker has rarely appeared. While yet young he became tutor to the nephews of the Bishop of Ischia, where he spent nine years in the lonely solitude of Vatolla, dividing his thoughts between poetry, philosophy, and jurisprudence. His chosen comrades, besides the great Roman jurisconsults, were Plato and Dante, with the last of whom his ardent and melancholy genius closely allied him. But even such mental companionship was more frequently deserted for the pursuit of his own absorbing reflections.

He saw that the history of mankind was no medley or phantasmagoria, at the mercy of individual leaders, but a grand march of humanity, a total evolution of humanity, governed by great laws and reducible to a science. He

trod the lonely path of this discovery with unwearied
patience, every day rising higher in unknown regions,
meeting no rival or companion, leaving his fellow-beings
below him as fast as he mounted, until at last, seating
himself on the summit, and looking around, he saw, spread
out far beneath his feet, all mankind and their history in
one view. "Unhappily," said Michelet, "he found himself quite alone. No one could understand him. He
was equally isolated by the originality of his ideas and
by the strangeness of his speech. The opposite of that
which happened to the Seven Sleepers befell him. He
had forgotten the language of the past and could speak
only that of the future, — a language then too early and
now too late, — so that for this grand and unfortunate
genius the time has never come."

He gave surprising examples of the vigor of his comprehensive and penetrating intellect by originating the
doctrine of myths, which, in its application to history,
has since proved so fruitful in valuable results; also in
first propounding, with as much precision and thorough-
ness as it has since received, the profound truth that the
record of much of the pre-historic experience of mankind
is locked up in the etymological structure of language.
It is only just now, in our own generation, that, in the
hands of the gifted masters of the science of comparative
philology, this deep discovery is amazing the world with
its brilliant revelations, forcing from the dark matrix of
each primitive word some crystallized secret of the forgotten life of the human race.

Vico has written his own life; and it is piteous to read
the account of the painful isolation in which he was left
by the careless and the envious. His rivals, chagrined
by his vast superiority, treated him with cruel injustice.
Some called him insane, others an obscure and paradox-
ical genius. He was traduced, satirized, pursued with
ironical eulogies. He says, "Vico blessed these adversi-
ties which ever drew him back to his studies. Retired in
his solitude, as in an impregnable fortress, he thought, he
wrote, he took a noble vengeance on his detractors. There
he found the new science. From that moment he believed

he had nothing to envy Socrates, in regard to whom the good Phedrus expresses the magnanimous sentiment 'Insure me his fame, and I will not shrink from his death. The envy which followed me living shall absolve me when dead.'"

At another time he says : "Since completing my great work I feel that I have become a new man. I am no longer inclined to declaim against the bad taste of the age, which, in refusing me the place I demanded, has given me occasion to compose the new science. Shall I say it? I may deceive myself, but I do not wish to deceive myself. The composition of this work has animated me with a heroic spirit which lifts me above the fear of death and all the calumnies of my rivals. When I think of the judgment of God which rewards genius with the esteem of the wise, I feel myself seated on an adamantine cliff." Frequently in his poems he opens his inmost heart, and consoles himself for lack of honor and love with the thought of his great discoveries, " penetrating in the abyss of wisdom to the eternal laws by which humanity is governed," "all nations together forming one city, founded and ruled by God himself"; and also with the thought of posthumous fame. How beautiful are these words, breathed in his pining solitude : " My dear country has refused me everything. But I respect and revere her. A severe mother, who never caresses her son or presses him to her bosom, is none the less honored. In the thought of the unrecognized benefits I have conferred on her, I already find noble consolation."

At last "the unfortunate Vico," to use his own words, "broken down by age, wrongs, fatigues, and physical sufferings," welcomed the grave as a sweet shelter from all storms. His fame is still growing brighter, as reflected in lofty minds, congenial with his own, from generation to generation. No gentle spirit who has learned to appreciate his quick and tender genius, the unkindness and desertion he suffered in his time, can read the following apostrophe from one of his earlier poems without a quickening heart-beat, without longing to call him back to receive now, so late, the meed he merited then. " Pure and

tranquil life, calm and innocent pleasures, glory and treasure won by merit, celestial peace of mind, and that which
s dearest to my heart, the love of which love is the price,
delicious reciprocity of sincere faith, sweet images of happiness, although but to aggravate my pain, still, come
again!"

## DESCARTES.

THE great Descartes, the pollen of whose thoughts,
borne on all the breezes of inquiry, fertilized the philosophy of Europe for two centuries, is a fine example of
one who, in spite of brilliant accomplishments, extended
reputation, strong affections, and courteous manners, was
made essentially a solitary man by his intense devotion
to the discovery of truth. He repudiated traditional authority and prejudice, and with a sublime force and heroism of soul threw himself back on common sense and a
sceptical openness and freedom of search for the reality
of things. There are four ways in which most persons
arrive at their various degrees of wisdom : self-evident
notions, the experience of the senses, the conversation
of other men, the reading of books. There have been,
Descartes says, in all ages, great minds who have tried
to find a fifth road to wisdom, incomparably higher and
surer than the other four, namely, the search of first
causes and true principles from which may be deduced
the reasons of all that can be known.

On this fifth road few mightier travellers than he have
ever trod. Those who have passed him since were indebted to his guidance. He dared to strip off all past
beliefs that he might not be encumbered or misled.
" But," he says, " like one walking alone and in the dark,
I resolved to proceed so slowly and with such circum-
spection that if I did not advance far, I would at least
guard against falling." Regarding " the supreme good as
nothing more than the knowledge of truth through its first
causes," he allowed nothing to interfere with his pursuit
of it.

But his kind temper, good taste and prudence did not

disarm the fears and foes awakened by the boldness of his speculations. Stratagems and dangers surrounded him. Cousin says, "After having run round the world much, studying men on a thousand occasions, on the battle-field, and at court, he concluded that he must live a recluse. He became a hermit in Holland." Eight years later he writes, "Here, in the midst of a great crowd actively engaged in business and more careful of their own affairs than curious about those of others, I have been enabled to live without being deprived of any of the conveniences to be had in the most populous cities, and yet as solitary and as retired as in the midst of the most remote deserts." At another time he says, "I shall always hold myself more obliged to those through whose favor I am permitted to enjoy my retirement without interruption than to any one who might offer me the highest earthly preferments." There is no reason to doubt the sincerity of this declaration; yet there is another side to the truth. For when Queen Christina paid him honoring attentions, and invited him to her court at Stockholm, he went thither and occupied an academic post. His first passion was the pursuit of truth; his second, a love of the esteem of his fellow men. His own frank words give a pleasing proof of this. "My disposition making me unwilling to be esteemed different from what I really am, I thought it necessary by all means to render myself worthy of the reputation accorded to me. This desire constrained me to remove from all those places where interruption from any of my acquaintances was possible, and to give myself up to studies." There was no misanthropic ingredient in his isolation. Yet, once or twice, a little soreness, a little petulance at the neglect of the public, at the lack of the co-operation he needed, escapes him. "Seeing that the experiments requisite for the verification of my reasonings would demand an expenditure to which the resources of a private individual are inadequate, and as I have no ground to expect public aid, I believe I ought for the future to content myself with studying for my own instruction, and posterity will excuse me if I fail to labor for them." But if, contrary to his own opinion, the ambition of Descartes in relation to society and mankind was superior to his fruition, so that some dissatisfaction resulted, it did not sour or exasperate him. On the whole, he kept his moral equipoise well and sweetly. He has himself indicated his three great reserves of happiness.

First, his employment itself. "The brutes, which have only their bodies to conserve, are continually occupied in seeking sources of nourishment; but men, of whom the chief part is the mind, ought to make the search after wisdom their principal care; for wisdom is the true nourishment of the mind." "Although I have been accustomed to think lowly enough of myself, and although when I look with the eye of a philosopher at the varied courses and pursuits of mankind at large, I find scarcely one which does not appear vain and useless, I nevertheless derive the highest satisfaction from the progress I conceive myself to have already made in the search after truth, and cannot help entertaining such expectations of the future as to believe that if, among the occupations of men as men, there is any one really excellent and important, it is that which I have chosen."

Second, intercourse with the highest minds of all times and countries. "The perusal of excellent books is, as it were, to enjoy an interview with the noblest men of past ages, who have written them, and even a studied interview in which are discovered to us only their choicest thoughts."

And thirdly, the subjection of his wishes to his condition. "My maxim was always to endeavor to conquer myself rather than fortune, and change my desires rather than the order of the world, and, in general, to accustom myself to the persuasion that, except our own thoughts, there is nothing absolutely in our power. Thus we learn to regret nothing which is unchangeable, desire nothing which is unattainable. I confess there is need of prolonged discipline and repeated meditation to accustom the mind to view all objects in this light; and I believe that in this chiefly consisted the secret of the power of such philosophers as in former times were enabled to rise

superior to the influence of fortune, and, amid suffering
and poverty, enjoy happiness which their gods might have
envied."

## HOBBES.

THE famous philosopher of Malmesbury is an example
of the difficulty a man of great intellect and proud sensitiveness experiences in reconciling himself to the disparity
between his own estimate of himself and the estimate set
on him by his unappreciative neighbors.   He was one of
the most independent and powerful thinkers, one of the
most clear and energetic writers, that have ever appeared
in England.   Macaulay even calls him "the most acute
and vigorous of human intellects."   He lived to the age
of ninety-two, devoting his great endowments to a course
of earnest thought resulting in unpopular conclusions embodied in unpopular works.   He was misunderstood, misrepresented, misvalued, ill treated.

Although haughty and irascible, he had many good
qualities, which drew the interest of numerous distinguished contemporaries to him.   Three successive Earls
of Devonshire patronized him, thought highly of him,
gave him a home, with slight duties and great leisure.
He was a good hater, and evidently relished despising
the ignorant herd and dealing bitter blows against his
enemies.   He comforted himself for his unpopularity by
cherishing friendly relations and correspondence with the
chief great men of his time, such as Bacon, Harvey,
Descartes, Ben Jonson, Aubrey, Clarendon ; and by nour
ishing his keen sense of his superiority to the vulgar
crowds of people.   "What proof of madness," he asks,
"can there be greater than to clamor, strike, and throw
stones at our best friends?   Yet this is less than the
multitude will do."   His writings frequently betray how
warmly he welcomed every notice from others calculated
to soothe and confirm his self-estimate, how angrily he
resented whatever ruffled or tended to lower it.   He
speaks of one of his lesser writings as little in bulk, and
yet "great enough if men count well for great."   Again

he says, "the clamorous multitude hide their envy of the present under a reverence of antiquity." He also said of his friend, the discoverer of the circulation of the blood, " Harvey is the only man I know, that, conquering envy, hath established a new doctrine in his lifetime." Likewise he wrote, when publishing his treatise on Human Nature, " I know by experience how much greater thanks will be due than paid me for telling men the truth of what men are. But the burden I have taken on me I mean to carry through, not striving to appease but rather to revenge myself of envy by increasing it."

He waged a fierce war for many years with Wallis on certain mathematical questions. When Wallis — whom he called " the pest of geometry" — taunted him with flattering himself and maligning others in his writings, he replied as follows : " A certain Roman Senator, having propounded something in the assembly of people which they misliking made a noise at, boldly bade them hold their peace, and told them he knew better what was good for the commonwealth than all they. And his words are transmitted to us as an argument of his virtue : so much do truth and vanity alter the complexion of self-praise."

His strong peculiarities of habit no less than his extraordinary powers marked Hobbes out as a man by himself. Yet, after all his lonely walking, lonely thinking, lonely living, and repelling quarrels, he clung warmly to his friends, had a horror of being left alone in his illness, bequeathed all his property to the faithful servant and friend who had been his amanuensis. He was not afraid of death, but said he should willingly " find some hole to creep out of the world at," and was wont to amuse himself with choosing for the epitaph to be graven on his tombstone, " *This is the true philosopher's stone.*"

His toughness of stock and copiousness of force enabled him to weather the storms of nearly a century. His colossal bulk of mind and earnest search for truth removed him from the crowd. He was turned in upon himself still more by the rivalry, envy, hate, slanders, aggravating attacks provoked by his genius, fame, disagreeable speculations, hot partisanship, and personal spleen. In half

philosophic, half angry solitude, he sought to foster and
defend that reflex idea of himself in whose extension and
firmness the essential comfort of life resides for such men,
and every assault upon which he naturally resented as a
blow at the very vitality of his soul.  His life, perforce,
was greatly solitary.  Yet friendship, for the same reasons,
was particularly needful and precious to him, so far as he
could get it.  The many high compliments he received
from the leading thinkers of his age must have thrilled
him with a fiery gladness impossible to colder and feebler
natures.  It is pleasant to think of the pleasure he took
in the dedication of Gondibert to him by Davenant; also
of what a luxury the flattering and eloquent ode addressed
to him by Cowley must have yielded to his sensibility.

> Nor can the snow which now cold age does shed
>     Upon thy reverend head
> Quench or allay the noble fires within :
>     But all which thou hast been,
> And all that youth can be, thou 'rt yet ;
>     So fully still dost thou
> Enjoy the manhood and the bloom of wit.
> To things immortal time can do no wrong ;
> And that which never is to die, forever must be young.

## LEIBNITZ.

ALTHOUGH Leibnitz for much of his life held an office
at court, and carried on an extensive correspondence with
diplomatists, mathematicians, and philosophers, he was a
lonely man from his boyhood to his burial.  He says :
" I always inclined less to conversation than to meditation
and solitude."  Referring to the time when, a youth of
fifteen, he was an academic student, he says : " I used to
walk to and fro in a little grove near Leipzig, called the
Vale of Roses, in pleasing and solitary meditation, con-
sidering the questions of the Schoolmen."  Prevented
from obtaining the degree of Doctor of Laws, he felt so
aggrieved and offended at the machinations of his rivals
that he at once left his native city, and never returned to
Saxony again, excepting for brief visits.

He ever had a high opinion of his own mind and worth, and was easily irritated, though generous and forgiving in his temper. His secretary and intimate associate, Eckhart, says it was characteristic of him "to speak well of every one, put the best construction on the actions of others, and spare his enemies when having it in his power to dispossess them of their places." He was never married, but lived by himself absorbed in gigantic toils. Courtiers and people for the most part neglected him. He was a superior being whom they could not understand. The clergy hated him, because he looked down with pity on their superstitions; and they publicly assailed him as a contemner of the ecclesiastical creed. He said he had a great many ideas which he held back, because the age was not ripe for them, and also because he extremely disliked being misunderstood and misrepresented. In a letter to Burnet he says, in reference to his own situation : "There are many things which cannot be executed by a single isolated individual. But here one hardly meets with anybody to speak to." In connection with this statement it is pleasant to remember the beautiful and impressive incident, that, when he was thirty years old, Leibnitz paid a visit to Spinoza, at the Hague. When these two vastest and loneliest intellects of their century met in that little poor Dutch chamber, did the two brains there together hold more mind than all the rest of Europe? Two years before he died he had formed a distinct plan of a universal language ; but, aged, over-occupied, solitary, he failed to complete and publish it. He writes to Remond de Montfort, that he had spoken of it to several persons, and gained no more attention than if he had related a dream. He adds : "I could easily work it out if I were younger, or less busy, or enjoyed the conversation of men who would encourage me, or had by my side young men of talent." This great man died in the midst of as much local indifference as he had lived. His friend Ker, who happened to arrive in Hanover on that day, was grieved, not only by the event, but by the slight notice taken of it. The funeral, Ker testifies, "was more like that of a highwayman than of one who had been the ornament of his country." The faithful Eckhart says, that although the whole court were invited to attend the solemnities, no one appeared but himself. The Royal Academy of Sciences in Berlin, of which he was the founder and the first president, remained silent. The Royal Society at London, of which he was one of the oldest and most distinguished members, took not the slightest notice of the death of the great rival of Newton. The French Academy alone paid a tribute becoming its own chivalrous character, and worthy of one of the great-est of human minds.

But if Leibnitz was neglected by the conspicuous ob-scurities about him, in the high tasks of his genius he had his place among the most illustrious heroes of humanity. Pilgrims from far-away lands, who stand in the aisle of the church at Hanover, and read beneath their feet the laconic inscription, *Ossa Leibnitii*, thrill with reverence in memory of him whose powerful thought is vibrating to the ends of the earth, and whose fame will penetrate the remotest future.

### MILTON.

MANY of the chief conditions of spiritual solitude met in a high degree in Milton. A proud and pure mind, devotion to learning, a passion for liberty, a passion for truth and virtue, a passion for lasting fame, a deep and bold dissent from the prevalent theological doctrines and religious forms about him, general neglect, repeated dan-ger, and, at last, blindness. Numerous expressions of this experience are to be found in his writings. He says, referring to the text in Genesis, " Loneliness is the first thing which God's eye named not good." While yet a young man he wrote to his friend Diodati, " As to other points, what God may have determined for me I know not ; but this I know, that if he ever instilled an intense love of moral beauty into the breast of any man, he has instilled it into mine. Ceres, in the fable, pursued not her daughter with a greater keenness of inquiry, than I, day and night, the idea of perfection. Hence, wherever

I find a man despising the false estimates of the vulgar, and daring to aspire to what the highest wisdom through every age has taught us as most excellent, — to him I unite myself by a sort of necessary attachment."

The sonnet in which Wordsworth addresses him, and describes his holy seclusion and his noble services, is a household word.

> Thy soul was like a star, and dwelt apart ;
> Thou hadst a voice whose sound was like the sea :
> Pure as the naked heavens, majestic, free,
> So didst thou travel on life's common way,
> In cheerful godliness ; and yet thy heart,
> The lowliest duties on herself did lay.

The outrageous warfare waged against him by such foes as Du Moulin, Salmasius, and More, must have given him a keen relish for the refuge of a peaceful privacy. And there are repeated passages in his poems which plainly reveal a temperament fitted for the benefits of loneliness, a mind accustomed to enjoy the delights of it. Thus he makes Adam say to Eve, —

> If much converse perhaps
> Thee satiate, to short absence I could yield ,
> For solitude sometimes is best society,
> And short retirement urges sweet return.

And a perception full of the heartiest feeling of reality pervades the following lines : —

> In sweet retiréd solitude
> He plumed his feathers and let grow his wings,
> That in the various bustle of resort
> Were all too ruffled, and sometimes impaired.

Johnson wrongfully accuses Milton of a dark revengefulness, a bitter envy. His nature was profoundly sweet, gentle, and regal. His passionate retorts and invectives are not proofs of gall or hate, but either oratoric heats of battle, or weapons wielded in self-defence. He was a man of noble poetic angers, not of mean brooding enmities. When his foes assailed him with ignorance and wrong he repelled their slanderous insolence with contemptuous in dignation.

I did but prompt the age to quit their clogs
By the known rules of ancient liberty,
When straight a barbarous noise environs me
Of owls and cuckoos, asses, apes, and dogs :
As when those hinds that were transformed to frogs
Railed at Latona's twin-born progeny,
Which after held the sun and moon in fee.
But this is got by casting pearls to hogs,
That bawl for freedom in their senseless mood,
And still revolt when truth would set them free.

This retaliation shows more disturbance of spirit than is
becoming ; but to infer from such language the existence
in the writer of "a malignity at whose frown hell grows
darker," is absurd.   These jealous incompetents had, in
their judgment, hurled him down into a muddy pit of
error and wickedness, from the glorious peak of truth
and greatness on which, in his own judgment, he was
perched : and the vehemence of his scorn simply measures the intensity with which he resents their injustice
and replaces himself on his height.   The true Milton is
less expressed in his rousing polemic invectives, the trumpet blasts of his embattled spirit, than in the melodious
passages of meditative reminiscence and description in
which the affections of his natural character exhale.

As one who long in populous city pent,
Where houses thick and sewers annoy the air,
Forth issuing on a summer's morn to breathe
Among the pleasant villages and farms
Adjoined, from each thing met conceives delight,
The smell of grain, or tedded grass, or kine,
Or dairy, each rural sight, each rural sound ;
If chance with nymph-like step fair virgin pass,
What pleasing seemed, for her now pleases more,
She most, and in her look sums all delight.

Deserted, blind, old, even harshly treated by his ungrateful daughters, composing his immortal poem, he
depicts himself as singing, —

With mortal voice, unchanged
To hoarse or mute, though fallen on evil days,
On evil days though fallen, and evil tongues,
In darkness, and with dangers compassed round,
And solitude.   Yet not alone while thou

> Visitst my slumbers nightly, or when morn
> Purples the east — still govern thou my song,
> Urania, and fit audience find, though few.

In this spirit he made his age sublime as he had made his manhood heroic. Such steady approving respect had he for himself as he grew lonelier in his age, such grand memories of his bygone deeds, such high imaginative communion with the present and the future, that we cannot hesitate to apply to him his own words descriptive of the Saviour : —

> And he still on was led, but with such thoughts
> Accompanied of things past and to come
> Lodged in his breast, as well might recommend
> Such solitude before choicest society.

And so he died. And when strangers from distant lands linger in the chancel of Saint Giles at Cripplegate, as they read the inscription on his tomb they forget the surrounding roar of London. They find it difficult to think of him as sleeping there. They feel that he is truly interred in a monument which is co-extensive with the civilized world.

### PASCAL.

PASCAL was a personality apart, with ideas proud as his intellect, with faith apparently humble and sincere as his heart, but in reality more wilful than natural, and under-arched by a scepticism awful to himself. This sceptical character of his mind is conclusively shown by Cousin in his celebrated report to the French Academy on the " Thoughts " ; and later editors of his posthumous writings have brought to light the unscrupulous changes and suppressions practised by the first editors. Dr. Lélut has demonstrated, in his instructive treatise, " L'Amulette de Pascal," the deeply diseased condition, in his later years, of both the body and the mind of this great unfortunate genius.

With a nervous system overcharged with force and out of equilibrium, the brain expending an abnormal share

of his vitality, his strange precocity deprived him of boyhood.   While others of his age were nappy at their sports,
he was by himself, earnestly grappling with the deepest
questions, now wresting brilliant secrets from science with
joy and glory, now musing over the darker problems of
human nature, pale, weary, sad, hopelessly baffled in rea
son, imperiously remanded by his education and his heart
to faith.   He early became an invalid, and was scarcely ever
after free from pain.   As time wore on his state grew
worse.   His excessive mental labors shattered his constitution.   A morbid depreciation of the worth of all worldly aims gradually possessed him.   He became extremely
unhappy, not merely in his outward relations but also in
his speculations.   His vast genius, out of tune and balance, saw disproportion, misery, and frightful mystery
everywhere.   He furnished another exemplification of the
truth that great men, unless blessed with health, are more
unhappy than others, because their transcendent powers
are intrinsically less harmonized with their earthly conditions.   Their faculties overlap the world, and the superfluous parts, finding no correspondent object, no soothing
returns, are turned into wretchedness.  Pascal asks, " Shall
he who alone knows nature alone be unhappy?"   Yes,
if knowledge of nature be the pioneer of discord and rebellion against nature.   Only let love for what is known
and conformity to it keep even pace with knowledge, and
the more one knows the happier he will be.   But there is
danger with great genius, especially if there is any dis-
balancement in it, that perception will generate undue
feeling, feeling out of tune with the facts, and therefore a
source of irritable wretchedness.

    This is clearly to be seen in the case of Pascal himself.
He wore an iron girdle stuck full of steel thorns, which
he pressed into his side whenever worldly thoughts allured
him.   " Seek no satisfaction on earth," he said ; " hope
nothing from men ; your good is in God alone."   A true
religious philosophy would rather say, Seek a relative satisfaction in every normal fact of nature, every finite manifestation of the will of God ; never despair of your fellow
men.   Whatever the God of nature has made is good ;

whatever the God of grace does is well. The sound master of moral insight labors to ennoble human nature and life by every possible imaginative aggrandizement and exaltation. The school represented by Pascal strives to demean human nature and life by every possible imaginative impoverishment and degradation. This direful mistake is committed in the imagined interest of a supernatural antidote for the bane of a ruined world. It aggravates the evils it seeks to cure, by exciting what needs to be soothed, namely, the friction of man with his fate.

The noble but overstrung sensibility of Pascal is shown by the fact that once, when Arnauld seemed to prefer peace to truth, the shock of grief and pain was so great that he fainted away. To read his meditations on the nature and state of man, is like wandering through some mighty realm of desolation, where gleams of light fall on majestic ruins, lonely columns, crumbling aqueducts, shattered and moss-grown temples. His logic, his vigor, his irony, did shining and permanent service to morality in the Provincial Letters. But in his "Thoughts" a dark tinge of disease, a perverse extravagance, vitiate the unquestionable originality, and give the whole strain of argument an unsoundness as gloomy and pervading as the intellect is powerful and the rhetoric brilliant. He sees man suspended between the two abysses of infinity and nothingness. He never wearies of varying the melancholy antithesis of the sublimity and the contemptibleness of man, the grandeur and the misery of our nature and lot. Man is a chimera, an incomprehensible monster, a contradiction, a chaos, judge of all things, victim of all, depositary of truth, sewer of error, the brother of the brutes, the equal of the angels, the glory and the scum of the universe. He is a closed and inexplicable enigma, — unless we accept the scheme of Christianity in the dogmatic exposition of the Catholic Church. Original sin is the key to the otherwise incomprehensible riddle. The violence of its fall in Adam crushed human nature into a mass of piteous and venerable ruins, an incongruous collection of suns and dungheaps.

The genius of Pascal is displayed in the magnificence

of his lamentations, the gorgeous ornaments with which he enhances the degeneracy he describes. His disease is revealed in the dismal melancholy he throws over all, and in the perverse factitiousness of his remedial devices. "Vulgar Calvinism," Hallam says, "exhibits man as a grovelling Caliban, Pascal paints him as a ruined arch-angel." But both endeavor to exaggerate the evils of our nature and deepen the darkness of our state, in order to lend increased preciousness and splendor to the super-natural remedy. This method surely violates the mod-eration of nature, the sanity of reason. Imagination is given us to secure equilibrium in our powers and con-ditions, to bring in ideal compensations for actual defects, to harmonize our nature and lot. It is a dreadful abuse to employ it to multiply incongruities and annoyances, enlarge existing disbalancements, and intensify the dis-cords already experienced. To see the truth and conform to it what is out of proportion, is the final cure for every human ill. To aggravate a malady, half supposititious, so as to give imaginary value to some artificial panacea, is the method of quacks and dupes.

The soul of Pascal was a lonely battle-field, — the scene of a struggle between opposite tendencies, which must sometimes have been as terrible as it was noiseless and hidden. His logical acuteness and intrepidity penetrated sophisms, and exposed the innumerable difficulties and perplexities of human life in their most formidable array; while his fears, affections, weakness, made him cling to the Catholic creed. His sublime imagination pictured man as a grain of dust on the earth, the earth itself as a grain of dust in the bosom of nature, the eternal silence of whose boundless spaces was frightful to him. Disease made deeper encroachments on his digestive organs and on his brain. He was weary of the struggles of the few for glory, sick of the insincerity and frivolity of the many; and often said, "I shall die alone." Although he was never personally a misanthrope, his fearful insight of the frenzied self-love, the folly, vanity, flippancy, and false-hood of common men painfully alienated him from them. He says, "If all men knew what others say of them, there would not be four friends in the world." And his desire for friendly fellowship, his feeling of his own loneliness, appear clearly enough from the following paragraph : — "The little communication that we can have in the study of the abstract sciences disgusted me with them. But I expected to find many companions in the study of man. I was deceived. There are still fewer who study man than who study geometry."

He had suffered two attacks of partial paralysis in his limbs, attacks which seemed, however, not to touch his mind. While in this weak condition driving across one of the bridges over the Seine, the horses took fright and leaped off, leaving the carriage poised on the edge. The shock affected him so severely that he from that time frequently had the hallucination of an abyss yawning at his side. In this state, his sister, a devoted nun, persuaded him forever to abandon the world. Disappointed, sick, excited, his capacious mind hungering for truth and peace in the infinite, he turned with a morbid eagerness to the seclusion and austerity of Port Royal. Here he gave the last eight years of his life — from his thirty-first to his thirty-ninth year — almost exclusively to religious meditation, and to literary labors for the cause of Christianity.

The hypochondriacal state of Pascal betrays itself through all his poetic sophistry and glorious declamation. Health, without denying the evil of the world, enjoys its good, and tries, by making the best of it, to rise into something better. It is disease that reacts against it in disgust and horror, and paints it a thousand times worse than it is, in order to lend a keener relish to some theoretic good. The incongruities of human nature are better explained by the doctrine of a tentative progress towards our destiny, an advance still incomplete, with complicated faculties not yet harmonized, than by the doctrine of the Fall, which simply adds a new problem more fearful than the one it professes to solve. If man stands midway between infinity and nothing, — which is an oratoric, not a philosophic, expression, — his desires allying him to that, his attainments to this : if he clasps hands on one side with the ape, on the other side with the angel, it is that he has risen thus high and his passions are not yet equilibrated with his conditions, rather than that he has fallen thus low, and in his plunge been caught there by grace, and is now torn by the contradictory attractions of salvation and perdition. The facts of the problem are far more satisfactorily solved by the idea that the exorbitant faculties and demands of man are the pre paratory rudiments of the divine estate he is to inherit, than by the idea that they are the discordant fragments of a celestial state from which he has been expelled. Man is a child of nature sensuously chained to the earth, but ideally scaling the heights of immensity ; not a lord of heaven tumbled in ruins, mourning over what he has lost while clutching at what is within his reach. But Pascal took the latter view. He fought down his doubts, or thrust them out of sight, and clung frantically to the traditional theology as a shipwrecked man to a spar. We see his face look out at us as he drifts, a white and piteous speck of humanity, in the black flood. He regarded the soul as the convulsed ground of a supernatural conflict between the fiends of nature and the ministry of grace. He called man a reed that thinks. His soul was alone, a geometric point of thought in the infinitude of space. Impelled by the grandeur of his soaring mind and the wretchedness of his tortured body, both aggravated by the theological scheme reciprocally ministering to them and ministered to by them, he was constantly darting to and fro between the two poles of imagination, All and Nothing, and constantly associating one of these monstrous extremes with everything human. Shocked and lacerated by such a tremendous vibration, no wonder his strength so early gave way, no wonder his view of life was so overwrought. The disease which the surgeons laid bare in his gangrened vitals and brain, is equally revealed by a psychological autopsy of his writings, gangrenous blotches interspersing the splendid and electric pages. Thus, in his self-depreciating unhappiness and solitude, he affirmed that it was sinful for any one to love a creature so unworthy as he, and so soon to perish.

Perhaps the most pathetic passage from his pen, — when we view it in the light of his pure character, transcendent talents, and sad biography, — is the following : "Man has, springing from the sense of his continued misery, a secret instinct that leads him to seek diversion and employment from without. And he has, remaining from the original greatness of his nature, another secret instinct, which teaches him that happiness can exist only in repose. From these two contrary instincts there arises in him an obscure propensity which prompts him to seek repose through agitation, and even to fancy that the contentment he does not enjoy will be found, if, by struggling yet a little longer, he can open a door to rest." He had known long ages in thought and feeling, but not forty years in time, when death kindly opened for him the door to rest.

## ROUSSEAU.

ROUSSEAU was so lonely a man that the ground of his life was one long soliloquy, interrupted only as its surface now and then broke into distasted dialogues. He had singularities which made sustained companionship extremely difficult ; singularities for the understanding of which few of his critics have had the only available key, namely, sympathetic insight. The connection of heart and brain in him was wonderfully intimate, the quantity and obstinacy of emotion extraordinary. His states of consciousness had greater impulsive force in their origin, greater vascular diffusion in his system, greater persistency in his nerves, than those of other people. In his youth he sought to avoid the other boys who wanted him to join them in their sports. "But," he says, "once really in their games, I was more ardent and went further than any. Difficult to start, and difficult to restrain — such was ever my disposition." Again he says : "If no better than others, I am at least different from them. I am made unlike any one I have ever seen." Tragedies lie latent in those simple, daring words.

A truer picture of the shaded glory and prominent

wretchedness of the youth of genius has never been drawn than that painted in the "Confessions" of Rousseau. At sixteen, discontented. with himself and everything about him, devoured by desires of whose object he could form no conception, weeping without any cause for tears, sighing for he knew not what, given up to caressing the creations of his own fancy, he was happy only when he could escape, alone, among the lonely charms of nature, and abandon himself to impassioned visions, ideas, and dreams. "My delight in the world of imagination and my disgust with the real world gave rise to that love of solitude which has never since left me. This disposition, apparently so misanthropic and so melancholy, in reality proceeds from a heart all too fond, too loving, too tender ; a heart which, failing to find real beings with whom to sympathize, is fain to feed on fictions."

In the case of Rousseau, the sensitive pride imbedded in his constitution, too deep and constant for his own recognition, his fiery and persistent consciousness of his own soul, of objects, ideas, and emotions, required a soothing, deferential sympathy more pronounced and sustained than men were willing to give. The failure to receive what he wanted cut him to the quick. He regarded the disappointment as a cruel injustice, and retreated for solace into his own fancy and into seclusion. "The beings of my imagination," he declares, "disgust me with all the society I have left." Yes ; because the reactions towards him of the beings of his imagination were under the domination of his own will, while the persons he met in society exercised their own opinions and feelings towards him, however much they fell below his. It is plain that for many years he disliked society because he did not shine in it as he thought he ought. In his first letter to Malesherbes he repudiates this conclusion, even for the past, when it was true, because it was now no longer true He claims as the real reason, "An indomitable spirit of liberty, arising less from pride than from my incredible dislike of effort. The slightest duty laid on me in social life is insupportable. Therefore is ordinary

intercourse with men odious to me ; but intimate friendship is dear, because it imposes no duty ; one follows his
heart, and all is done." A deeper analysis would have
shown him that this "indomitable spirit of liberty" was
itself based on a perversity of pride. Friendship was
dear because it reflected himself to himself; ordinary
intercourse was odious because it asked him equally to
reflect others to themselves. He would not have been
unhappy in society if he had been either humble in himself or indifferent as to the opinions others entertained
of him. To have a high idea of self and to need the
sympathy of others to sustain it, is to be miserable in all
society except the most congenial. The only cure is to
be found either in self-sufficingness or in self-renunciation.

The fervid quickness and strength of Rousseau's feelings keyed him on so high a pitch that he could hardly
sink into contentful unison with others. Finding every
response distressingly inadequate to his craving he almost ceased to ask a response. "The evening," he
writes, "when I have spent the day alone, finds me
happy and gay ; when I have passed the time in company it finds me taciturn and depressed." Accordingly
he took refuge in imagination and solitude. Once, when
his friend Diderot had long been confined in the Bastile,
he obtained permission to visit him. On being admitted
he rushed forward in convulsive joy, fell on his neck, and
covered his face with kisses and tears. Diderot, instead
of returning the demonstration of affection, coolly said to
his jailer : "You see, sir, how my friends love me." The
ice that fell on his heart in that moment poor Rousseau
never forgot. And when, at a later time, Diderot, with
direct reference to him, said that no one but a bad man
could love to live in solitude, it was no wonder that with a
deep sense of injury he indignantly repelled the assertion.
One of his most characteristic works, published after his
death, was entitled "Reveries of a Lonely Walker." The
first reverie begins with these words : "Behold me, then,
alone on the earth, with no brother, neighbor, friend, society, save myself." And on a later page he writes : "I

Q

was born with a natural love for solitude, which has grown the more I have known men." His pride was incomparably greater than his vanity, making it difficult for him to sympathize with those socially above him. His democratic dislike for royalty and the nobility was vehement. But his sympathy with his inferiors was easy and strong. He even says : "My dog is not my servant, but my friend." When the pent lava-flood rose too high in his breast, it forced a ravaging vent in literature, setting whole countries aflame, and fevering successive generations. His electric words, as fresh to-day as a hundred years ago, as fresh a hundred years hence as to-day, compel sighs and tears to answer them, and can strike no inflammable soul without kindling it into a blaze. From the nature of the case it was scarcely to be expected that such a man would achieve peaceful and sufficing intercourse with his contemporaries. Thirsting for an ocean of love and admiration, when but a cupful was offered he turned away disappointed, chagrined, unhappy, and fled into the wild and lovely retreats of the country, to soothe his fever with the grass and rocks, the snows, the woods, the waters, and the stars. The expression he gave to his impassioned sensibility for the charms of romantic scenery and the sweets of loneliness has impregnated much of subsequent literature, and has given him a high rank among the missionaries of the love of nature and solitude.

When staying at Vevay, on the borders of Lake Leman, he experienced overpowering expansions of sentiment. " My heart poured itself out in a thousand innocent and ardent joys. Melting into tenderness I sighed and wept like a child. How often, stopping to indulge my feelings, seated on some projecting piece of rock, I occupied myself with watching my tears drop into the stream." This was morbid sentimentality, beyond a question. But it is unjust to stigmatize it thus, as if that were all. This wealth of soul, this passionate mental sensibility so much above the common endowment, was the hiding-place of his genius.

Theodore Parker wrote in his diary, " I have just been reading the Confessions of Rousseau. A thief ! a liar !

a great knave! — how I abhor him!" And to give this as the proper valuation of the man whose immense fire-soul inspired whole nations and ages with the love of nature, the love of liberty, the love of man, and the love of nobleness! The estimate is as fair and adequate as the estimate pronounced by Sir Isaac Newton on the art of sculpture, when he called the statuary in the gallery of the Duke of Devonshire "stone dolls." The purloining of the ribbon and the dastardly falsehood he told to conceal the act, were not chronic traits of the character of Rousseau, but violent deviations from it; and he atoned for them by the heroic confession of his guilt before the whole world. His foibles and vices, gross as they were, would appear venial in contrast with those of the most of his compeers, had these unveiled their shame as fully as he, for a high moral purpose, did his. It is to be remembered that he has made us know the very worst of him, so stripping his soul and life in the "Confessions" that he might dare, as he startlingly affirmed, present himself before the judgment-seat of God with his book in his hand! And, on the other side, the mass and keenness of his love for mankind and for moral beauty were superlative, in spite of his frequent personal defections. Despite those defections, too, he defended the cause and served the interests of freedom, virtue, and humanity. with an eloquence never equalled, and with a practical effectiveness rarely surpassed. One of the best read and wisest writers of recent days says: "The modern fashion of gentleness in feeling and manners was introduced mainly through the influence of Rousseau." He recalled mothers to the nursing of their own children. He vindicated with peerless energy the simplicities of nature against the cruel corruptions of luxurious conventionality. He was the first French gentleman to put off the wearing of gold lace and a sword, and adopt a simple costume. He inaugurated the most invaluable reforms in education, and in behalf of the equal rights of all men struck blows that made the old despotisms of Europe reel on their thrones. And as to his jealousy, quarrels, moral extravagances, it is to be borne in mind that the sufferings of disappointment and disease, together with the various peculiar trials to which he was subjected, had repeatedly wrought him into a state of virtual insanity : " His mind had grown suspicion's sanctuary."  To name the men of illustrious worth, from Kant to Channing, from Schiller to Whittier, who, well aware of his vices, have yet loved and honored him for his matchless merit, would be answer enough to an abusive condemnation of him by wholesale.  His genius is a glory and inspiration to his race ; his services claim grateful homage ; his errors and sins plead for forbearance.

Schiller says : —

> In Rousseau, Christians marked their victim — when
> Rousseau enlisted Christians into men !

Banned by church and state, hated by bigots, feared by good men, unloved by the common people whose wrongs had so deeply moved him, he lived apart and misunderstood.  He reminds us herein of the remark made by Zimmermann: " There is always something great in the man against-whom the world exclaims, at whom every one throws a stone, on whose character all attempt to fasten a thousand crimes without being able to prove one."  Rousseau rejected not the sanctities and authorities of nature and conscience, only the base counterfeits substituted for them.

> For peace or rest too well he saw
> The fraud of priests, the wrong of law ;
> And felt how hard between the two
> Their breath of pain the millions drew.
> A prophet-utterance strong and wild,
> The weakness of an unweaned child,
> A sun-bright hope for human kind,
> And self-despair — in him combined.
> The love he sent forth, void returned ;
> The fame that crowned him scorched and burned,
> Burning, yet cold, and drear, and lone —
> A fire-mount in a frozen zone.

Born with delicate organs and irritable nerves, developing a precocious sensibility, his early reading of Plutarch and romances joined with his native bent to blend in his soul the heroic ideality of Rome and Sparta with the poetic ideality of chivalry. Lacking the equilibrium of sober reason, at the frequent sight of individual instances of cruelty and meanness he reacted from his high-strung notions of absolute good and human perfectibility into a wretched despondency. Had he invariably turned from the special examples of wrong to the general laws of right, to the deep, steady, moral sanctions and tendencies in the nature of things and in the nature of man, he might have been happy. But he kept up a vibratory action between the thought of himself and the thought of those he disliked or suspected. And he had known but too many reasons for dislike and suspicion. Thrown into contact with many vile characters, he suffered base misinterpretations at their hands. He was so often belied, cheated, slandered, that his outraged mind took on a chronic impression of his wrongs and colored the whole world with them. He lost his health and became a hypochondriac, tremulously shrinking from contact with men. Henri Martin says : " He fancied himself surrounded with a universal conspiracy to degrade his character and blight his memory before posterity. Instead of exaggerating his influence he exaggerated his isolation. He disbelieved the sincerity of the disciples who flocked to him ; and did not taste the highest consolation, to a heart like his, of enjoying the good that he had done to his fellow-beings. This was doubtless a harsh expiation of the offences he may have committed in this world."

Unquestionably there is a basis for the severe judgment pronounced on him by Joubert, who yet ascribes to him a vast moral superiority over Voltaire. Joubert says that Rousseau was envious, vain, proud, voluptuous, irreligious in his piety, corrupt in his severity, dogmatic against authority, discontented with everything beyond himself, a beggar basking in the sun and deliciously despising the human race ! I cite this judgment that it may have its effect ; for there is a truth in it, though it is only one side of the truth, and that side most uncharitably heightened. He did voluptuously pamper the notion of himself, had the egoistic vices of excessive sentiment and

an over-heated brain.   He was his own Pygmalion.   **But**
this was a morbid reaction from an unworthy society, and
rather deserves a sigh than a curse.   It is the grand, unspotted Fichte who says : " Rousseau would have been as
modest and happy as his critics, had he been tormented
with as few noble aspirations."

The reader who is swift to blame the faults of Rousseau should remember his merits, pity his woe, and learn
to avoid his mistakes.   For our own part we like best to
leave him with the words of his grateful friend, the author
of " Paul and Virginia."   Bernardin St. Pierre says : " I
derived inexpressible satisfaction from his society.   What
I prized still more than his genius was his probity.   He
was one of the few literary characters tried in the furnace
of affliction, to whom you could, with perfect security,
confide your most secret thoughts.   Even when he de-
viated, and became the victim of himself or of others, he
could forget his own misery in devotion to the welfare of
mankind.   He was uniformly the advocate of the miserable.   There might be inscribed on his tomb these affecting words from that Book of which, during the last years
of his life, he carried always about him some select passages: 'His sins, which are many, are forgiven ; for he
loved much.' "

### ZIMMERMANN.

THE fame of his treatise on the " Influence of Solitude " has echoed the name of Zimmermann through the
world.   Born at Brugg, a little town on the banks of the
Aar, near Zurich, he received an elaborate education
covering the various provinces of history, science, philosophy, and poetry.   His masculine understanding made
him a good proficient in mathematics, politics, and statis-
tics, while his uncommon sensibilities and taste gave him
delighted range in the richer field of romantic literature.
He was familiar with the Greek and Latin poets, the best
German and French authors, and the English Shakespeare, Pope, Thomson, and Young.   He must have
had by nature not only a clear and powerful intelligence,
but also an unusually tender and noble heart.

He was greatly capable of enthusiastic admirations. When still a mere youth, studying his profession at Göttingen under the celebrated Haller, he felt the warmest love and reverence for this great physician, formed relations of charming intimacy with him, and afterwards wrote a glowing life of him. In his old age, the last flame of his hero-worship broke out with tenacious heat and brilliancy in connection with the king of Prussia, Frederick the Great. His soul was fitted to enjoy friendship in its most sacred delicacies. His writings and his life abound with the proofs. What a gracious charm of sincerity and fervor breathes in his numerous allusions to his friends in his literary works! This is especially the case when he refers, as over and over he does in his "Solitude," to Lavater, Hotze, Hirtzel, Tissot. He was highly esteemed by his friends, on whom his many noble qualities made their proper impression. We have a life of him written by Tissot, who does full justice to his renowned and lamented associate.

He was exceedingly fortunate, too, in his wife. A niece of Haller, she was lovely in person and mind, with the mildest temper, the softest voice, extreme cultivation and brightness, and fascinating manners. While she lived, she was his sweet and sure asylum from every care. When dying, she said, "O my poor Zimmermann! who will understand thee now?" The first shock of tangible affliction he had known was her death : the second, following a few years later, was the death, by consumption, of his only daughter, whose worth he has affectingly celebrated in his literary masterpiece. Some time later, he married, again, a beautiful and estimable lady, whose assiduous fondness alleviated as far as possible the miseries of his remaining years. For, gifted as Zimmermann was with talents and accomplishments, true and kind as his friends were, widely as his celebrity as author and physician extended, he was still, a great deal of the time, a wretchedly unhappy man.

Goethe, who had considerable intercourse with him, has left an incisive sketch of this vehemently impulsive nature, outwardly polished and self-controlled, inwardly untamed and exacting. Zimmermann was the precise opposite, he says, of those persons who dance about in frivolous delight over the vacant nothings which they are; he had great deserts with no inward satisfaction. His severity towards his children, — even towards the favorite daughter whom he so eloquently mourned, — was "a partial insanity, a continuous moral homicide, which, after having sacrificed them, he at last directed against himself." Yet Goethe generously says he was himself deeply indebted to this brave, rich-souled, most instructed and public-spirited man; and adds that all who understand the sad life he led will not condemn him, but pity him.

The unsocial side of Zimmermann was based in twofold disease: first, the mental disease of an excessively sharp and constant desire to be appreciated, to be noticed and admired; second, the bodily disease of hypochondria, — that sickly irritability which results from an over-tasking of the nervous system. Carlyle says: "He had an immense conceit of himself, and generally too thin a skin for this world. A person of fine, graceful intellect, high, proud feelings, and tender sensibilities, hypochondria was the main company he had." He suffered dreadfully from what may be called social hyperesthesia, a morbid over-feeling of the relations between himself and others. At twenty, while yet a student in the University, he wrote to his friend Tissot, "I pass every hour of my life here like a man who is determined not to be forgotten by posterity." Later, when established as a physician in his native village, the feeling of his own superiority to the rude people around him destroyed all comfort in intercourse with them. Still later, when promoted to a more courtly sphere, as physician to the king of Hanover, a keen perception of the neglect he received from some, of the envy and gall of others, of the innumerable foibles and vices of most, incessantly nettled and depressed him, and kept him in a ferment of misery. Had he possessed a stable self-complacency, contemptuous of foreign opinion, or calmly superior to it; could he have been content with the approval of his own conscience, trying himself by the fixed standard of duty, — his distress and melancholy would have been unknown. But the idea and desire of being thought highly of by all were nailed to his imagination and heart, and they fastened him in misery.

His impartial biographer says, "Many parts of his work betray the feebleness of his nerves, and the peevishness of his temper. But there was a striking difference between his manners and his writings. When with others, he was always generous, gentle and polite, incapable of saying an offensive word. He always made his patients his friends, by the unwearied complaisance of his attentions. But the moment he was alone, and at his desk, his urbanity left him, and he grew satirical : his natural energy, his vehement love of virtue and hatred of vice, carried him away, and he painted the worse characteristics of men in the liveliest colors." His very words seem to tingle with indignation, when he speaks of hearing dolts praised for their learning, and atrocious villains complimented on their well-known humanity. He is misanthropic because the glowing height of his ideal of humanity ironically condemns the base deviations from it which are so common. If he said, "Who lives with wolves must join in their howls," he also said, "He alone is fit for solitude who is like nobody, liked by nobody, and likes nobody."

The chief and chronic happiness of man ought to arise from himself and his own conduct. Feeling that his witness is on high, he ought to be satisfied with the approval of his own conscience, and not rise and fall in soul, like barometer, with the favors and frowns of other men. The misery of Zimmermann originated in his inability to secure this self-sufficing independence. It is astonishing to see how clearly he knew the truth he so grossly failed to practise. "It is not," he says, "my doctrine, that men should reside in deserts, or sleep like owls in the hollow trunks of trees ; but I am anxious to expel from their minds the excessive fear which they too frequently entertain of the opinion of the world. I would, as far as is consistent with their respective stations in life, render them independent. I wish them to break the fetters of prejudice, to imbibe a just contempt for the vices of society, and to seek occasionally a rational solitude, where they may so far enlarge their sphere of thought and action as to be able to say, at least for a few hours every day, 'We are free.'" Yet the writer of this fine paragraph was never free from the bondage against which he so well inveighs. Vanity was his colossal foible. His elation at the attentions of Frederick, the pomp of happiness with which he proclaims the flattering gift and letter sent him by the Russian Empress Catherine, are ludicrous. And the cleaving agony he suffered from every mark of opposition or undervaluation tends to provoke laughter in one class of spectators as much as pity in another. If the poetic side of his susceptibilities challenges admiration, their personal side is obnoxious to contempt. When he had been blamed for an article which he had published in some medical review, he says: "There arose against me a universal shrieking-combination, a woman-epidemic." It is obvious that much of his pain originated in the sore imagination that he occupied a greater space in the thoughts of others than he really did, and that he was less favorably judged by them than he desired to be and believed he deserved to be.

Zimmermann has been a prominent member of the Apostolate of Solitude. He experienced it so thoroughly, meditated on it so patiently, in its wretchedness and in its happiness, in its inspiring influence and in its blighting influence; he saw the truths on both sides of the subject so sharply, that, on the whole, he has handled the theme with remarkable fairness. He has not been frightened by the appearance of inconsistency, but has stated the facts of the case in its opposite aspects with energetic boldness. Here are some of his scattered aphorisms: " In the crowd we are impudent; in the closet, modest." " Genius stagnates in solitude : where merit shines, merit is kindled." " Those who are good alone should not be left alone." " We are most of us, even in our maturest age, infants : we cannot go alone." " Great characters are their own heralds, though they have thousands to announce them."

A few hours before expiring, Zimmermann uttered these words, the last that he spoke : "I am dying : leave me alone." He died. They left him alone in his coffin, un-derground, — at least, seemed to leave there such part of him as may be left in any material enclosure ; for the pow-er of God reclaims at once his returning child, the bosom of Nature soon her sundered elements.

Two generations had passed, when, at sunset, after a day of calm beauty, in the summer of eighteen hundred and sixty-five, the writer of this sketch, a pilgrim from America, stood in the burial-ground of the quiet and quaint old city of Hanover beside that deserted tomb. He pondered the lessons written and lived for the benefit of others by the silent slumberer beneath his feet. He meditated on the different forms of human loneliness, their causes, their accompaniments, and results. His musing ended in a peaceful thought, which blended the memory of the once popular author with the loneliness of death and oblivion, and with a fame glimmering swiftly over the nations of the earth to subside in the dark silence of the grave. Kneeling then, half-unconscious of what he did, he wrote with traceless finger on the stone, Here Zimmermann drinks his fill of solitude !

## BEETHOVEN.

THE personality and life of Beethoven were profoundly lonesome. His immense native power of mind and sen-sibility, early set askew with the world of men, made him peculiarly sensitive to exactions, slights, and irritations. The death or the fickleness of the maiden he loved in his youth apparently made a dark and sinister stamp on his social character, and left a permanent bitterness in his blood. His averseness to common intercourse was aggra-vated by his poverty, his devouring absorption in the sci-ence and art of music, and a singular combination in him of awkwardness and scorn, tender diffidence and titanic pride. The lack of popular favor, the incompetent con-demnation his wonderful compositions long suffered, must also have been a trial tending to sour him.  Furthermore, as in the case of every man of primal genius, his tran-scendent originality doomed him to a determined struggle with the past, an uncompromising insurrection against conventional authority and usage.  He defied the pre-scriptions of his predecessors, broke pedantic fetters, re-futed his teachers, made new rules for himself, upheaved a world dead in professional routine and tradition that he might inspire it with fresh freedom and fresh triumphs ; and thus, perforce, he stood alone, battling with obscurity, contempt, and hate, until he slowly conquered the recog-nition he deserved.  Finally, in addition to these previous causes, the sternness of his isolation was made complete by the dreadful calamity of a dense and incurable deaf-ness.

Dark indeed was his melancholy, bitter the revulsion of his capacious soul upon itself.  He says, " I was nigh taking my life with my own hands.  But Art held me back.  I could not leave the world until I had revealed what lay within me."  Resolved at any cost to be him-self, and express himself, and leave the record to pos-terity, he left behind opponents and patrons alike, and consecrated all to his genius and its ideal objects.  Occu-pying for a long time a room in a remote house on a hill, he was called the Solitary of the Mountain.  "His life was that of a martyr of the old legends or an iron-bound hero of the antique."  Poor, deaf, solitary, restless, proud, and sad, sometimes almost cursing his existence, some-'imes ineffably glad and grateful, subject now to the soft-est yearnings of melancholy and sympathy, now to tem pestuous outbreaks of wrath and woe, shut up in himself, he lived alone, rambled alone, created alone, sorrowed and aspired and enjoyed alone.

The character of Beethoven has many times been wronged by uncharitable misinterpretations.  He has been drawn as a misanthrope, a selfish savage.  His nature had attributes as glorious as the music born out of them.  He was a democrat, who earnestly desired that the rights of all men should be secured to them in the enjoyment of freedom.  Asked, in a law-suit before a

German court, to produce the proof of his nobility, he pointed to his head and his heart, and said, " My nobility is here, and here." He was a fond reader of Plato and of Plutarch. One of his biographers says, " The republic of Plato was transfused into his flesh and blood." He always stood by his republican principles stanchly. It was in the firm belief that Napoleon meant to republicanize France that he composed and inscribed to him his Heroic Symphony. On learning that the First Consul had usurped the rank of Emperor, he tore off the dedication and threw it down with explosive execrations. He sympathized intensely with that whole of humanity which to a genius like his ever reveals itself as a great mysterious being, distinct from individuals, yet giving the individual his sacredness and grandeur. His uncertain and furious temper was an accident of his physical condition, the unequal distribution of force in his nervous centres. He once suddenly quitted a summer retreat, where he was supremely happy, because his host persisted in making profound bows whenever he met him in his walks. Such an incident makes his nervous state clear enough. An idea which to a man of stolid health and complacency would be nothing, entering the imagination of the rich and febrile Beethoven, was a terrific stimulus. To judge such an one justly, discriminating insight and charity are needed.

In his lofty loneliness his mislikers considered him as " a growling old bear." Those who appreciated his genius thought of him as the mysterious " cloud-compeller of the world of music." Nearly all regarded him as an incomprehensible unique into whose sympathetic interior it was impossible to penetrate. Carl Maria von Weber once paid him a visit, of which his son, Max Weber, has given a graphic description full of interesting lights. Himself kept scrupulously clean by an oriental frequency of bathing, he sat in the disorderly desolate room, amidst the slovenly signs of poverty, his mass of lion-like face glowing with the halo of immortality, his head crowned with a wild forest of hair. He was all kindness and affection to Weber, " embracing him again and again, as though he could not part with him."

When he produced his mighty opera, Fidelio, it failed. In vain he again modelled and remodelled it. He went himself into the orchestra and attempted to lead it; and the pitiless public of Vienna laughed. To think now of the Austrian groundlings cackling at the sublimest genius who has ever lifted his sceptre in the empire of sound, making him writhe under the torturing irony of so monstrous a reversal of their relative superiorities! After suffering 'his cruel outrage, he fled more deeply than ever into his cold solitude. As Weber says, "He crept into his lair alone, like a wounded beast of the forest, to hide himself from humanity." Nothing can be sadder in one aspect, grander in another, than the expression this unapproachable creator, this deaf Zeus of music, has given of his isolation. "I have no friend; I must live with myself alone; but I well know that God is nearer to me than to my brothers in the art."

Of course this is no entire picture either of the soul or the experience of Beethoven. He had his happy prerogatives and hours. Life to him too was often sweet and dear. He knew the joy of a fame which before he died had slowly grown to be stupendous. Almost every one of the musical celebrities who arose in his time, from 'the author of Der Freischütz to the author of Der Erlkönig, with pilgrim steps brought a tributary wreath to him as the greatest master. Above all, he had a sublime consciousness and fruition of his own genius. At one time he says, "Music is like wine, inflaming men to new achievements, and I am the Bacchus who serves it out to them." At another time he says, "Tell Goethe to hear my symphonies, and he will agree with me that music alone ushers man within the portals of an intellectual world ready to encompass him, but which he can never encompass." If he suffered hunger, loneliness, the misunderstanding of the vulgar and conventional, he kept himself free, and felt himself supreme in his sphere. An anonymous critic has well written of him, "He gained what he sought, but gained it with that strain of discord in his finer nature which is to the soul of the artist what the shadow of a cloud is to a landscape. The desire to make the world different from what it was in kind as well as degree was the error which ruined his earthly peace ; for he persisted in judging all relations of life by the un attainable ideals which drew him on in music. Yet it was out of this opposition to the reality, which was to him a sorrow and bitterness known to but few beside, that there came the final victory of his later creations." He also knew that his strains would sound his name and worth down the vista of future ages with growing glory. " I have no fear for my works. No harm can betide them. Whoever understands them shall be delivered from the burdens that afflict mankind."

But despite all these alleviations Beethoven was pre eminently a lonely nature. He was extremely fond of taking long walks by himself through beautiful scenery — as Petrarch, Rousseau, and Zimmermann were. One hardly knows where to look for a more pathetic outbreak of a loving and disappointed heart than is given in the follow ing expressions in the will he left for his two brothers. The thoughts in that passage of his Heroic Symphony wherein, as he said, he prophesied the melancholy exile and death of Napoleon, are not charged with a more penetrative sadness or immense grief than is in the strain of these pleading, parting words : — " O ye who consider me hostile, obstinate, or misanthropic, what injustice ye do me ! Ye know not the secret causes of what to you wears this appearance." " My deafness forces me to live as in an exile." " O God ! who lookest down on my misery, thou knowest that it is accompanied with love of my fellow creatures, and a disposition to do good. O men ! when ye shall read this, think that ye have wronged me. And let the child of affliction take comfort on find ing one like himself, who, in spite of all obstacles, did everything in his power to gain admittance within the rank of worthy artists and men." " I go to meet death with joy. Farewell, and do not quite forget me after I am dead."

## SHELLEY.

THE cruel injustice with which Shelley was hunted, in the abused names of morality and religion, by persons immeasurably beneath him in every attribute of nobleness, is one of the bitter tragedies of our century. Few men have existed so brave, thoughtful, disinterested and affectionate as he. Hundreds of passages in his poems, and in his letters, make the heart of the sensitive reader bleed. He writes to his wife, "My greatest content would be utterly to desert all human society. I would retire with you and our child to a solitary island in the sea, would build a boat, and shut upon my retreat the flood-gates of the world." At another time this exquisite child of intellect and sensibility says, "I feel myself almost irresistibly impelled to seek out some obscure hiding-place, where the face of man may never meet me more." The sorrow of the case is that he so passionately loved his kind all the while, and passionately longed for love in return. He made a transcript from his own heart when he wrote, "In solitude, or in that deserted state when we are surrounded by human beings, and yet they sympathize not with us, we love the flowers, and the grass, and the waters, and the sky." The poor people in Florence who saw him wandering through the neighborhood and in the galleries there, called him "the melancholy Englishman."

In his "Stanzas written in dejection near Naples," occur the lines :

> I sit upon the sands alone ;
> The lightning of the noontide ocean
> Is flashing round me, and a tone
> Arises from its measured motion,
> How sweet did any heart now share in my emotion !

One can hardly help recognizing in the following passage of his Julian and Maddolo, a description partly copied from his own experience.

> There are some by nature proud,
> Who, patient in all else, demand but this, --

> To love and be beloved with gentleness ;
> And being scorned, what wonder if they die
> Some living death ?

Shelley comforted himself with high studies and works, with the deep love of a few chosen intimates, with doing good to every poor sufferer who came within his reach, with the loftiest ideal philanthropy, and with as intense a communion with nature as ever blessed the soul of a poet. But, with his transcendent capacities of imaginative feeling, he walked ensphered in a mystic loneliness. His words are :

> I love all waste
> And solitary places, where we taste
> The pleasure of believing what we see
> Is boundless, as we wish our souls to be.

During his stay in Rome, himself almost as lonely as the glorious Titan he describes, he wrote his Prometheus Unbound. He composed it, in his own language, " on the mountainous ruins of the baths of Caracalla, among the flowery glades and thickets of odoriferous blossoming trees which are extended in ever-winding labyrinths upon its immense platforms and dizzy arches suspended in the air." If the objects of nature were ever made the beloved playmates of a mortal, that mortal was Shelley. But it is his great triumph, his profound medicinal lesson for other men, that extreme as was his suffering from wrong and obloquy, and deep as were his resources elsewhere, he never sank to misanthropy, but always continued to love and always sought to bless those who hated and strove to injure him. The scornful Landor, after an eloquent eulogium of the rare virtues of Shelley, adds, " This is the man against whom such clamors have been raised by bigots and cowards, and by those who live and lap under their tables." In contrast with this frank and galling contempt, how divine is the strain in which the outcast poet himself addressed his persecutors : —

> Alas ! good friend, what profit can you see
> In hating such a hateless thing as me ?
> There is no sport in hate where all the rage
> Is on one side. In vain would you assuage

12 *                                                    R

> Your frowns upon an unresisting smile,
> In which not e'en contempt lurks to beguile
> Your heart by some faint sympathy of hate.

Of all the expressions of the mind and heart of Shelley, perhaps the most wonderful and sustained in intensity
of richness is his " Alastor, or the Spirit of Solitude." In
the prose preface to it he explains its purpose to depict
a poet of the rarest gifts, who, after a devoted pursuit of
the choicest ends of life, thirsts for the sympathy of an
intelligence like his own, and, failing to find it, is blasted
by the disappointment, and droops into an untimely
grave. It is an allegory of the loneliness of genius, describing how " the pure and tender-hearted perish through
the intensity and passion of their search after the communities of human sympathy." The lesson that " the
self-centred seclusion of genius will be avenged by darkness, decay, and extinction," and that " the selfish, blind,
and torpid multitudes constitute, together with their own,
the lasting misery and loneliness of the world," is taught
in this poem with vivid power by one who

> In lone and silent hours,
> When night makes a weird sound of its own stillness,
> Like an inspired and desperate alchemist,
> Staking his very life on some dark hope,
> Had mingled awful talk and asking looks
> With his most innocent love and his strong tears,

until, conqueror of his enemies in the loving conquest of
himself, he could say,

> Serenely now,
> And moveless as a long-forgotten lyre
> Suspended in the solitary dome
> Of some mysterious and deserted fane,
> I wait thy breath, Great Parent, that my strain
> May modulate with murmurs of the air,
> And motions of the forest and the sea,
> And voice of living beings, and woven hymns
> Of night and day, and the deep heart of man.

Few men, indeed, have either loved solitude better or
had keener experience of wrong at the hands of society
than Shelley. Yet he knew well, and well teaches others,

how profound is the need of a loving fellowship with his kind, for the health, force, and joy of every man, for the highest as well as the lowest. He says, with reference to self-centred seclusion, "The power which strikes the luminaries of the world with sudden extinction by awakening them to too exquisite a perception of its influences, dooms to a slow and poisonous decay the meaner spirits that dare to abjure its dominion. Their destiny is more abject and inglorious, as their delinquency is more contemptible and pernicious. They who, deluded by no generous error, instigated by no sacred thirst of doubtful knowledge, duped by no illustrious superstition, loving nothing on this earth, and cherishing no hopes beyond, yet keep aloof from sympathies with their kind, — have their appointed curse. They are morally dead."

The magnificent scorn which Shelley felt for every form of meanness or cruelty breathes throughout his works, especially in the burning wrath with which, in his "Adonais," he blasts the author of the brutal attack in the Quarterly on Keats. But the terrible contempt with which he swooped down on the "miserable calumniators," the "nameless worms," the "viperous murderers," the "carrion-kites" of men, was only a passing ideal anger, never a chronic hatred or personal revengefulness. The comparison of his own extraordinary mind with the dwarfish intellects around him, the perception of the vast superiority of his own power and passions to those of ordinary men, never, as it did in the case of Byron, fed a devouring pride in him ; never filled him with disdain for his race or with disgust at the worthlessness of the prizes of life. He nobly practised the precept he so nobly urges : —

> There is one road
> To peace, and that is TRUTH, which follow ye !
> LOVE sometimes leads astray to misery.
> And some perverted beings think to find
> In scorn or hate a medicine for the mind
> Which scorn or hate hath wounded. O, how vain !
> The dagger heals not, but may rend again.

No man familiar with the writings of Shelley, who has any appreciation of the scale of ranks in human character, can roam in the haunts wonted to his feet; muse on the landscapes his eyes loved to drink; stand in the halls where he dwelt; pause on the beach of the bay where the sea, with late remorse, gave up the drooping marble of his form; recall the scene of his friends restoring his limbs to dust in fire mixed with wine and frankincense; linger in votive thought, the soul of the dead poet transfusing the conscious soul of the pilgrim, before the grave in Rome holding his heart, and read through tears that tenderest of all inscriptions, *cor cordium*, heart of hearts, without emotions of pity, reverence, love, and wonder, which words can hardly convey.

It is impossible more fitly to end this sketch of one who, in proportion as he is appreciated, will be the darling of gentle and generous hearts, than by quoting the words of his enthusiastic friend, Leigh Hunt: "He was like a spirit that had darted out of its orb and found itself in another world. I used to tell him that he had come from the planet Mercury. When I heard of the catastrophe that overtook him, it seemed as if this spirit, not sufficiently constituted like the rest of the world to obtain their sympathy, yet gifted with a double portion of love for all living things, had been found dead in a solitary corner of the earth, its wings stiffened, its warm heart cold; the relics of a misunderstood nature slain by the ungenial elements."

### COLERIDGE.

THE opinion has been expressed by De Quincey, that the intellect of Coleridge was "the subtilest and the most spacious that has yet existed among men." His heart was not inferior to his mind. Yet how profoundly lonely he was! The rich fire of his fancy and the fatal faintness of his will made the world a dream peopled with phantoms. He once characterized himself as "through life chasing chance-started friendships." Among the lines he wrote after spending a night in the house once occupied by the Man of Ross, we read with strong emotion the following: —

But if, like mine, through life's distressful scene,
Lonely and sad thy pilgrimage hath been,
And if, thy breast with heart-sick anguish fraught,
Thou journeyest onward tempest-tossed in thought,
Here cheat thy cares ; in generous visions melt,
And dream of goodness thou hast never felt.

His dear friend Charles Lamb, who almost idolized him, said "he had a hunger for eternity." No doubt, in the immensity of his spiritual isolation from ordinary minds, when he turned back from baffled efforts after a competent communion, he often felt

So lonely 't was that God himself
Scarce seeméd there to be.

Aubrey De Vere, in the fine poem he wrote after the death of the Seer of Highgate, says : —

And mighty voices from afar came to him ;
Converse of trumpets held by cloudy forms,
And speech of choral storms.
Spirits of night and noontide bent to woo him.
He stood the while, lonely and desolate
As Adam when he ruled a world yet found no mate.

Though it is true that Coleridge had a few dear friends, he appeared to live in a spell, with an enchanted barrier about him. His existence was a long soliloquy of wondrous richness, weirdly remote from contact, which other men seem to overhear as unseen listeners.

He said himself : "Perhaps never man whose name has been so often in print for praise or reprobation had so few intimates as myself." When he died at Highgate, after a residence of twenty years, a biographer says " he was a stranger in the parish, and therefore was interred alone ! "

## WORDSWORTH.

SOLITUDE is to different persons what their characters, habits, and aims make it. To one and another it is variously a covert, a prison, a sanctuary, a studio, a forge, a throne. To Wordsworth, that grand and peaceful spirit, patriarch of the intimate muse, it was a bower, a chamber, a library, and a temple, — his place of joy, rest, work, and worship. Here he retreated from the distressful medley of popular whims, from the deteriorating strain of common ambitions, to the intrinsic standards of truth and good, and the authoritative companionship of greatness and worth. He retreated hither not to brood over woes, indulge disdain, and meditate revenge ; but to enjoy thought, nature, and God, and impart the enjoyment to mankind.

Wordsworth was fitted for solitude by his informing and overpowering ideality; by his brooding, interior tenderness ; by his heroic originality, self-respect, and independence. His impassioned imagination turned things to thoughts and thoughts to things, and frequently made absorbing emotions suffice in place of sights and sounds, deeds and words. He said himself that he was often so rapt into the world of ideas that the external world seemed not to be, and he had to reconvince himself of its existence by clasping a tree or some other object that happened to be near him. When Sir George Beaumont had made him a munificent gift, and he had for many weeks neglected to acknowledge the favor, he apologetically says : "I contented myself with thinking over my complacent feelings, and breathing forth solitary gratulations and thanksgivings, which I did in many a sweet and many a wild place during my late tour." The winter he spent at Goslar, in Germany, he walked daily on the ramparts by a pond. "Here," he writes, "I had no companion but a kingfisher, a beautiful creature that used to glance by me. I consequently became much attached to it."

He was too occupied and grave and continuous, too wealthily sensitive and devoutly dedicated, for that flippant, fragmentary jocosity, that free and easy intercourse on the level of little nothings, in which average natures take pleasure. His microscopic studies of himself and his states ; his steadfast sympathies with the simplest and poorest objects ; his telescopic sweeps of the sublimities of nature, history, and philosophy, — insulated him equally from the vulgar and the proud. In his own words, —

He was retired as noontide dew,
Or fountain in a noonday grove ;
And you must love him, ere to you
He would seem worthy of your love.

With such souls as his sister Dorothy, Coleridge, and Charles Lamb, he maintained a glorious community of mind and heart in a friendship of rare beauty ; but to the multitude he was uninteresting, and positively repul sive to conventional and conceited critics. The multitude neglected him ; the critics made contemptuous war on him.

This was on account of the peculiar offence of his originality. Dismounting from the traditional stilts of poetry, this great poet and true man sought to portray, in the simple language of genuine insight and passion, the permanent and universal elements of beauty, dignity, and joy in the outward works of God, in the structure of human nature, and in the experience of human life. The inspiration of loveliness, worth, and sublimity had hitherto been chiefly sought in the most imposing outer aspects of life and nature ; in kings, courts, conquerors, philosophers ; in the romantic, the exceptional. He sought to unveil it in the commonest places and forms ; to show the grandest materials of wisdom, poesy, and religion, tragedy and happiness, in huts and laborers, in the most ordinary lot and landscape of man. He had the moral courage, love, and perseverance to do this, and genius to succeed. But, until he had educated a public to appreciate his originality, he had to pay the penalty of his superiority in the suffering of a long series of insults and incompetent scorn. With reference to his " Idiot Boy," he was called the hero of his own story. His " Peter Bell " was saluted with a chorus of jeers. His books were little read, while volumes of trash had a large circulation, and were praised by all the reviews. Was not here enough to make a man break down in despair, or recoil into misanthropy? He had an extraordinary passion for fame, knew himself worthy of it, but was attacked and despised for the very things for which he ought to have been admired and prized.

Without despondency or hate, he fell back on his gifts and call ; turned from actual men to ideal man, from the irritating society of fashion and emulation to the pacifying society of the landscape and the Infinite Spirit, and determined to conquer usefulness and renown by perfect ing himself and improving his productions. It is one of the noblest examples in history, — the example of " pre ferring to any other object of regard the cultivation and exertion of his own powers in the highest possible degree."

The peculiar experience of Wordsworth, united with his peculiar characteristics, made him a solitary, but not a lonesome, man. He enjoyed a happy tranquillity nursed by meditative sympathy ; whereas many a one, under less trying circumstances, has fallen victim to an irritable wretchedness nursed by restless antipathy. He was continually thinking of the things worthy to be loved and adored ; they, of the things worthy to be shunned and loathed. The conditions of his moral victory afford an example worthy of careful study.

In the first place, his self-respect and self-confidence never failed him ; and —

> Happy is he, who, caring not for pope,
> Consul, or king, can sound himself to know
> The destiny of man, and live in hope.

He certainly had dipped his pen in his deepest blood when he wrote, —

> Creative art
> Demands the service of a mind and heart
> Heroically fashioned to infuse
> Faith in the whispers of the lonely Muse,
> While the whole world seems adverse to desert.

He did not need to be "nourished with the sickly food of popular applause." Perceiving that men praise us only as they recognize in us some counterpart of what they are or wish to be, he saw that often there is no surer test of merit than obloquy. Taught to feel (perhaps too much) the self-sufficing power of solitude, invulnerable to the sleet of hisses and arrows, knowing himself divinely called to his work, assured that his place was with the great

and good of all ages, he wrote to Southey, "Let the age
continue to love its own darkness; I shall continue to
write, with, I trust, the light of heaven upon me"; and
to Bernard Barton, with reference to a bitter critique on
him, "I doubt not but that it is a splenetic effusion of
the conductor of that Review, who has taken a perpetual
retainer from his own incapacity to plead against my
claims to public approbation." Again, he writes to Lady
Beaumont, that, knowing the absolute ignorance in which
worldlings of every rank and situation must be wrapt as
to the feelings, thoughts, and images on which the life
of his poems depended, the envy and malevolence which
stand in the way of a work of any merit from a living poet,
he had only the lowest expectations concerning the immediate effect of his writings on the public. But, he adds,
as to the assaults of the critics, "My ears are stone-deaf
to this idle buzz, and my flesh as insensible as iron to
these petty stings. I have an invincible confidence that
my poems will co-operate with the benign tendencies in
human nature and society wherever found; and that they
will, in their degree, be efficacious in making men wiser,
better, and happier."

His joy in nature was as great as in his vocation.
However isolated, he never felt lonesome, as his personality rather blended with objects that stood relieved
against them.

> They flashed upon that inward eye
> Which is the bliss of solitude,

and were either taken up with imaginative enrichment
and transfusion into himself, or himself blent and lost in
them. Many a time and oft he sat amidst his own
thoughts, and amidst the scenes of nature, in such entrancement that "even the motion of an angel's wing
would have interrupted the intense tranquillity." In the
silent faces of things he could read unutterable love.

> Sound needed none,
> Nor any voice of joy; his spirit drank
> The spectacle; sensation, soul, and form
> All melted into him; they swallowed up

> His animal being : in them did he live,
> And by them did he live : they were his life.

No wonder he was fond of solitude ; and, though he "wandered lonely as a cloud that floats on high over vales and hills," he could not be lonesome.   In the heart of the mist ; on the bare moor, or the top of the moun tain, or under the cope of midnight ; in the haunt of the heron, by the shy river ; or the remoter nook, where the pelican sits on the cypress spire and suns himself, — he was at home, with things in which he joyed, and which seemed to love him.   He says well, —

> I learned betimes to stand unpropped :
> And independent musings pleased me so
> That spells seemed on me when I was alone :
> Yet could I only cleave to solitude
> In lonely places ; if a throng was near,
> That way I leaned by nature.

Thirdly, his reverential and joyous communion with himself, and his reverential and joyous communion with nature, keeping him pure in heart, content with modest pleasures, removed from little enmities and low desires and every malignant passion, enabled him also to maintain a reverential and joyous communion with man.   He gave no harbor to suspicions and envies, but wholesomely threw them off.

> He kept,
> In solitude and solitary thought,
> His mind in a just equipoise of love.

When he thought of the cruelties and miseries of men, it was to pity and try to cure them.   When he thought of the oppressions and degradations of men, it was not weakly to despond, or to give way to hate or scorn ; but, with generous indignation, to denounce them, and aspire to liberty and nobleness for all, with a firm reliance on the great laws and principles tending to realize the predestined order of the Creator.   He contemplated prevailingly the diviner qualities and sublime connections of human nature, the glorious facts and hopes of human life, until he recognized —

A grandeur in the beatings of the heart.

He saw man in his own domain of the earth, —

As a lord or genius, under God,
Presiding ; and severest solitude
Had more commanding looks when he was there.

Man rose on his sight, set in the most beautiful and im·
posing scenery of the world, encompassed with august
powers of virtue and bliss and tragic troops of woe. The
imagination and passion which many embittered geniuses
have used to darken and degrade the image of man,
Wordsworth employed to depict him in dazzling lights,
endowed with godlike attributes, not as a mere dusty
brother of the worm, but as a being first in every capa-
bility of wisdom, goodness, and rapture, through the divine effect of truth and love. Thus, when he left his
lonely mountains, and, in the tribes and fellowships of
men, was begirt by shapes of vice and folly, bustling
greeds of manners, objects of sport, ridicule and scorn ;
when he " heard humanity in fields and groves pipe solitary anguish, or hung brooding above the fierce storm of
sorrow, barricaded evermore within the walls of cities,"
— he was not downcast or forlorn. He turned to the
true ideal man, ennobled by associated connection with
nature and the presence of God, with past and future,
with history, science, and philosophy. There he found
unfailing comfort and inspiration.

But, besides the happiness Wordsworth had in his undisturbed self-respect, in the forms and motions of nature,
and in his ardent sympathy with the human race, he knew
a rarer and perhaps keener happiness in the profound
presentiment of his own benignant and illustrious fame
in the future. How well he knew his own place !

If thou indeed derive thy light from heaven,
Then, in the measure of that heaven-born light,
Shine, poet ! in thy place, and be content.

Since his estimate was the simple truth, and not unac-
companied with devout humility, it is a shame to call it
egotism. Hundreds of the best men and women of the

generation succeeding, and still more of generations yet to come, will echo the truth of his anticipations. "Of what moment is the present reception of these poems," he wrote to Lady Beaumont, "compared with what I trust is their destiny? To console the afflicted; to add sunshine to daylight, by making the happy happier; to teach the young and the gracious of every age to see, to think, and feel, and therefore to be more actively and securely virtuous, — this is their office, which I trust they will faithfully perform long after we are mouldered in our graves." Not many men have ever been better entitled to feel and to say, —

> There is
> One great society alone on earth, —
> The noble living and the noble dead.

Lastly, the happiness of Wordsworth in all his solitudes was completed by his unaffected communion with God, — not the dead God of tradition, not the abstract God of verbal formulas; but the living God, who is the Lord of all that is. By purity, holiness, humility, waiting sympathy, his mind became a conscious temple for "the Prophetic Spirit that inspires the human soul of universal earth dreaming on things to come." He became wonderfully aware of the significance of those awful incumbencies under which human thoughts creep; and recognized them as

> Visitings
> Of the Upholder of the tranquil soul,
> That tolerates the indignities of Time,
> And, from the centre of Eternity
> All finite motions overruling, lives
> In glory immutable.

No man, without the utmost sincerity and intensity of unborrowed religious experience, could have written passages that abound in the poems of Wordsworth, particularly in his "Excursion," "Tintern Abbey," and "Ode on the Intimations of Immortality."

> In such access of mind, in such high hour
> Of visitation from the living God,
> Thought was not; in enjoyment it expired.
> No thanks he breathed, he proffered no request;

> Rapt into still communion that transcends
> The imperfect offices of prayer and praise,
> His mind was a thanksgiving to the Power
> That made him.

This religious originality ranks him among the inspiring prophets of the race. To know this, must, amidst the abuse and the neglect he suffered, have administered exalted consolation to him. "It shall be my pride," he says, —

> That I have dared to tread this holy ground,
> Speaking no dream, but things oracular.
> . . . . . . . . . . This I speak,
> In gratitude to God, who feeds our heart
> For his own service ; knoweth, loveth us,
> When we are unregarded by the world.

In Wordsworth, the transitions of consciousness were ever from the insignificant to the august, from the ugly to the fair, from individuals to humanity, from the transient and exceptional to the permanent and universal, from the finite creation to the Infinite Spirit. Thus he avoided the rasping shocks of disappointment, neutralized exasperating vexations, healed grief and despondency, rested serenely on sublime supports of peace and happiness. In this manner, he so informed his mind with quietness and beauty, so fed it

> With lofty thoughts, that neither evil tongues,
> Rash judgments, nor the sneers of selfish men,
> Nor greetings where no kindness is, nor all
> The dreary intercourse of daily life,
> Could e'er prevail against him, or disturb
> His cheerful faith, that all which we behold
> Is full of blessings.

Who was ever better rewarded than Wordsworth in the realization of what, in one of his earliest poems, he had distinctly seen and coveted as the highest earthly prize?

> A mind, that, in a calm, angelic mood
> Of happy wisdom, meditating good,
> Beholds, of all from her high powers required,
> Much done, and much designed, and more desired, —
> Harmonious thoughts, a soul by truth refined,
> Entire affection for all human kind.

It is a beautiful thing, too, to know, that, before he died, the meeds he had so grandly earned were poured at his feet in lavish tributes from abroad, not alone in the silent honor and love of the best minds in the world, but also when old Oxford twined her lofty laurel round his head, while her children made her arches shake above their shouted welcome ; and, still more, when the wronged and glorious Shelley said to him, —

> Thou wert as a lone star whose light did shine
> On some frail bark in winter's midnight roar ;
> Thou hast like to a rock-built refuge stood
> Above the blind and battling multitude.

Wordsworth is still more a teacher than a poet, — one of the very deepest and soundest moral teachers of the world. His muse, indeed, often reminds us rather of Academus than of Parnassus. But more than is lost in art and beauty is gained in guidance and edification. He incarnates for us the endless lesson that to promote and fortify the general welfare of our own being is the paramount end, — a fact which almost all forget in a distracted pursuit of externalities. He would call us home into the possession of ourselves, not for any egotistic pampering, but in order that we may lose ourselves in fruition and worship of the whole. "We live too little within," sighs poor Maurice de Guérin. "What has become of that inner eye which God has given us to keep watch over the soul, to be the witness of the mysterious play of thought, the ineffable movement of life, in the tabernacle of humanity? It is shut; it sleeps."

The special value of Wordsworth is as an exemplifying teacher and contagious imparter of certain habits of thought and feeling. He is an original apostle of the enthusiasm of nature, the enthusiasm of principles, and the enthusiasm of humanity. There is a deep and vital philosophy in his creed, —

> To her fair works did Nature link
> The human soul that through me ran ;

as life itself is primarily "an adjustment of inner relations with outer relations," a moving reflection within us of something originally without us. He has in this direction given an invaluable new impulse both to literature and to direct experience. He teaches us to recognize humanity not as the mere sum of existing men, but, in addition to this, as a spirit diffused through time and space over the whole world from its beginning to its end, enriched with all history and all hope, — a conception under whose influence our entire existence becomes thronged with inspiring impregnations of reflection and sentiment. Who truly receives this instruction will learn

> To prize the breath we share with human kind,
> And look upon the dust of man with awe.

He teaches us that

> By love subsists
> All lasting grandeur, by pervading love ;
> That gone, we are as dust.

He would make us

> Know that pride,
> Howe'er disguised in its own majesty,
> Is littleness ; that he who feels contempt
> For any living thing hath faculties
> Which he has never used.

With unwearied earnestness of conjoined example and precept, he illustrates how —

> Unelbowed by such objects as oppress
> Our active powers, those powers themselves become
> Strong to subvert our noxious qualities ;
> They sweep distemper from the busy day,
> And make the big, round chalice of the year
> Run o'er with gladness.

He leads us to be

> Studious more to see
> Great truths than touch and handle little ones.

He teaches us pre-eminently the lesson which alienated moroseness so constantly inverts : —

> To enfeebled power,
> From clear communion with uninjured minds
> What renovation may be brought, and what
> Degree of healing to a wounded spirit,

> Dejected, and habitually disposed
> To seek in degradation of her kind
> Excuse and solace for her own defects.

He exemplifies in all his life, and in all his works, the habit of seeing the great in the small, the sublime in the vulgar, the strange in the common, the awful authority and charm of humanity in the poorest and most ignorant men, and God everywhere, — a habit invaluable alike for wisdom, for virtue, for dignity, for peace, and for happiness. Fortunate every one who learns the secret !

He teaches us, finally, the restorative efficacy and charm of solitude, like a prophet familiar with all her secrets. To turn from a Heine to a Wordsworth is like changing attention from the roar and blaze of brothels, groggeries, and hells, to a nightingale warbling on a moonlit bough in heaven. What a strain he pours on the ears of the fops, loungers, gladiators, and slaves of time ! —

> When from our better selves we have too long
> Been parted by the hurrying world, and droop,
> Sick of its business, of its pleasures tired,
> How gracious, how benign, is solitude !
> How potent a mere image of her sway !
> Most potent when impressed upon the mind
> With an appropriate human centre, — a hermit,
> Deep in the bosom of the wilderness ;
> Votary, in vast cathedral, where no foot
> Is treading, where no other face is seen,
> Kneeling at prayers ; or watchman on the top
> Of lighthouse beaten by Atlantic waves ;
> Or as the soul of that great Power is met
> Sometimes embodied on a public road,
> When, for the night deserted, it assumes
> A character of quiet more profound
> Than pathless wastes.

The immortal fame of Wordsworth is secure with the immortal benefits he will render his docile readers. While Winander, Fairfield, and Rydal remain, to all visionary minds his wraith will haunt them ; and as long as Derwent runs, it will murmur his name to the pilgrims on its banks. Men will have a more blessed and mysterious communion with nature, a more constant and pervading sense of the presence of God, a more firm and tender love of their fellow-beings, because he has lived and sung. Fitly does Lowell say, referring to him : " Parnassus has two peaks : the one where improvising poets cluster ; the other, where the singer of deep secrets sits alone, — a peak veiled sometimes from the whole morning of a generation by earth-born mists and smoke of kitchen fires, only to glow the more consciously at sunset, and after nightfall to crown itself with imperishable stars."

## BYRON.

ANY list of the great solitary spirits of the world, not to be strikingly defective, must contain the name of Byron. He has written the best lines in the English language on the subject of solitude. His personality, full of fascinating interest, stands in relief from the mass of men. His experience furnishes instructive illustration of many of the conditions and consequences of spiritual isolation. We may in his example trace the dark secrets of unhappiness more clearly than almost anywhere else.

Byron was marked out from average humanity from his very birth. He inherited from both parents a " blood all meridian," on one side rich with voluptuous sensibility, on the other side tingling with vehement irritability. He had a dark, tempestuous passionateness of temperament, combining in the most singular manner a remarkably keen and abiding sense of himself with a remarkable freedom from the meannesses of selfishness, and an unusual susceptibility to noble thoughts and sentiments. The deceitfulness, fickleness, coldness, meanness which his sharp intelligence, aggravated by his morbid consciousness, taught him to trace in the characters and deeds of most of his associates, — the great disparity between what he craved and what he found, — very early gave a stronger warp and an intenser tinge to his natural bias towards loneliness and a melancholy brooding over his own thoughts.

The brain of Byron, physiologically considered, was a wonderful organ. It was at once uncommonly powerful and uncommonly small. What fineness and firmness of

fibre, what compactness and vigor of cells, what profuse
ness of polarity it must have had! And in addition to
the original concentrated strength of his highly charged
nervous structure, everything in his circumstances and life
contributed to heighten his genius by intensifying his
mental polarities and disturbing their equilibrium. One
distinguishing element in his self-consciousness was his
rank. He was heir to a title giving him prominence
among his fellows, yet without the better accompaniments
of respectful deference and tenderness which should attend such a birth. His intellectual pride kept him from
obtruding his titular supremacy. His generous democratic
impulses and his contempt for the illustrious mediocrities
of the peerage and the throne, made him disdainful of
inherited insignia ; yet in him the feeling of the peer ever
lay underneath the feeling of the poet and fast by the
feeling of the man. The poverty and neglect which
shrouded his childhood lent a new acuteness to his feeling of his social and personal claims. His sufferings
when first sent from home to school,—poor, proud, shy,
affectionate, unknown, unfriended, unnoticed in the herd of
boys, — were pitiable. When by the death of a relative he
succeeded to his ancestral honors, as the master called his
name in school, for the first time with the prefix of lord,
he burst into tears in the midst of his staring mates. A
dark slough of mortified pride, created by his young
school experience at Harrow, hung over his mind for
years. There is a tomb in the churchyard at Harrow,
commanding a view over Windsor, which was such a favorite resting-place with him that the boys called it Byron's
tomb. Here he would sit for hours wrapt in thought,
"brooding lonelily over the first stirrings of passion and
genius in his soul, perhaps indulging in those forethoughts
of fame, under the influence of which, when little more
than fifteen years of age, he wrote " : —

> My epitaph shall be my name alone :
> If that with honor fail to crown my clay,
> O may no other fame my deeds repay !
> That, only that, shall single out the spot,
> By that remembered or with that forgot.

Another heightener of self-consciousness in Byron was his superb personal beauty. The romantic charm of his noble features, with the mystery of his genius, drew all eyes on him wherever he went, as soon as he had become known, and secured for him that flattery of attention and curiosity which cannot fail to react on its object. That he was fully conscious of this, and that it wrought on him with a keen force, is obvious from many particulars; among others from the pains he took with his toilet, shaving the hair off his temples, setting the fashion of the turned-down collar. This over-consciousness of himself was raised to a painful pitch by the slight malformation and lameness of one of his feet. "The embittering cir cumstance of his life," Moore says, "which haunted him like a curse, and, as he persuaded himself, counterbalanced all the blessings showered on him, was the trifling deformity of his foot." He once said mournfully to his friend Becher, who was trying to cheer him by the assurance of his great gifts, "Ah, my dear friend, if *this* (laying his hand on his forehead) places me above the rest of mankind, *that* (pointing to his foot) places me far, far below them." He said the "horror and humiliation" which came over him once in childhood, when his mother called him "a lame brat" were unutterable. He also once, on overhearing Mary Chaworth, of whom he was desperately enamored, say to a female friend, "Do you suppose I could love that lame boy?" darted out of the house in a state of frenzy, and fled into the solitude of the forest, where he stayed until late in the night.

The boyish sensibility of Byron was strangely empassioned, easily piqued, resentfully retentive of wrongs, slights, and pains. When his favorite schoolmate, young Lord Clare, expressed regret at the departure of another friend, Byron was tortured with jealousy. It affords the skilled psychologist a deep glimpse into the secrets of his bosom to know that he was bashful even to the end of his life, and had the habit of blushing. When staying in his boyhood at Mrs. Pigot's, if he saw strangers approaching the house, he would leap out of the window to avoid meeting them. The same trait is ascribed to himself by

Rousseau, with whom Byron had much in common, in spite of his elaborate disclaimer of the asserted likeness. It is also recorded of Virgil that his diffidence often caused him to beat a sudden retreat into shops, to escape the gaze of those who met him in the streets of Rome. Such a union of qualities always makes its possessor fond of seclusion, and gives him at least a superficial twist of misanthropy. The youngest muse of Byron sang to one of his earliest friends : —

> Dear Becher, you tell me to mix with mankind :
> I cannot deny such a precept is wise :
> But retirement accords with the tone of my mind,
> And I will not descend to a world I despise.

It is affecting to see how soon a half sad, half angry soreness towards the world mingled with his strong and haughty boldness of self-assertion. On the death of his mother, nearly at the same time with that of his two friends, Matthews and Wingfield, he wrote to Hodgson, " I am solitary, and I never felt solitude irksome before."

During his first journey in Greece, he said, his chief delight was to climb to some high rock above the sea, and remain there for hours, gazing on the sky and waters, lost in reverie. There are immortal passages in his poems which demonstrate how often and how sincerely he must have enjoyed this sombre luxury. When he testified, " My nature leads me to solitude, and every day adds to this disposition," his words expressed the simple truth, and no freak of affectation. His mind was cast in a deep and gloomy mould. Few could adequately sympathize with him. Conscious of this he strove to exaggerate it, with an emphatic liking for whatever emphasized his unlikeness from the human commonalty. He drew himself in such characters as a Childe Harold, — " Apart he stalked in joyless reverie," — a Conrad, a Lara — " lord of himself, that heritage of woe," — to exaggerate the more his contrast with other men, to make them wonder and tremble, to give a stronger charge to their awe and curiosity pertaining to him.

He took a dark delight in cherishing tragic ideas and

looking on objects of terror from which other persons would shrink horrified. His stormy soul felt most at home with the storm. Amidst the awful revelry of the elements on the benighted Alps, hearing the live thunder leap from crag to crag, seeing the lake lit into a phosphoric sea, he longs to fly abroad in the carnival of natural horrors, a disembodied portion of the tempest and the night. His passion for images of terror, his passion for female beauty, his passion for all lonely and savage scenes of nature,—almost exclusively gratified in seclusion from the distractions of company, — fed his great passion for solitude, because there he felt himself lifted into distinct prominence from other men, saved from what he regarded as the profane vulgarity of being sunk and confused in the mass of humanity. He loved solitude too because it set his faculties free. Men of his style of mind are intolerably restive under any external restraint. Their intractable self-will cannot bear a yoke, or a foreign direction, but must obey its own impulse alone.

> I have not loved the world, nor the world me ;
> I have not flattered its rank breath, nor bowed
> To its idolatries a patient knee,
> Nor coined my cheek to smiles, nor cried aloud
> In worship of an echo. In the crowd
> They could not deem me one of such ; I stood
> Among them, but not of them ; in a shroud
> Of thoughts which were not their thoughts.

The morbid bias in the soul of Byron received a darker and more decisive turn from his distress at the marriage of Mary Chaworth. His mad love for her, his grief on her wedlock with another, the permanent influence it exerted on his character and fortunes, are depicted in his marvellous poem, " The Dream." His genius spread the wretched workings of this bitter disappointment over all society. He felt himself singled out for desolation : —

> As some lone bird without a mate,
> My weary heart is desolate ;
> I look around and cannot trace
> One friendly smile or welcome face,
> And even in crowds am still alone.

He sought relief in foreign travel. Taking the page, Robert Rushton, in his train, he said, "I like him, because, like myself, he seems a friendless animal." The scornful review given to his juvenile "Hours of Idleness" had already stung him to the production of a vigorous satire, which had made him suddenly famous, the object of attention from all quarters.

The centres of contrast or sets of poles in his mind furnished by the opposite ingredients of his nature and experience, acquired still greater power when, returning home, he published the first two cantos of "Childe Harold." His reputation and popularity rose to an extreme height. He became the most observed and courted of all observers. Then followed, in swift shocks of succession, the envy of mortified rivals, the hatred of jealous inferiors, the sneers of incompetent critics, the attacks of the religionists whose convictions both his precepts and his practice offended, his marriage, and the separation of his wife from him. The outbreak of odium, obloquy, calumny, horror which came upon him, in suddenness and intensity and extent surpasses belief. It would have overwhelmed any one whose pride was less colossal, whose strength less obstinate, whose resources less rich, than his. Moore says, "Such an outcry was raised against Lord Byron as in no case of private life perhaps was ever before witnessed." Bigots and the galled enviers of his success gave vent to their dislike of him in all sorts of libels, hints, and innuendoes; and good people who believed him the foe of morality and religion, swelled the current with dark interpretations and reports. He was cut in the street, and excluded from all but the fewest houses. His acquaintances deserted him; of all the obsequious crowd only half a dozen friends stood by him. His previous recluseness and cynicism took on a sullener hue. His isolation from the average fellowships of life was carried to its climax. He left his country, never more to return until brought with sword and laurel on his bier.

> Adieu, adieu! My native shore
> Fades o'er the w ers blue.

But the tempest of hate, slander, and depreciation was beating behind him, and chased him with every post. A poem published at that time referred to him thus : —

> Wisely he seeks some yet untrodden shore,
> For those who know him less may prize him more.

And in a rhyming pamphlet, addressed to him, these lines occur : —

> Shunned by the wise, admired by fools alone,
> The good shall mourn thee, and *the Muse disown.*

This to the author of the " Hebrew Melodies," " Manfred," and the " Prophecy of Dante," the object of the admiring homage of a Shelley and a Goethe, the electric shaker of his age ! A writer in Blackwood called Venice " the lurking-place of his selfish and polluted exile." Insult and injury alight on a spirit like that of Byron as a whirlwind on the waters. Sinful and faulty as he was, he deserved not a tithe of the penalty inflicted. Few can know how sincere and how fearful were his sufferings. It was only the gigantic power of his personality which enabled him to surmount the convulsing experiences heaped on him. He said : " I felt that if what was whispered and muttered and murmured was true, I was unfit for England ; if false, England was unfit for me. I withdrew ; but this was not enough. In other countries, in Switzerland, in the shadow of the Alps, and by the blue depth of the lakes, I was pursued and breathed upon by the same blight." How painfully he felt every fresh hurt appears from the pains he took to avoid them. For five years in his exile he never once looked into an English newspaper. Those unsympathetic critics who sneer at Byron for breaking his heart in public once a month, and making melodramatic capital out of his woes, are unjust to him. They do not feel the fearful severity of his trial ; do not see that it belonged to his tenacious, associative, creative genius to accumulate magnifying materials around every seated pang, and that a God gave to him, as to Goethe, power to find some relief from his sufferings by *singing* what he suffered. His self-agony was no fiction which he played with to catch attention, but a fact which he had to deal with as best he could. It became the central string of his lyre and tone of his voice. He is to be pitied for it, not scorned. Those literary critics who ridicule Petrarch, Rousseau, Byron, for making an exhibition of their personal sorrows, implying that they themselves have similar or severer trials of wrongs, noble griefs and loneliness, only they are too great and modest to expose them to the public, — these critics seem not to be conscious that they reveal in double refinement the very trait they complacently condemn, the same egotism raised to the third power. Byron virtually says : " O world ! I am most unhappy. I will bear my undeserved fate with defiant fortitude. Look on my grief with sympathy, on my heroism with admiration ! " His critic virtually says, half to the public, half in self-pampering soliloquy : " The afflictions proper to exalted genius have fallen on me too; but the compensating triumphs of exalted genius, its strong curative reticence, are also mine still more. I cannot stoop to expose the secrets of my soul to the vulgar world. I leave them to the insight of kindred greatness." The latter pride, which repels sympathy, is more aggravated and less amiable than the former, which seeks sympathy.

So absorbing and acute were the melancholy and wretchedness of Byron, that it is rather a wonder, as he said, that he did not, in some of their crises, burst a blood-vessel or blow out his brains. He writes in his private journal one day : " O God ! I shall go mad " ; relieving his anguish with the words of Lear. With *his* nature, after what he had gone through, the habit of an exaggerating and revengeful recollection of his wrongs and miseries was inevitable. His feeling towards those who were instrumental in his domestic ruin, desertion, and banishment, was fiendish. His letter after the suicide of Romilly is absolutely diabolical. " Do you suppose," he says, " that I have forgotten or forgiven the deliberate desolation piled upon me when I stood alone upon my hearth with my household gods shivered around me ? It has comparatively swallowed up in me every other feeling,

and I am only a spectator upon earth till a tenfold opportunity offers." The workings of this habit in generating
unhappiness were as pernicious as its moral quality was
bad. Yet he has undoubtedly expressed it with poetical
extravagance, for he was essentially generous, though not
religiously forgiving. A truer reflection of him is seen in
the verse, —

> Here 's a sigh to those who love me,
> And a smile to those who hate ;
> And, whatever sky 's above me,
> Here 's a heart for every fate.

He had a diseased sensitiveness to affronts, and could
not endure being looked down upon. In his unhappy
departure from his country, he spread the distempered
feeling of hostility over all his countrymen, and shunned
them with a tremulous revulsion poorly coated with arrogance. "Rome is pestilent with English." " I abhor the
nation, and the nation me." " In two or three years this
tribe of wretches will be swept home again, and the Con
tinent will be roomy and agreeable." " As to the estimation of the English, let them calculate what it is wor'h
before they insult me with their insolent condescension."
The isolating and embittering influences of Byron's quarrel with his country, and of his long exile, were envenomed
by his habits of life and by the character of many of his
associates. Plunging deeply into dissipation, spending
much time in company with persons of depraved tastes,
loose principles, and bad experience, his irregularities
preyed on both his health and his heart, familiarized him
with the baser and sadder phases of human life and human nature, and tended strongly to make him sick ot
man and weary of existence. We have his own mature
testimony as to dissipation, that

> It is a sad thing, and not only tramples
> On our fresh feelings, but — as being participated
> With all kinds of incorrigible samples
> Of frail humanity — must make us selfish,
> And shut our souls up in us like a shellfish.

In melancholy answer to the question why he played,
drank, rode, wrote, he said : " To make some hour less

13 *

dreary." "I have had a devilish deal of tear and wear of mind and body in my time." "As I grow older, the indifference, not to life, but to the stimuli of life, in creases." "My heart is as gray as my hair."

> Count o'er the joys thine hours have seen,
>   Count o'er thy days from anguish free,
> And know, whatever thou hast been,
>   'T is something better not to be.

"Selfishness is always the substratum of our damnable clay." He was alienated and set apart from the conventional sympathies of society by his contemptuous rejection of the theological beliefs of his time, not less than by his reckless mode of life. The hue and cry against him was greatly swelled by the angry votaries of the creed he repudiated. The puissance of independent reason was a chief current in the inspiration of power by which he stood against the world and shaped his enduring verse.

But among the slaves of vice, in the revels of abandoned indulgence, in the academy of mockery and scoffing, he could never be at home and at peace. He was an eagle, not a buzzard. Still his smitten lyre vibrates, —

> Though gay companions o'er the bowl
> Dispel awhile the sense of ill ;
> Though pleasure fires the maddening soul,
> The heart — the heart is lonely still.

He longed for sympathy in his best thoughts and feelings; he wanted sincere love and praise ; but his pride, stung by frequent repulses, would not let him simply own the want. He gave it indirect expression in his dark lamentations : —

> The fire which on my bosom preys
> Is lone as some volcanic isle : —

and perverse expression in proud assertions of self-suffic-ing solitude : —

> I stood and stand alone, — remembered or forgot.

In an eloquent letter he dedicated the fourth canto of Childe Harold to his friend Hobhouse, "to relieve a

次 t which has not elsewhere, or lately, been so much
accustomed to the encounter of good-will as to withstand
the shock firmly." Despite his sport and sarcasm he was
earnest at bottom, and his heart was ever flying to his
head. He said,

> Though hymned by every harp, unless, within,
> Your heart joins chorus, fame is but a din.

Though Byron had fits of disgust and anger with the
world of mankind, he was never a settled misanthrope.
He was of too rich and high an order for chronic malignity. Those appearances and expressions thought to
imply the contrary of this are sometimes poetic extravagances, and always rest on grounds quite distinct from a
deliberate hate of his kind. It is true that he says, with
cool literality, "The more I see of men the less I like
them": true, that he sings in fervid verse, —

> We have had our reward — and it is here ;
> That we can yet feel gladdened by the sun,
> And reap from earth, sea, joy almost as dear
> As if there were no man to trouble what is clear.

These, however, are vents of his bubbling resentments,
not of his deepest tendencies and sentiments. He loved
other things more, rather than man less. His love of
solitude was glorious, not sullen. He used it not to
fabricate plots of vengeance, but for creations of beauty
and emotional reveries. He asks and answers, —

> Why do they call me misanthrope? Because
> They hate me, not I them.

Misanthropes are not liberals in politics, nor profuse in
relieving misery with their money. Byron, notwithstanding his rank, was a republican, a deadly hater of every
form of despotism, a martyr to Greek liberty, and always
munificently kind in helping the distressed. It was in
indignant vexation at the restoration of the Bourbons that
he dashed in his journal, "To be sure I have long despised myself and man, but I never spat in the face of my
species before." Undoubtedly he had an undue amount

of anger and contempt, the result of his flaming sensibility, no sign of stagnant affection. His thrilling flesh, electric blood, thrilling nerves, must perforce render him impatient of obtuseness and affectation and incompetency. He demanded action in others on the height of his own. "I remember at Chamouni, in the very eyes of Mont Blanc, hearing an Englishwoman exclaim to her party, 'Did you ever see anything more rural?' As if it was Highgate or Hampstead. 'Rural!' quotha? Rocks, pines, torrents, glaciers, clouds, and summits of eternal snow far above them — and 'rural'!"

Byron loved to be alone because he keenly resented the hurts and hostilities he had experienced, and because he shrank in all his better moods from the comparatively ignorant, frivolous, and insensible average of society. He said that "the reason why he disliked society was that the follies and passions of others excited the evil qualities of his own nature." When alone, he put on his royal prerogatives, and rose along the blue wilderness of interminable air, to coast the crystal worlds of infinity, and sound the mysterious treasures of truth and beauty with the palpitating doubt and terror in which he delighted. The overwhelming mind that produced "Cain" must have loved to retire apart with its forces, which strike stars into chaos and mould chaos into stars ; its thoughts, which compress immensity into a point and expand a point into immensity ; its events, which are rare in time though frequent in eternity ; its vision of God

> On his vast and solitary throne
> Creating worlds to make eternity
> Less burdensome to his immense existence
> And unparticipated solitude.

Well indeed might he say that could he have *kept* his spirit to such flights he had been happy ; but in the restraints of human dwellings he became a thing of worn and weary restlessness. It was his last redemptive resource to keep his mind in frequent hours free from the dominion of worldly intercourse, and find or make

> A life within himself, to breathe without mankind.

He loved loneliness because in it he possessed the un-
trammelled enjoyment of his own powers, and particularly
because he could indulge the delicious and unfathomable
tenderness of his ideal affections.   It is impossible to
read the various poems he addressed to his sister Au
gusta, especially that wondrous autobiographic piece
written at Diodati in his twenty-eighth year, and not be
drawn to him with a mingled yearning of love, pity, and
admiration.

> O that thou wert but with me !   But I grow
> The fool of my own wishes, and forget
> The solitude which I have vaunted so,
> Has lost its praise in this but one regret.

Soured as he was on the surface, the depths remained
sweet, and his misanthropy was only lip-deep.   One day
seeing a kid vainly trying to get over a fence and piteous-
ly bleating, he dismounted, and with much pains helped
it over.   He loved little children, as the savage and depraved do not.   And he was incapable of ingratitude

> The heart must
> Leap kindly back to kindness, though disgust
> Hath weaned it from all wordlings : thus he felt.

The quiet sail of his boat on Lake Leman was as a
noiseless wing to waft him from distraction.   Surely a
gentle and noble heart speaks in the wish : —

> O that the desert were my dwelling-place,
> With one fair spirit for my minister,
> That I might all forget the human race,
> And, hating no one, love but only her !

His shrinking from ungenial fellowship in the higher
moments of his consciousness, his thirst for the restorative nutrition of solitude, and his magnanimous recoil
from the detestable acridity of experience, are all finely
shown in the following lines : —

> There is too much of man here, to look through
> With a fit mind the might which I behold :
> But soon in me shall loneliness renew

> Thoughts hid but not less cherished than of old,
> Ere mingling with the herd had penned me in their fold.
> To fly from need not be to hate mankind ;
> All are not fit with them to stir and toil.

With all his pride and power Byron lacked the complacency which is the base of happiness.  He was at war
with himself, torn between contending energies.  He was
not his own friend.  Milton, with his interior unity and
peace, reverenced and loved himself with an august
steadiness.  He was his own friend in calm and uniform
companionship.  But Byron, seeing much within to honor
and much to condemn, much to applaud and much to
deplore, rather admired and pitied, than reverenced and
loved, himself.  His interior division and perturbation
were full of unhappiness.  Half dust, half deity, he alternately lauded and loathed himself, now haughtily skirring
extinguished worlds and gazing on eternity, now sensible
of grovelling wants and littleness.  At one moment, he
sonorously cries,

> I may stand alone,
> But would not change my free thoughts for a throne ;

at another, he mourns that he was ever born, and medi-
tates suicide.  Such scorn and dislike as he had for his
kind were a reflection of his scorn and dislike for himself.

> Fain would I fly the haunts of men, —
> I seek to shun, not hate mankind :
> My breast requires the sullen glen
> Whose gloom may suit a darkened mind.

He bore a melancholy bosom and was of a moody texture from his earliest day, believing himself predestined
to woes.  " I never could keep alive even a dog that I
liked or that liked me," he wrote, when the Countess
Guiccioli fell sick.  His peculiar trials confirmed this
fatal proclivity.  Once, in the streets of Venice, with
Moore, late at night, he saw a poor destitute creature
moaning in pain ; he gave her money and soothing words ;
when she rose up, and, walking before him, mocked the
sidling motion his lameness gave his gait.  He passed on
without a word.  To appreciate his sufferings we must

understand his exquisite susceptibility, together with the noble impulses native to his soul. " Tell me that Walter Scott is better. I would not have him ill for the world," he wrote to Murray. The gorgeous and solemn elo- quence of his Monody on Sheridan is the expression of a generous and mighty heart. His valet in Ravenna saw him kneel on the pavement before the tomb of Dante, and weep. His emotions were so violent on seeing a representation of Alfiéri's Mirra in the theatre at Bologna, that he was seriously indisposed for several days from the effects of " the convulsions, the agony of reluctant tears, and the choking shudder." With such a temperament, a nature so at variance in itself, and sent through a dire ordeal, it is not strange that at times he thought

> Too long and darkly, till his brain became,
> In its own eddy boiling and o'erwrought,
> A whirling gulf of phantasy and flame.

There was also in Byron an element of perversity, the result of his injured pride pouring from above upon the teeming tenderness below. The obscure working of these two elements in combination gave him a perverse liking to invert the demands of others, baffle their expec- tations, and appear worse than he was. He liked to sur- round himself with mystery, even with dread, for the sake of the curiosity it provoked. He sought to blacken him- self beyond the truth, for two reasons. First, it distin- guished him from other men, who wish to seem better than they are ; *he* would seem worse than he was : it was the implicit satire in which he clothed his scorn of hypoc- risy. Secondly, it was a comfort to him, it sustained him in his own eyes, to react from other people's unjust esti mate of him to his own knowledge of the truth. Giving fifty guineas to an unfortunate Venetian, — when asked what he had done, he would say : " I told him to go about his business "; and then take pleasure in turning from the mistaken verdict of "cruel " to the approval of his own conscience.

In practice Byron longed for the esteem and love of his fellows, loved and praised the richness of the world.

In experience he was wretched, in consequence of the discord of his faculties and aims : because he had not attained to inner unity.   In theory, desiring to reconcile the incongruity, and justify himself, he asked, Why am I so unhappy?   The answer he gave was, Because men are bad, and the world is poor.   How much sounder the aphorism of the strong, wise Goethe : —

> Wouldst lead a happy life on earth ?
> Thou must, then, clothe the world with worth !

Byron would have outgrown his unhappiness if he had resolutely labored with clear purpose to suppress his too sharp and constant consciousness of himself and of his distracted relationships.   A pampered and tyrannical idea of self, or a despised and scourged idea of self, is irreconcilable with happiness.   An objective treatment of self in the light of truth, as any other object is treated, will gradually adjust it so that the truth itself will be agreeable to it and attune it to a firm concord.   He did not live long enough to work himself clear of the feculence of his will, the slag of his passion, and become a pure intelligence, a serene and joyous force.   So mightily endowed was he it seems a few years more must have brought him the religious victory all the supreme masters have won.   He would have conquered the lesson that detachment, self-renunciation, is the only path for the morbid and moody individual into the free, glad, healthy life of nature and humanity.

That fiery breast is cold now, that titanic spirit at rest It is well.   If his was the pain, be the moral ours.


### BLANCO WHITE.

An impressive exemplification of the cruel treatment and isolation consequent on an abandonment of conventional opinions and usages in obedience to personal convictions of truth, is seen in the life of that beautiful type of Christian character, the tender-hearted, self-sacrificing and heroically truthful Joseph Blanco White.   Born in

Spain, reared a Catholic, his conscientious inquiries led him to become successively — after he had taken up his abode in England — an Episcopalian, a Unitarian, a free theist ; and, as the penalty of his disinterested search for truth and adherence to his conclusions, he died, in the purest spirit of martyrdom, poor, obscure, sadly solitary. As could not but be the case in such a country as England, a few noble friends loved him as he deserved to be loved, and never forsook him. One of them has given us the story of his life, — a precious legacy for the spirits who are pure enough, lofty and devoted enough, to appreciate it.

A friend whom nature had exempted from doubt on subjects which habit and feeling had sanctioned to him, once found Blanco White " bathed in tears, lamenting that his faith had vanished without the least hope of recovering it." The constitution of his mind made it impossible for him to stop inquiry, and he was often torn with pain on discovering that his fancied belief of doctrines had really arisen from sympathy with persons whom he loved, and whose esteem he could not preserve without subscribing their creed. After hearing how sore his Episcopalian friends were because he had found himself obliged to leave their church, he writes in his jour nal : — " I have taken, as usual, a walk in the cemetery. Sitting, very tired, among the tombs, the following thought occurred to me. He who deceives, injures mankind ; by not separating myself from the Church of England, I should deceive ; therefore, by not separating, I should injure mankind. Kind and excellent friends seem to take a delight in saying to me that *I have given a mortal stab to my usefulness*. Secret feeling does not allow them to perceive that what leads them to say so is the desire of giving me a stab ; for I have already taken a decided step, and that observation can have no effect but that of adding to my sufferings. Do they think that I have acted according to my conscience, or against it ? The latter is inconceivable ; but if I have acted according to the dictates of my conscience, do they wish I had acted against them.? Do they wish that the

T

*stab* should be given my conscience instead of my usefulness?" A year and a half later he wrote, "The violence
of party feeling and the selfish worldliness prominent
around, make me shrink more and more from all contact
with society. I feel that I must wait for death in this
perfect moral solitude, without a single human being near
me to whom I may look up for that help and sympathy
which old men that have walked on the beaten paths of
life expect when their dissolution approaches." On attending the funeral of a clergyman, he says : "I could
not prevent a tear from rolling down when the coffin was
lowered. There is, indeed, much of my sensibility
which is nervous ; yet a mind so stored with baffled affections and regrets as mine, may be excused for its weakness. My efforts to suppress external marks of feeling
are very great, but not equal to the object. My tear,
however, was not for the deceased personally, with whom
I was not at all intimate. It was for *humanity*, suffering,
struggling, aspiring, daily perishing and renewed humanity. It is not death that moves me ; but the contemplation of the rough path and the darkened mental atmosphere which the human passions and interests, disguised
as religion, oblige us to tread and cross on our way to the
grave." After this touching glimpse into the depths of
his soul, it is piteous to read such expressions in his journal as follow : "I felt so oppressed by solitude in the
afternoon, that I desired Margaret to sit in the room,
that I might see a human being. My solitude in this
world (I do not mean the absence of company) increases
in a most melancholy degree. Intellectual convictions,
at least with me, are powerful in the regulation of conduct, but very weak in regard to the feelings." What affecting pathos and nobleness of spirit mingle in this epit-
ome of his life, written by him in the album of one of
his dearest friends !

> Reader, thou look'st upon a barren page :
> The blighting hand of pain, the snows of age,
> Have quenched the spark that might have made it glow.
> Long has the writer wandered here below,
> Not friendless, but alone. For the foul hand
> Of Superstition snapped every band

That knit him to his kindred : then he fled ;
But after him the hideous monster sped
In various shapes, and raised a stirring cry :
" That villain will not act a pious lie."
Men, women, stare, discuss, but all insist,
" The man must be a shocking Atheist."
Brother or sister, whatsoe'er thou art !
Couldst thou but see the fang that gnaws my heart,
Thou wouldst forgive this transient gush of scorn,
Wouldst shed a tear, in pity wouldst thou mourn
For one who, spite the wrongs that lacerate
His weary soul, has never learned to hate.

Much later he said, " How vehemently I long to be in the world of the departed ! " And again, " My bodily sufferings are dreadful, and the misery produced by my solitude is not to be described. But trusting in God's Spirit within me, I await my dissolution without fear. Into thy hands, O Eternal Lord of life, of love, of virtue, I commend my spirit." And then at last, the glad hour, so long waited for, came ; and that divine soul sped to its infinite release, no longer to be an exile for truth's sake, to pine for love no more, never again to know what it is to be lonely.

## LEOPARDI.

PERHAPS no one of all the men of genius who have lived in recent times has had so lonely a soul and led so lonely a life as Leopardi, the Italian philologist, thinker, and poet, whose name is growing into fame, as his character and fate are becoming known and winning more of love and pity. His intellect, imagination, and heart alike were remarkable for their scope and fervor. He dared to think without checks, and to accept as truth whatever he saw as such. Consequently he rejected the common notions prevalent around him, and was pointed at as a sceptic. He loved his country with a burning patriotism ; her bondage and torpor, and the supine degradation of her children, alternately aroused his indignation and oppressed him with the deepest sadness. His sense of his own powers was high, enkindling a grand ambition which his unfortunate circumstances combined to irritate, thwart and baffle. "Mediocrity frightens me," he says, "my wish is to love, and become great by genius and study." His intense susceptibility to beauty, his impassioned and exacting sympathy, created in him the deepest necessity for love ; but his deformity, poverty, and sickness, prevented the fulfilment of this master desire. Opposed by a hostile fate within and without, disappointed at every turn, without health of body or peace of mind, accepting in its direst extent that philosophy of despair which denies God, Providence, and Immortality ; surrounded for the most part by tyrannical bigots and ignorant boors, possessed by an inexpressible melancholy, alleviated only by the activities of his own genius and the occasional attentions of one or two friends and correspondents, — the unhappy Leopardi lived in the deepest and saddest of solitudes. Knowing how great his intellectuality and his sensibility were, it makes one's heart ache to read his recorded wish that he might become a bird, in order, for a little season, to experience their happiness and peace. He has partially described the hopeless monotony of his life, in the dilapidated old town of Recanati, in his poem, "La Vita Solitaria." In the poem on the "Recollections of Youth," he paints the dismal and trying lone·liness of his maturity with painful power : —

> Condemned to waste and pass my prime
> In this wild native village, amid a race
> Unlearned and dull, to whom fair Wisdom's name,
> And Knowledge, like the names of strangers sound,
> An argument of laughter and of jest ;
> They hated me and fled me. Not that they
> Were envious ; of no greater destiny
> They held me than themselves ; but that I bore
> Esteem for my own being in my heart,
> Though ne'er to man disclosed by any sign.
> Here passed my years, recluse and desolate,
> Without or love or life. Bitter and harsh
> Among the unkindly multitude I grew.
> Here was I robbed of pity and of trust,
> And, studying the poor herd, became of men
> A scorner most disdainful. Ah, at times
> My thoughts to you go back, O hopes, to you,
> Blessed imaginations of my youth !
> When I regard my life, so mean, and poor,

And mournful, and that death alone is all
To which so much of hope has brought my days,
I feel my heart stand still, and know not how
To be consoled for such a destiny.

The soul of Leopardi was too powerful — surpassingly affectionate and terribly disappointed as he was in life — to permit him usually to express his misanthropy, his grief and wretchedness, either in sentimental sighs or in wails of despair. His dark views and unhappy feelings vented themselves rather in forms of smiling irony, philosophic satire, and a quiet humor, wherein tender melancholy and bitter force of thought are equally mixed. His writings are marked by classic finish and repose. The manly courage and fortitude that breathe in them are not less obvious than the plaintiveness — not lackadaisical, but heroic — which betrays how constant and deep his pain was. The cause of his spiritual isolation and misery was not merely his rare genius and earnestness, absorbing thought and study, not merely his profound unbelief, not merely his yearning and regurgitating affection, but also his chronic ill-health and nervous exhaustion. Nearly all his life he was the victim of depressing physical disease. He says, " It appears to me that weariness is of the nature of air, which fills all the space intervening between material things, and all the voids contained in them. When anything is removed and the room is not filled by another thing, weariness takes its place immediately. Thus all the interstices of human life between the pleasures and misfortunes are filled up with weariness."

From the bleaker climate and more inhospitable society of Recanati, Leopardi wandered to Florence, Bologna, Rome, and lastly to Naples. Here he died in the arms of his good and dear friend Ranieri. He had written in his fine poem of " Love and Death," — " the two sweet lords, friends to the human race, to whom fate gave being together,"— at the close of this poem he had said, "Lovely Death ! bow to the power of unaccustomed prayers, and shut my sad eyes to the light. Calm, alone, I await the time when I shall sleep on thy virgin breast."

Rarely has death been more welcome to a mortal,

rarely has one lived capable of a keener or vaster happiness, had his fellow-creatures but come up to the standard
his genius exacted, and answered his cravings. In the
suburbs of Naples, in the little church of San Vitale,
stands the monument reared by the loving friend and
biographer on whose bosom "he gave up his soul with
an ineffable and angelic smile." The traveller who lingers to read the inscription, traced by the pen of Giober-
ti, draws a deep sigh, and hopes that the great hapless
spirit whose clayey part sleeps there, is now, in a higher
form, under fairer conditions, enjoying the harmony and
love he so vainly longed for here.

### FOSTER.

THE author of the essay on "Decision of Character,'
John Foster, was always distinguished for his separateness
of life and soul. His capacious, earnest, sombre, extremely sensitive and tenacious cast of mind unfitted him
to herd with society. The greater his need the less his
fitness. Contrasting himself with a lady whose "habit
was so settled to solitude that she often felt the occasional
hour spent with some other human beings tedious and
teasing," he says : "Why is this being that looks at me
and talks, whose bosom is warm, whose nature and wants
resemble my own, more to me than all the inanimate objects on earth and all the stars of heaven ? Delightful
necessity of my nature ! But to what a world of disappointments and vexations is this social feeling liable, and
how few are made happy by it in any such degree as I
picture to myself and long for !" Expressing his sympathy with his friend Mrs. Mant, who had complained of
feeling desolate and solitary among uncongenial neighbors,
he says : "Shall you be sorry that your mind is too
thoughtful and too religious to suit their society ? Could
you be willing to humble yourself to a complacent agree-
ment with their levity or their oddity ? You ought to feel
your superiority, and dismiss the anxious wish for a companionship which you cannot purchase but by descending

to a level where you would never feel happy if you *did* descend to it." After spending an hour with a handsome but ignorant and unsocial woman and a cat, Foster said he felt he could more easily make society of the cat than of the woman. He characterized fashionable worldlings, the hardened habitués of society, as "people who worship Indifference and are proud of their religion." One of his sharpest and saddest aphorisms is this : " We are interested only about self or about those who form a part of our self-interest. Beyond all other extravagances of folly is that of expecting or wishing to live in a great number of hearts." It was but natural that he should, as he did, fall back on himself, nature, and God, and spend the time in solitude, revolving the sombre and massive meditations out of which his writings grew.

## CHANNING.

WILLIAM ELLERY CHANNING is one of the most exalted and influential characters of modern time. He is a character as distinctly American as Washington, and worthy to be compared with him. For, if less commandingly conspicuous and imposing, he is far finer, sweeter, more spiritual, ideal, and religious. The average multitude of mankind live, by mechanical habit, on tradition. There are two classes of great men whose mission is by their original power and fire to redeem common men from their deathly passivity, and inspire them to newness of living. First, the creative minds who audaciously cast off the bonds of old authority, break through the limits of routine, and lay bare unheard-of regions of life. Second, those less endowed, but equally inspired natures, who, staying for the greater part by the traditions and authorities honored in their time, cannot abide anything lifelessly formal, but must vitalize all they touch, repudiating torpid conformity, making the old as good as new by stripping off its bandages and breathing a soul under its ribs of death. Channing belonged rather to this latter class, though not excluded from the other one. His was

more the greatness of balanced faculties, sincerity, patience, earnestness, consecration, than that startling greatness which, goaded to unparalleled deeds by a strange
fire shut up in its bones, despises trembling prudence and
leaps into the unknown to pluck its prizes.

Channing was great by the translucency of his large
and lofty mind, and by the permeating morality of his
character. No mechanical conformity could satisfy him.
He must see for himself, and vitalize all his views. He
sought with patience, by many-sided comparisons and
tests, with the aid of the minds around him, to understand
subjects justly. He fought as a divine champion to drive
from his own soul the conceit, ignorance, delusion, dead
traditionality of opinion he saw infesting ordinary souls,
and always to live as far as possible at first hand, in genuine perception, faith, and love. His sincerity and earnestness fused his powers in every expression, so that he
acted as a unit flowing with irresistible fervor and momen-
tum. No accompaniments of his utterance created any
obstacle to its effect : his impression was therefore integral, without those contradictions and jerks which take so
much away from the influence of many speakers. The
pillars of his being went down to the basis of primal
truths, and rested, naked, alive, electric, on the moral
foundations of things, in contact with the original sources
of inspiration. Thus, although he was not a great scholar,
nor a discoverer of any important thoughts or methods,
he had great originality of character. The intensely sustained action of his faculties lent to the best thoughts and
sentiments which he derived from his time new fire and
importance, and gave a fresh impulse towards their dominion in the breasts of others. If he did not with penetrative intellect uncover new principles or provinces, his
inspired contemplation made the commonplace burn unto
the kindling of souls indifferent before. He is at this
moment a creative moral influence, breathing in the literature and life of America. He is also at this moment,
through translations of his works, a high ethical, educational, liberalizing influence in France, Germany, and
Russia.

Channing was impressively separated throughout his life from the bulk of those around him by the manifold superiority of his soul, the greater quickness and richness of his sensibility, the greater vitality and breadth of his reason, the greater keenness and gravity of his conscience, the greater force and constancy of his aspiration after internal harmony and public usefulness. In early youth he was much given to lonely rambles, secret self-communion, romantic reveries, "seeking in unreal worlds what the actual world could not give." "Much of my time," he writes, "is thrown away in pursuing the phantoms of a disordered imagination. Musing wears away both my body and my mind. I walk without attending to the distance." He suffered severely from home-sickness. Recalling it long after he said : "I remember how my throat seemed full, and food was tasteless, and the solitude which I fled to was utter loneliness."

When he was eighteen, brooding over enthusiastic dreams at once glorious and sad, he wrote to his dear classmate Shaw, "I am sensible that my happy days are passed, and I can only weep for them. My walks now are solitary ; no friendly voice to cheer me, no congenial soul to make a partner of my joy or sorrow. I am, indeed, in the midst of my family, with the best of mothers, brothers, and sisters. But alas ! I have no *friend*." Soon after this expression he went as a tutor to Richmond in Virginia. Here is a glimpse of his experience, given in a letter to Shaw written at the time. "I have a retired room for my study, a lonely plain to walk in. I often look towards the North with a sigh, and think of the scenes I have left behind me. But I remember that cruel necessity has driven me from home, and wipe away the tear which the painful recollection has wrung from my eyes. O heaven ! what a wretch should I be, how wearisome would existence be, had I not learned to depend on myself for enjoyment. Society becomes more and more insipid. I am tired of the fashionable nonsense which dins my ear on every side, and am driven to my book and pen for relief. Nature or education has given this bent to my mind, and I esteem it as the richest bless

ing Heaven ever sent me.    I am independent of the
world."    Despite this brave rally, however, his isolation,
absorption, stern abstemiousness, and over-toil, often depressed his spirits ; and when, on a Christmas day, he
found himself too meanly clad to join the gay party in
another part of the mansion, he felt a bitter blow of heart-
break.

Forty-three years afterwards, he thus reverted to those
hard, yet most fruitful days : " I lived alone, too poor to
buy books, spending my days and nights in an outbuild-
ing, with no one beneath my roof except during the hours
of school.    There I toiled as I have never done since.
With not a human being to whom I could communicate
my deepest thoughts and feelings, I passed through intellectual and moral conflicts so absorbing as often to banish
sleep and to destroy almost wholly the power of digestion.    I was worn wellnigh to a skeleton.    Yet I look
back on those days and nights of loneliness and frequent
gloom with thankfulness.    If I ever struggled with my
whole soul for purity, truth, and goodness, it was there.
There, amidst sore trials, the great question, I trust, was
settled within me, whether I would obey the higher or
lower principles of my nature, — whether I would be the
victim of passion, or the free child and servant of God.
It is an interesting recollection that this great conflict
was going on within me, and my mind receiving its impulse towards the perfect, without a thought or suspicion
of one person around me as to what I was experiencing."

When twenty-seven years of age, four years after his
ordination, he wrote to one of his friends : " I have a
strong propensity to lead the life of a recluse and a book-
worm ; and perhaps, if I were able to study all the time,
I should neglect the active duties of my profession.    My
life is very tranquil.    I *will* not mingle with the conten-
tions of the world.    Angry politicians and theologians are
raging around me, but I try not to hear."    Dewey, after
living with Channing in his family, in his mature age,
when he had acquired a great fame, has thus described
him : " He stood alone.    I found him embosomed in
reverence and affection, and yet living in a singular iso-

lation. No being was ever more simple, unpretending, and kindly-natured than he ; and yet no such being, surely, was ever so inaccessible. Not that he was proud, but that he was venerated as something out of the earthly sphere." In his sixtieth year, Channing said : " I try in solitude to keep up my interest in my fellow-creatures ; and my happiness, when alone, is found in labors for their improvement." And he wrote to a young friend, only a few weeks before his death : " At the end of life, I see that I have lived too much by myself. I wish you more courage, cordiality, and real union with your race." Yet he who said this was a most celebrated preacher and writer, with wealth, an extensive correspondence, a high social position, and crowds of admirers. Even in the midst of an upgazing world, a mind of unusual strength, tenderness, earnestness and consecration, is likely to be alone.

With the prostration and pain of chronic ill-health, Channing had a nervous system in which ideas distributed thrills of emotion with as much energy as in others objects distribute shocks of sensation. While others were passionately absorbed in the pursuit of money and outward rank, or slothfully abandoned to pleasure and ease, he was heroically studying to know the truth, examining all sorts of opinions to sift out the false and effete, retain the real and vital. Others were generally content to tread a lifeless routine of conventionality and self-ignorance ; he toiled with burning devotion to advance steadily towards perfection, and to set before others the methods of such a progress, both by example and by precept. With reference to this end he strove with unwearied patience to understand himself, human nature, the good and ill of human life, the laws of duty, as they are, — neither fanatically exaggerating the defects and misery, nor idolatrously heightening the gifts and deserts, that met his gaze.

Besides the distance resulting from these traits, he was further separated by the misunderstanding and opposition he experienced from unworthy judges of his character. His immense self-respect, his deliberate setting of his

own conscientious conviction above every other authcr-
ity, the firmness with which he fell back on his own
perceptions and feelings, the sincerity with which he
recognized the voice of God in the sovereign ideas and
sentiments of his soul, constraining him to express them
and serve them, made his personality not less odious to
some than fascinating to others. The least worthy among
his associates, failing to distinguish from an ignoble egotism his grand esteem for himself as an accredited representative of the supreme interests of truth and humanity,
were offended at his saying so much of his " mission," his
" great thoughts," his " sacred ideas." The enemies, provoked by the high-toned boldness with which he rebuked
the sins and wrongs he recognized, assailed him with
anonymous letters, with outrageous imputations of bad
motives. And when he poured forth in inspired strains
the fulness of his soul on the godlike worth of human
nature, and the future glory of human destinies, he was
laughed at as a wild dreamer. "These ideas are treated,"
he remonstrates, " as a kind of spiritual romance ; and
the teacher who really expects men to see in themselves
and one another the children of God, is smiled at as a
visionary. The reception of this plainest truth of Christianity would revolutionize society, and create relations
among men not dreamed of at the present day. A union
would spring up, compared with which our present friendships would seem estrangements. Men would know the
import of the word Brother." Yet, notwithstanding the
combined force of these influences to chill, alienate, and
depress, he grew ever calmer and happier in his faith in
men, and love of nature, and enjoyment of life, and said
his last year was the dearest of all he had known.

With the isolating characteristics by which he was
marked, characteristics which have led so many superior
souls to scornful withdrawal and bitter wretchedness, it is
an important inquiry how Channing managed to keep
himself interested and happy. In the first place, he early
formed the blessed habit of meditating on the divine
aspects of nature, man, and society, in preference to dwelling on their dark and distressful aspects. The glory of

the attributes of God, the inspiration of disinterestedness, the privilege of existing in a universe of progressive order and beauty, the blissful freedom and grandeur of self-sacrifice, the vision of a perfect society yet to be realized on earth, the boundless possibilities in the destiny of the soul, — an assimilating communion with such themes as these fed the fountains of his life always with strength, and often with rapture. At one time he was so wrought up by his convictions of the dignity of human nature and the promises of universal good, that, as he afterwards described the experience to a friend, "I longed to die, feeling as if heaven alone could give room for the exercise of such emotions; but when I found I must live, I cast about to do something worthy of these great thoughts."

That he suffered much from lack of satisfying fellowship, cannot be concealed. Hear his own confession, as romantic as though written by the fiery enthusiast of Lake Leman, instead of by a Puritan of cold New England: "My whole life has been a struggle with my feelings. I walk and muse till I can walk no longer. I sit down with Goldsmith or Rogers in my hand, and shed tears — at what? At fictitious misery. Ask those with whom I have lived and they will tell you that I am a stoic. But I only smothered a fire which will one day consume me. I sigh for tranquil happiness; but still continue sanguine, ardent, inconstant. One reason why I now dislike the rapture and the depression which I formerly encouraged, is probably this: I find none to share them with me." Again, he writes: — "I often want faith in the sympathy of individuals with whom I converse, and shrink from expressing the truth, lest it should meet no response. This I am trying to overcome." We cannot help suspecting that his own isolation quickened his sympathetic perception of instances of solitude, when we find him, during his voyage to Europe, noting in his journal: — "The sight of the sea-bird struck me with its loneliness. I thought of its spending its night on the ocean. But I remembered that it had no home to forget."

Channing, with all the sickness, pain, ideal sorrow, extreme sensitiveness and intellectuality, which have led a host of gifted men to misanthropy and despair, never became conceited or sour, but won the victory because he paid the price of the victory in perseveringly observing its conditions. He kept a holy watch over his own tendencies, and adjusted them to the sober standards of virtue. He was, in his own words, "too wise to waste in idle lamentations over deficiencies the energy which should be used in removing them." How weighty in moral wisdom and valor is the following passage : — " I can remember the days when I gloried in the moments of rapture, when I loved to shroud myself in the gloom of melancholy. But I have grown wiser as I have grown older. I now wish to do good in the world, and must throw away these ridiculous ecstasies, and form myself to habits of piety and benevolence. The other day I handed to a lady a sonnet of Southey's, which had wrung tears from me. 'It is pretty,' said she, with a smile. 'Pretty !' echoed I, as I looked at her. 'Pretty !' I went home. As I grew composed I could not help reflecting that the lady who had made this answer was universally esteemed for her benevolence. I knew that she was goodness itself; but still she wanted feeling. And what is feeling? I blushed to find, when I thought more on the subject, that the mind was just as passive in that state which I called 'feeling' as when it received any impressions of sense. One consequence immediately struck me ; that there was no moral merit in possessing feeling, and, of course, no crime in wanting it."

The growing depth and serenity of Channing's happiness were the reward of his wisdom and virtue both in relation to society and to himself. The great standard of association ready to link itself with every state of consciousness that arose in him, was the divine set of principles inwrought by God with the structure of humanity and destined at last to harmonize all things with themselves. This was an unfailing spring of comfort and power. He says, referring to this mode of thought, " I feel a noble enthusiasm spreading through my frame ; every nerve is strung, every muscle is laboring ; my

bosom pants with a great, half-conceived and indescribable sentiment; I seem inspired with a surrounding
deity." He associated the idea of the race with the
person of the individual, saw the essence and glory of
the whole in each of the parts. He felt this with such
extraordinary vividness, spoke and wrote it with such an
iterated eloquence of sincerity, as really to diffuse around
him a new impulse to the philanthropy of the age ; — as
high a moral benefit, as pure a religious service, as can
be rendered to men by a man. This is the noble and
healthy opposite of the habit exemplified by such morbid
characters as Pascal, Leopardi, Schopenhauer, whose
trains of meditations always fly to the dark ideas of the
wickedness, weakness, wretchedness of our nature and
state, and thus cause each successive thought to deposit a
new layer of melancholy.

Channing illustrates another felicitous contrast to the
habit of these unhappy men, in his treatment and regard
of *himself*. No one can rationally expect other men to
think as much of him, estimate him as highly, love him
as warmly, as he desires to be thought of, estimated, and
loved. Each one is too much occupied at home, has not
time or force enough, even if he had inclination, to do
this. Each can give nearly all to the whole ; the whole
could not, without universal destruction, give more than a
little to each. Accordingly those whose happiness depends on their seeing themselves reflected in the minds
of others, in lights sufficiently flattering to minister to
their vanity and ambition, must be, as a general rule,
prevailingly unhappy. He alone can have a stable and
increasing happiness, who, trying faithfully to do his duty,
is content with the approval of God and his own conscience, by the intrinsic standards of what is right and
good. All men who, like Petrarch, Rousseau, Byron,
sensitively refer to their self-reflections in the judgments
of others, are miserable ; for their pride topples, their
complacency is destroyed by stinging disappointments,
either real or fanciful. But men of the Dante, the Fenelon, or the Wordsworth stamp, who esteem themselves,
not indirectly through the figures they make in other peo-

ple's imaginations, but directly by the divine authorities to which they bow, the sublime ideals to which they are loyal, the boundless and everlasting good which they appreciate, are possessed of an immovable content. Channing was of these. He was devoted, whether in company or alone, not to pampering, but to perfecting, himself. On this base his self-respect stood firmly. " The true tone of virtue," he says, " is the tone of conscious superiority, calm, expressive of unaffected dignity, strong in itself, and therefore not disturbed by clamors." "My own opinion of what I publish is not at all affected by the general reception it meets with ; but if no souls are reached, there is cause of distrust." " I like to know the evil that is said of me, because much of it may be traced to misapprehension, and because sometimes part of it has a foundation in real defects of character, and may be used for self-knowledge and self-reform." From all painful wrongs, injurious reports, sympathetic woes, he had delicious retreats and cures in his Christ-like ideas, sentiments, and efforts. He says in one of his own choice sentences, sweet and high, " We visionaries, as we are called, have this privilege from living in the air, that the harsh sounds from the earth make only a slight impression on the ear." To lose self-respect is to touch the bottom at the same time of degradation and of misery. He who is upheld by a sound self-respect may be calm and happy even in the midst of a thousand trials.

Channing was a lonely man, but he wisely shunned most of the evils and nobly gained all the benefits of solitude. Laboring to perfect himself, he disinterestedly served his fellow-men, resolutely sought truth, and humbly worshipped God. In this manner he neutralized misery and was happy to the end, exerting a noble influence and setting a redemptive example, an influence and example which the expansion of his pure fame promises to diffuse and perpetuate. He is one of the few men who make us mend our idea of man.

## ROBERTSON.

ONE of the bravest confessors of our age was the noble English preacher, Frederick William Robertson. Although his eloquence brought him much publicity, and his charming qualities of character made him warmly beloved by many friends, and his extreme fidelity to his professional duties kept him busy both in his library and with people, he lived, as to the inner man, in a trying solitude. The rare tenderness of his spirit, the uncommon capacity and · earnestness of his mind, his heroic loyalty in the pursuit of truth, his extraordinary breadth of perception and catholicity of temper, removed him quite out of the range of vulgar natures, and made him an object of suspicion and hate to partisans and bigots. He paid in sorrowful irritation and suffering the penalty of his exquisite sensitiveness and his unflinching courage. What a glimpse the following sentence opens into his life : " I am sometimes tempted to doubt whether any one who tries to open people's eyes in science, politics, or religion is to be reckoned as a martyr or a fool. The cross ? Or the cap and bells ? "

When in Switzerland he wrote home to his wife, " I cannot tell you how the love of solitude has grown upon me. I can enjoy these mountains, with their sombre pine-woods and their wild sights and sounds, only when I am alone." At another time, with a noble depth of pathos and of thought, he writes : " I am alone, lonelier than ever, — *sympathized with by none, because I sympathize too much with all.* But the All sympathizes with me. A sublime feeling of a Presence comes upon me at times, which makes inward solitariness a trifle to talk about." Yet, despite such divine compensation, to such a soul a loving society, and not a compulsory solitude, was the genuine atmosphere of enjoyment. He said, " Sympathy is too exquisitely dear to me to resist the temptation of expecting it ; and then I could bite my tongue with vexation for having babbled out truths too sincere and childlike to be intelligible. But as soon as the fit of misanthropy is

14 *                                                          U

passed, that absurd human heart with which I live, trusts
and confides again.   Yet, yet, say what I will, — when
any one soothes me with the semblance of sympathy, I
cannot for the life of me help baring my whole bosom in
gratitude and trust."

It was the too natural sequel of such extreme fondness
that, after repeated experiences of unfairness and unkindness, from natures so far inferior as to be incapable of
appreciating and responding to him in his own kind, he
should, in final revulsion of pain, say : " I am resolved
now to act, and feel, and think, alone."   It is doubly
melancholy to remember the cruel wrong he suffered
when we recall the revelation of its painfulness which
he left on record.   "Unless a man," he says, "has a
skin like a rhinoceros, and a heart like a stone-fruit, it is
no easy thing to work alone.   The bad feelings of pride
or vanity get as little to feed them in such a struggle as
the better ones of sympathy and charity.   Elijah, stern
and iron as he was, should be a warning to any common
man to expect that many a day he will have to sit under
his juniper-tree in despondency and bitter sense of isolation and uselessness."

Robertson was a man of the extremest refinement and
purity of heart.   In the most vivid sense of the phrase,
he was a soldier of Christ.   His loyalties and reverences
were surpassingly quick and deep.   He was a true incarnation of chivalry ; his elastic vigor of nerve making his
steps spurn the earth as he walked, his inspired imagination spreading over all the moral interests of humanity a
web of associations sensitive to pleasure and pain through
its whole extent.   His noble courage, both physical and
moral, was as supreme as his consecration.   Whenever
he spoke of battle his lips quivered, his eyes flashed, his
voice shook.   A soldier's son, he was rocked and cradled
amidst military sights and sounds ; and, to the last, as he
himself said, he could not see a review without being impressed to tears, nor look on the evolutions of cavalry
without a choking sensation.   He turned with loathing
from the crooked policies, petty reticences, clinging scandals, and bigoted denunciations of Evangelical and Tract-

arian controversies, of High and Low Church parties, and, flinging off their constraint, with a sense of measureless relief, wished "to die sword in hand against a French invader."

He was unhappy. The loss of health, the development of a morbid sensitiveness, together with his experience of the ignobleness of many men, the deceitfulness, envious hate and vulgarity forced on his notice, made him unhappy. Had he been less pure and holy, had he been meaner and colder, he would not have suffered as he did. The essence of chivalry, the *honor* of the true soldier, existed in him with extraordinary power and keenness. Chivalry is the ordering of conduct and the judging of ourselves by the highest standards of duty and sentiment reflected by our imagination in the minds of others. We act and estimate our acts, when we are truly chivalrous, not in the direct light of our own conscience alone, but by the most disinterested code of moral right and beauty, which we conceive as enthroned in the minds of our fellow men. Ideally subjecting himself to this exalted tribunal, Robertson found judgments continually pronounced, whose foul selfishness on the one side, whose cruel injustice on the other, disgusted and distressed him. One of his constitutional traits was a habit of self-depreciation. He underestimated himself and his deserts. This was owing to the strength of his perception of the standards of perfection, the models of success, and to the sharpness of his feeling of his shortcomings when tried by these. Superiority to the averages of attainment gives common natures an assured self-complacency that makes them happy.

To a fine and lofty soul the failure to reach what it aspires to is constantly depressing, and when such a soul looks down on the inferior averages, the superiority to them which would elate vulgar aspirants only stings it with double shame. It looks up to its baffling ideal, and despairs; looks down on the degraded contentment, and loathes. Such a soul was that of Robertson. He was high-spirited, with an immense self-respect, yet modest, sensitive, easily depressed. He always regarded the ten-

dency to sink back in a self-complacent peace on a feeling
of superiority to the moral exactions of public opinion
as the worst temptation of the Devil.  He resisted this
temptation as he would a profanation of the very shrine
of conscience, the deadliest contamination of the soul.
This is the key which explains, in a manner perfectly
consistent with his sweet sincerity of Christian charity,
those violent reactions and expressions which seem at
first glance to be almost bitterly misanthropic, — his dislike of popularity, his abhorrence of the reputation of
being a popular preacher, his resentful condemnation of
some styles of English orthodoxy.  "Would to God I
were not a mere pepper-cruet, to give a relish to the
palates of the Brightonians!"   "The popular religion
represents only the female element in the national mind,
at once devotional, slanderous, timid, gossiping. narrow,
shrieking, and prudish."

In spite of such apparent self-reliance and severity,
the strength and genuine catholicity of his sympathies
made him solitary, made him feel a pining lack of the co-
operative esteem and the kindred aspirations of others.
"Friendly looks and kind deeds," he says, "stir into
health that sour, rancid film of misanthropy apt to co-
agulate on the stream of our inward life if we live in
heart apart from our fellow-creatures."  He knew many
and many an hour of painful humility, loneliness, and
melancholy.  Alas! like so many gallant and spotless
souls before, while he lived he was persecuted, his experience embittered, and his days shortened by misrepre-
sentation, contumely, and hardship.  He died under a
cloud of excruciating pain, but with a clear conscience
and triumphant trust.  After his death came a harvest
of admiration, grateful love, and fruitful influence.


### CHOPIN.

THE Life of Chopin, by his friend Liszt, is a work of
rare interest, as an example of a noble friendship, as an
acute and powerful psychological portraiture of an ex-

traordinary genius, and as a revelation of that wonderful world of emotion in which the souls of great musicians live. The intense fineness and ardor of Chopin's imagination, the violence of his feelings, his sickly and irritable constitution, his exiled lot, his secretive pride, his subtle originality of mind and sentiment, the lofty earnestness of his aims, and his fastidious purity, made his experience one of bitter contrasts, unhappy and lonely. Sheathed in manners of kind and tranquil courtesy, which covered his convulsive soul as slopes of verdure and vine cover a volcano, he moved among men separate from them, reading the secrets of all, never baring his own.

He veiled his sufferings under the impenetrable calmness of a proud resignation that scorned either to utter complaints or to make demands. He strictly excluded from conversation all subjects relating to himself, carefully keeping others in the circle of their own interests lest they should intrude into his. He was apparently so free from self-occupation that people thought him absorbingly interested in them. Accordingly, he gave much pleasure but awakened little curiosity. "His personality remained intact, unapproachable under the polished surface on which it was impossible to gain footing." Excluded by his infirm health from the ordinary arena, where "a few bees with many wasps expend their strength in useless buzzing, he built a secluded cell for himself, apart from all noisy and frequented ways." He never suffered the world to suspect the secret convulsions that agitated him, never unveiled the shudder caused by the contact of more positive and reckless individualities with his own. His caustic perception caught the ridiculous both on the surface and in the depth, and he could easily hide within or repel without whatever he wished to hide or repel, by gay mystification or satirical raillery. No *ennui* annoyed him, because he expected no interest. Yet this unsuspected absence of his soul from the outward scene, this dense concealment of his real life, arose not from any shallow apathy or poverty of being, but in truth from the haughty royalty of his wants, the inconceivable susceptibility of his soul to hurts.

"He constantly reminded us," Liszt says, "of a convol-
vulus balancing its heaven-colored cup on an incredibly
slight stem, the tissue of which is so like vapor that the
slightest contact wounds and tears the misty corolla."
Conscious of the uselessness of his vivid indignation and
vexation, and too jealous of the mysteries of his emotions to betray them, he sought strength in isolation and
self-control, and, "by dint of constant effort, subjected
his sensibilities, in spite of their tormenting acuteness, to
the rule of what ought to be, rather than of what is."
Shrinking from the world and the crowd, with the mystic
richness of his fancy, and a bleeding sensitiveness of forlorn feeling, he had one charmed resource, — music. "In
his compositions he collected, like tears in a lachrymatory,
the memories of his youth, the passions and dreams of
his country, the affections of his heart, the mysteries
of his desires, the secrets of his sorrows." "What the
pious never say except on their knees, in communion with
God, he said in his palpitating compositions, uttering in
the language of tones those mysteries of emotion which
man is permitted to understand without words, because
no words can utter them. He was a tone-poet. He
seemed to live upon music, the moody food of imagination. All the elegiac tenderness, passionate coquetry,
martial heroism, and profound melancholy of the Polish
nationality, echoed from his soul, breathe in his strains.

He knew that he could not warm and move "the multitude, which is like a sea of lead." The public intimi-
dated and paralyzed him. But his magic performance
electrified the select audiences to whom he revealed the
secrets his delicate genius had caught from "those reserved yet impassioned hearts which resemble that plant
so full of burning life that its flowers are always surrounded by a subtle and inflammable gas." Liszt com-
pares the ineffably poetic fascination of Chopin's playing
to the perfume of the Ethiopian calla, which refuses to
diffuse its aroma in the breath of crowds, whose heavy air
can retain only the strong odor of the tuberose, the incense of burning resin. His friendly biographer thinks
his abnegation of popular applause veiled an internal

wound. He was perfectly aware of his own superiority; it did not receive sufficient reverberation to assure him that he was appreciated.

A gnawing discontent, scarcely understood by himself, secretly undermined him. "The praise to which he was justly entitled not reaching him in mass, isolated commendations wounded him. This was evident from the polished phrases with which he shook such commendations off, like troublesome dust," making it clear that he preferred to be left undisturbed in the enjoyment of his solitary feelings. "The joys, the consolations which the creations of true art awaken in the weary, suffering, believing hearts to whom they are dedicated, are destined to be borne into far countries and distant years by the sacred works of Chopin. He could not labor to attract auditors and to please them at whatever sacrifice." He aimed to leave a celestial and eternal echo of the emotions of his soul. "What are the fading bouquets of an hour to those whose brows claim the laurel of immortality?" If he could not have from men all he deserved and wanted, he would have nothing from them, — nothing except love and kindness from his chosen friends. He would build his hopes in God, wreak his soul in art, and leave his fame to time.

Repeatedly, Chopin seemed for months to be in a dying state, when he would rally, as by some surprising volition. In such ethereal natures imagination is almost omnipotent, and through its fixed ideas, its magnetic centres of association, works miracles. Twelve years before his death he started for Italy in such a condition that the hotel-keepers demanded pay for the bed and mattress that he used, that they might be burned. Yet the winter that he then spent on the Island of Majorca, under the ministrations of natural beauty and a sleepless love, wrought on him with a strange efficacy of restoration. His biographer becomes a poet in describing this enchanted oasis in the existence of the Polish composer. "In this solitude, shaded by groves of oranges, and surrounded by the blue waves of the Mediterranean, he breathed that air for which natures unsuited to the world and never feeling themselves happy in it long with such a painful homesickness; that air which may be found every-where, if we can find the sympathetic souls to breathe it with us, and which is to be met nowhere without them, — the air of the land of our dreams, of the country of the ideal." The story of this bewitching residence is described by Madame Sand in "Lucrezia Floriani," with all the empassioned gorgeousness of her art : she herself is La Floriani; Chopin is Prince Karol, and Liszt is Count Albani.

At length, after a fatal rupture of affection, an agony worse than death, by a lingering decline not fuller of pain and sadness than of beauty and majesty, the long tragedy of life drew to a close ; the lacerating conflict of the outer and the inner life, so successfully shrouded under that de-meanor of tranquil politeness, was to find relief. The noblest of his Polish countrymen, the loveliest of his countrywomen, idolatrous friends, were unremitting in their attentions. One evening near his end, at sunset, he saw the beautiful Countess Potocka, draped in white, weeping, at the foot of his bed. " Sing," he murmured. Amidst the hushed group of friends, the rays of the set-ting sun streaming upon them, she sang with her own ex-quisite sweetness the famous canticle to the Virgin which once saved the life of Stradella. " How beautiful it is ! My God, how beautiful !" sighed the dying artist. None of those who approached the dying Chopin "could tear themselves from the spectacle of this great and gifted soul in his hours of mortal anguish." Whispering "Who is near me," he was told, Gutman, — the favorite pupil who had watched by him with romantic devotion. He bent his head to kiss the faithful hand, and died in this act of love.

They buried the room in flowers. The serene loveli-ness of youth, so long dimmed by grief and pain, came back, and he lay there smiling, as if asleep in a garden of roses. At the farewell service in the Madeleine Church, his own Funeral March and the Requiem of Mozart were performed. Lablache, who had sung the supernatural *Tuba Mirum* of this Requiem at the burial of Beethoven, twenty-

two years before, now sang it again. In the cemetery of
Père la Chaise, under a chaste tomb surmounted by his
own marble likeness, between the monuments of Bellini
and Cherubini, where he had asked to be laid, sleeps the
hapless musician, whose weird and solemn strains are
worthy to carry his name into future ages as long as men
shall continue to contemplate the mysterious changes of
time and the mute entrance of eternity.


## THOREAU.

IF any American deserves to stand as a representative of
the experience of recluseness, Thoreau is the man. His
fellow-feelings and alliances with men were few and feeble;
his disgusts and aversions many, as well as strongly pronounced. All his life he was distinguished for his aloofness, austere self-communion, long and lonely walks. He
was separated from ordinary persons in grain and habits,
by the poetic sincerity of his passion for natural objects
and phenomena. As a student and lover of the material
world he is a genuine apostle of solitude, despite the taints
of affectation, inconsistency, and morbidity which his writings betray. At twenty-eight, on the shore of a lonely pond,
he built a hut in which he lived entirely by himself for
over two years. And, after he returned to his father's
house in the village, he was for the chief part of the time
nearly as much alone as he had been in his hermitage by
Walden water. The closeness of his cleaving to the landscape cannot be questioned : "I dream of looking abroad,
summer and winter, with free gaze, from some mountain
side, nature looking into nature, with such easy sympathy
as the blue-eyed grass in the meadow looks in the face of
the sky." When he describes natural scenes, his heart
lends a sweet charm to the pages he pens : " Paddling up
the river to Fair-Haven Pond, as the sun went down, I
saw a solitary boatman disporting on the smooth lake.
The falling dews seemed to strain and purify the air, and
I was soothed with an infinite stillness. I got the world,
as it were, by the nape of the neck, and held it under, in

the tide of its own events, till it was drowned ; and then
I let it go down stream like a dead dog. Vast, hollow
chambers of silence stretched away on every side ; and
my being expanded in proportion, and filled them."

In his little forest-house, Thoreau had three chairs,
"one for solitude, two for friendship, three for society."
"My nearest neighbor is a mile distant.  It is as solitary
where I live as on the prairies.  It is as much Asia or
Africa as New England.  I have, as it were, my own sun
and moon and stars ; and a little world all to myself."
"At night, there was never a traveller passed my door,
more than if I were the first or last man."  "We are
wont to imagine rare and delectable places in some re
mote and more celestial corner of the system, — behind
the constellation of Cassiopea's Chair, far from noise and
disturbance.  I discovered that my house actually had its
site in such a withdrawn, but forever new and unprofaned,
part of the universe."  "I love to be alone.  I never found
the companion that was so companionable as solitude."
In this last sentence we catch a tone from the diseased
or disproportioned side of the writer.  He was unhealthy
and unjust in all his thoughts on society ; underrating the
value, overrating the dangers, of intercourse with men.
But his thoughts on retirement, the still study and love
of nature, though frequently exaggerated, are uniformly
sound.  He has a most catholic toleration, a wholesome
and triumphant enjoyment, of every natural object, from
star to skunk-cabbage.  He says, with tonic eloquence,
"Nothing can rightly compel a simple and brave man to
a vulgar sadness : while I enjoy the friendship of the seasons, I trust that nothing can make life a burden to me."
But the moment he turns to contemplate his fellow-men,
all his geniality leaves him, — he grows bigoted, contemptuous, almost inhuman : "The names of men are of course
as cheap and meaningless as Bose and Tray, the names
of dogs.  I will not allow mere names to make distinctions for me, but still see men in herds."  The cynicism
and the sophistry are equal.  His scorn constantly ex-
hales : "The Irishman erects his sty, and gets drunk, and
jabbers more and more under my eaves ; and I am re-

sponsible for all that filth and folly. I find it very unprofitable to have much to do with men. Emerson says
that his life is so unprofitable and shabby for the most
part, that he is driven to all sorts of resources, and, among
the rest, to men. I have seen more men than usual, lately;
and, well as I was acquainted with one, I am surprised to
find what vulgar fellows they are. They do a little business each day, to pay their board; then they congregate
in sitting-rooms, and feebly fabulate and paddle in the
social slush; and, when I think that they have sufficiently
relaxed, and am prepared to see them steal away to their
shrines, they go unashamed to their beds, and take on a
new layer of sloth." Once in a while he gives a saner
voice out of a fonder mood : "It is not that we love to
be alone, but that we love to soar; and, when we soar,
the company grows thinner and thinner, till there is none
at all." But the conceited and misanthropic fit quickly
comes back : "Would I not rather be a cedar post, which
lasts twenty-five years, than the farmer that set it; or he
that preaches to that farmer?" "The whole enterprise
of this nation is totally devoid of interest to me. There
is nothing in it which one should lay down his life for, —
nor even his gloves. What aims more lofty have they
than the prairie-dogs?"

This poisonous sleet of scorn, blowing manward, is
partly an exaggerated rhetoric; partly, the revenge he
takes on men for not being what he wants them to be;
partly, an expression of his unappreciated soul reacting
in defensive contempt, to keep him from sinking below
his own estimate of his deserts. It is curious to note the
contradictions his inner uneasiness begets. Now he says,
"In what concerns you much, do not think you have companions; know that you are alone in the world." Then
he writes to one of his correspondents, "I wish I could
have the benefit of your criticism; it would be a rare help
to me." The following sentence has a cheerful surface, but
a sad bottom: "I have lately got back to that glorious
society, called solitude, where we meet our friends continually, and can imagine the outside world also to be
peopled." At one moment, he says, "I have never felt

lonesome, or the least oppressed by a sense of solitude, but once; and then I was conscious of a slight insanity in my mood." At another moment he says, "Ah! what foreign countries there are, stretching away on every side from every human being with whom you have no sympathy! Their humanity affects one as simply monstrous. When I sit in the parlors and kitchens of some with whom my business brings me—I was going to say—in contact, I feel a sort of awe, and am as forlorn as if I were cast away on a desolate shore. I think of Riley's narrative, and his sufferings." That his alienation from society was more bitter than sweet, less the result of constitutional superiority than of dissatisfied experience, is significantly indicated, when we find him saying, at twenty-five, "I seem to have dodged all my days with one or two persons, and lived upon expectation"; at thirty-five, "I thank you again and again for attending to me"; and at forty-five, "I was particularly gratified when one of my friends said, 'I wish you would write another book,—write it for me.' He is actually more familiar with what I have written than I am myself."

The truth is, his self-estimate and ambition were inordinate; his willingness to pay the price of their outward gratification, a negative quantity. Their exorbitant demands absorbed him; but he had not those powerful charms and signs which would draw from others a correspondent valuation of him and attention to him. Accordingly, he shut his real self in a cell of secrecy, and retreated from men whose discordant returns repelled, to natural objects whose accordant repose seemed acceptingly to confirm and return, the required estimate imposed on them. The key of his life is the fact that it was devoted to the art of an interior aggrandizement of himself. The three chief tricks in this art are, first, a direct self-enhancement, by a boundless pampering of egotism; secondly, an indirect self-enhancement, by a scornful depreciation of others; thirdly, an imaginative magnifying of every trifle related to self, by associating with it a colossal idea of the self. It is difficult to open many pages in the written record of Thoreau without oeing confronted with examples of these three tricks. He is constantly, with all his boastful stoicism, feeling himself, reflecting himself, fondling himself, reverberating himself, exalting himself, incapable of escaping or forgetting himself. He is never contented with things until they are wound through, and made to echo himself; and this is the very mark of spiritual disturbance. "When I detect," he says, "a beauty in any of the recesses of nature, I am reminded, by the serene and retired spirit in which it requires to be contemplated, of the inexpressible privacy of a life." In the holiest and silentest nook his fancy conjures the spectre of himself, and an ideal din from society for contrast. He says of his own pursuits, "The unchallenged bravery which these studies imply is far more impressive than the trumpeted valor of the warrior." When he sees a mountain he sings : —

Wachuset, who, like me,
Standest alone without society,
Upholding heaven, holding down earth, —
Thy pastime from thy birth, —
Not steadied by the one, nor leaning on the other,
May I approve myself thy worthy brother !

This self-exaggeration peers out even through the disguise of humor and of satire : "I am not afraid of praise, for I have practised it on myself. The stars and I belong to a mutual-admiration society." "I do not propose to write an ode to dejection, but to brag as lustily as chanticleer in the morning, standing on his roost." "The mass of men lead lives of quiet desperation." But he, — he is victorious, sufficing, royal. At all events he will be unlike other people. "I am a mere arena for thoughts and feelings, a slight film, or dash of vapor, so faint an entity, and make so slight an impression, that nobody can find the traces of me." "I am something to him that made me, undoubtedly, but not much to any other that he has made." "Many are concerned to know who built the monuments of the East and West. For my part, I should like to know who, in those days, did not build them, — who were above such trifling." "For my part, I could easily do without the post-office. I am sure that I never read any memorable news in a news-paper." This refrain of opposition between the general thoughts and feelings of mankind and his own, recurs until it becomes comical, and we look for it. He refused invitations to dine out, saying, " They make their pride in making their dinner cost much ; I make my pride in making my dinner cost little." One is irresistibly re-minded of Plato's retort, when Diogenes said, " See how I tread on the pride of Plato." — " Yes, with greater pride."

But he more than asserts his difference ; he explicitly proclaims his superiority : " Sometimes when I compare myself with other men, it seems as if I were more favored by the gods than they." " When I realize the greatness of the part I am unconsciously acting, it seems as if there were none in history to match it." Speaking of the scarlet oaks, he adds with Italics : " These are *my* china-asters, *my* late garden-flowers ; it costs *me* nothing for a gardener." The unlikeness of genius to mediocrity is a fact, but not a fact of that relative momentousness entitling it to mo-nopolize attention. He makes a great ado about his ab-sorbing occupation ; his sacred engagements with himself ; his consequent inability to do anything for others, or to meet those who wished to see him. In the light of this obtrusive trait the egotistic character of many passages like the following becomes emphatic : " Only think, for a moment, of a man about his affairs ! How we should respect him ! How glorious he would appear ! A man about *his business* would be the cynosure of all eyes." He evidently had the jaundice of desiring men to think as well of him as he thought of himself ; and, when they would not, he ran into the woods. But he could not es-cape thus, since he carried them still in his mind.

His quotations are not often beautiful or valuable, but appear to be made as bids for curiosity or admiration, or to produce some other sharp effect ; as they are almost invariably strange, bizarre, or absurd : culled from ob scure corners, Damodara, Iamblichus, the Vishnu Purana, or some such out-of-the-way source. He seems to take oddity for originality, extravagant singularity for depth and force. His pages are profusely peppered with pungent paradoxes and exaggerations, — a straining for sensation, not in keeping with his pretence of sufficing repose and greatness : " Why should I feel lonely? is not our planet in the Milky Way?" "All that men have said or are, is a very faint rumor ; and it is not worth their while to remember or refer to that." He exemplifies, to an extent truly astonishing, the great vice of the spiritual hermit ; the belittling, because he dislikes them, of things ordinarily considered important ; and the aggrandizing, because he likes them, of things usually regarded as insignificant. His eccentricities are uncorrected by collision with the eccentricities of others, and his petted idiosyncrasies spurn at the average standards of sanity and usage. Grandeur, dissociated from him, dwindles into pettiness ; pettiness, linked with his immense ego, dilates into grandeur. In his conceited separation he mistakes a crotchet for a consecration. If a worm crosses his path, and he stops to watch its crawl, it is greater than an interview with the Duke of Wellington.

It is the wise observation of Lavater, that whoever makes too much or too little of himself has a false meas ure for everything. Few persons have cherished a more preposterous idea of self than Thoreau, or been more persistently ridden by the enormity. This false standard of valuation vitiates every moral measurement he makes. He describes a battle of red and black ants before his wood-pile at Walden, as if it were more important than Marathon or Gettysburg. His faculties were vast, and his time inexpressibly precious : this struggle of the pismires occupied his faculties and time ; therefore this struggle of the pismires must be an inexpressibly great matter. A trifle, plus his ego, was immense ; an immensity, minus his ego, was a trifle. Is it a haughty conceit or a noble loftiness that makes him say, " When you knock at the Celestial City, ask to see God, — none of the servants "? He says, " Mine is a sugar to sweeten sugar with : if you will listen to me, I will sweeten your whole life." Again, " I would put forth sublime thoughts daily, as the plant puts forth leaves." And yet again, " I shall be a benefactor if I conquer some realms from the night, — if I add to the domains of poetry." After such manifestos, we expect much. We do not find so much as we naturally expect.

He was rather an independent and obstinate thinker than a powerful or rich one. His works, taken in their whole range, instead of being fertile in ideas, are marked by speculative sterility. "He was one of those men," a friendly but honest critic says, "who, from conceit or disappointment, inflict upon themselves a seclusion which reduces them at last, after nibbling everything within reach of their tether, to simple rumination and incessant returns of the same cud to the tongue." This unsympathetic temper is betrayed in a multitude of such sentences as this : "O ye that would have the cocoanut wrong side outwards! when next I weep I will let you know." Thoreau is not the true type of a great man, a genuine master of life, because he does not reflect greatness and joy over men and life, but upholds his idea of his own greatness and mastership by making the characters and lives of others little and mean. Those who, like Wordsworth and Channing, reverse this process, are the true masters and models. A feeling of superiority to others, with love and honor for them, is the ground of complacency and a condition of chronic happiness. A feeling of superiority to others, with alienation from them and hate for them, is the sure condition of perturbations and unhappiness.

Many a humble and loving author who has nestled amongst his fellow-men and not boasted, has contributed far more to brace and enrich the characters and sweeten the lives of his readers than the ill-balanced and unsatisfied hermit of Concord, part cynic, part stoic, who strove to compensate himself with nature and solitude for what he could not wring from men and society. The extravagant estimate he put on solitude may serve as a corrective of the extravagant estimate put on society by our hives of citizens. His monstrous preference of savagedom to civilization may usefully influence us to appreciate natural unsophisticatedness more highly, and conventionality more lowly. As a teacher, this is nearly the extent of his narrow mission. Lowell, in a careful article, written after reading all the published works of Thoreau, says of him: " He seems to us to have been a man with so high a conceit of himself, that he accepted without questioning, and insisted on our accepting, his defects and weaknesses of character, as virtues and powers peculiar to himself. Was he indolent, — he finds none of the activities which attract or employ the rest of mankind worthy of him. Was he wanting in the qualities that make success, — it is success that is contemptible, and not himself that lacks persistency and purpose. Was he poor, — money was an unmixed evil. Did his life seem a selfish one, — he condemns doing good, as one of the weakest of superstitions."

In relation to the intellectual and moral influence of solitude, the example of Thoreau, with all the alleviating wisdom, courage, and tenderness confessedly in it, is chiefly valuable as an illustration of the evils of a want of sympathy with the community. Yet there is often a deep justice, a grandly tonic breath of self-reliance, in his exhortations. How sound and admirable the following passage : " If you seek the warmth of affection from a similar motive to that from which cats and dogs and slothful persons hug the fire, because your temperature is low through sloth, you are on the downward road. Better the cold affection of the sun, reflected from fields of ice and snow, or his warmth in some still wintry dell. Warm your body by healthful exercise, not by cowering over a stove. Warm your spirit by performing independently noble deeds, not by ignobly seeking the sympathy of your fellows who are no better than you self."

Though convinced of the justice of this sketch, the writer feels rebuked, as if it were not kind enough, when he remembers the pleasure he has had in many of the pages of Thoreau, and the affecting scene of his funeral on that beautiful summer day in the dreamy town of Concord. There was uncommon love in him, but it felt itself repulsed, and, too proud to beg or moan, it put on stoicism and wore it until the mask became the face. His opinionative stiffness and contempt were his hurt self-respect protecting itself against the conventionalities and scorns of those who despised what he revered and revered what he despised.  His interior life, with the relations of thoughts and things, was intensely tender and true, however sorely ajar he may have been with persons and with the ideas of persons.  If he was sour, it was on a store of sweetness ; if sad, on a fund of gladness.

While we walked in procession up to the church, though the bell tolled the forty-four years he had numbered, we could not deem that *he* was dead whose ideas and sentiments were so vivid in our souls.  As the fading image of pathetic clay lay before us, strewn with wild flowers and forest sprigs, thoughts of its former occupant seemed blent with all the local landscapes.  We still recall with emotion the tributary words so fitly spoken by friendly and illustrious lips.  The hands of friends reverently lowered the body of the lonely poet into the bosom of the earth, on the pleasant hillside of his native village, whose prospects will long wait to unfurl themselves to another observer so competent to discriminate their features and so attuned to their moods.  And now that it is too late for any further boon amidst his darling haunts below,

There will yet his mother yield
A pillow in her greenest field,
Nor the June flowers scorn to cover
The clay of their departed lover.


### MAURICE DE GUÉRIN.

MAURICE DE GUÉRIN, born in Southern France, in 1810, of an ancient and noble but impoverished family, was graced with such personal gifts as to attract extreme interest from his associates, and endowed with literary talents which have gained him an enviable fame by the few exquisite works bequeathed when he died, at the early age of twenty-nine.  His sister Eugénie, and his friends, Trebutien, La Morvonnais, Marzan, Sainte-Beuve, George Sand, and others, have secured the publication of his brief compositions, drawn attention to their singular charm, and

p.iid tributes to his memory not soon to be forgotten.
Nothing of the kind can be more interesting than the
peculiarities of constitution and experience which made
the character of this gifted young man so shy and lonely,
his career so unhappy, his death so pathetic, the image
of him left behind so strangely attractive and sad.

At twelve, the tender boy, "poor bird exiled from his
native turrets," went to Toulouse to study at a seminary
there ; afterwards to the College Stanislas in Paris.   At
a later period, he returned home, and tarried in the midst
of domestic love and the stillest seclusion.   But, inwardly
wounded, unhappy, uncertain, he was drawn in heart and
fancy alternately to a brilliant career in the world, and to
the mystic life of a religious retreat.   The following striking passage is a transcript from his own soul ; " Which is
the true God ?   The God of cities, or the God of deserts ?
To which to go ?   Long-cherished tastes, impulses of the
heart, accidents of life, decide the choice.   The man of
cities laughs at the strange dreams of the eremites : these,
on the other hand, exult at their separation, at finding
themselves, like the islands of the great ocean, far from
continents, and bathed by unknown waves.   The most to
be pitied are those who, flung between these two, stretch
their arms first to the one, then to the other."   The last
sentence describes his own state for a long time.   He
at length came under the influence of the renowned
Lamennais, whose disciples he joined at La Chênaie.
Amidst the wild scenery of Brittany, with a group of
enthusiastic young men of genius and devotion, under
the eye of the fascinating master whose combination of
Catholicism and Democracy, whose electric words, whose
conflict and subsequent rupture with the papacy, caused
such a sensation in that day ; whose soul was so torn,
and whose end so tragic, — Maurice remained for nine
months.   But he was made for a poet rather than for a
devotee.   The attraction of nature and letters overpow-
ered that of faith and the cloister.   And one day, with
deep emotion, he said farewell to his venerated master,
parted from his beloved comrades, and heard the gates
of the little paradise of La Chênaie shut behind him.

He paused on his way to Paris at the romantic home
of his friend La Morvonnais.   " Behold how good Providence is to me !   For fear the sudden transition from the
softly-tempered air of religious solitude to the torrid zone
of the world would try my soul too severely, it has drawn
me from my sanctuary into a house raised on the border
of the two regions, where, without being in solitude, one
still does not belong to the world ; a house whose windows open, on the one side upon the plain covered with
the tumult of men, on the other upon the desert where
the servants of God are singing ; there upon the ocean,
here upon the woods."   He went to the capital in which
the ambition, intellect, and pleasure of the world are concentrated.   His religious interest died down.   He drank
the cup which the senses are offered in that wondrous bewilderment of prizes, perils, delights, agonies, — the focus of the luxuries and excitements of the earth.   A hard
struggle with obscurity and poverty, interspersed with
ominous illness, with a few visits to dear Cayla, followed
by his happy marriage, crowned in less than a year by his
death, — and the bitter-sweet story of his outer life was
done.

Guérin was one of those natures gifted with vast
powers of intuition and sentiment, but small powers of
organization and execution, who exceedingly interest
others, but are unable to be sufficiently interested themselves, and therefore early become the victims of depression, weariness, sickness, and death.   His nervous system was of that ethereal and ravenous temperament,
which, not able to appropriate accordant and adequate
nutriment from without, preys upon itself.   Preternatu-
rally sensitive to ideal hurts and helps, he nursed those
delicious sadnesses which devour vitality while they feed
sentiment.   He felt his thoughts and emotions as though
they were material pictures, solid objects passing through
his imagination all alive, conscious atoms swimming in
the bosom of the soul.   Consequently, matters of the
inner life which would be to others only trifling impressions were colossal portents to him, — electrifying blisses
or overwhelming agonies.   He seemed to possess marvel-

lous modes of intellection and emotion of his own, sweeter and vaguer than are known by common mortals ; " an
intoxication of delicious monotony and languor ; a half
sleep, empty of thought, yet full of enchanting dreams
of beautiful things." Too rich to be insensible to the
wealth and loveliness of the universe, too poor to be able
to grasp and fix the divine shapes in solid forms of art, he
was torn between aspiration and weakness, will and want.
Few souls ever turned so lucid a mirror to the phenomena of nature, or were so intensely conscious of what occurred within them, as his. Musing on a fearful tempest, he said, " Strange and admirable, these moments of
sublime agitation joined with profound reverie, wherein
the soul and nature, arrayed in all their grandeur, lift
themselves face to face." At times, he said, he could
hear at the bottom of his being faint murmurs marking
the return of life from afar. " These rustling rumors are
produced by my thoughts, which, rising out of their dolorous torpor, make a light agitation of timid joy, and begin conversations full of memories and hopes." What
a delicate revelation of his poetic softness of soul in this
sentiment, " Happy who sits on the top of the mountain,
and sees the lion bound and roar across the plain, with
no traveller or gazelle passing near ! "
He was fascinating both by demonstrativeness and by
reticence, his frankness and his mystery. His father said
that "in childhood his soul was often seen on his lips
ready to fly." His writings show a spiritual unveiling,
wonderful in quality and quantity. Yet he says, " That
which every man of a certain choice nature guards with
the greatest vigilance, is the secret of his soul and of the
closest habits of his thoughts. I love this god Harpoc-
rates, his finger on his lip." And Sainte-Beuve says,
" He loved only on the surface, and before the first curtain of his soul : the depth, something behind, remained
mysterious and reserved." He was unlike those about
him, and the strange difference drew them, while it estranged him. The superlative tenderness of his spirit
was a weakness that disqualified him for happiness among
the coarse, noisy natures of the commonalty, and made

him in all things shrink from the vulgar, and yearn to the
select; detest the commonplace, and adore the sublime.
He said the reading of Chateaubriand's René dissolved
his soul like a rain-storm.  He complained greatly of the
loss in society of all simple and primitive tastes, the
sophistication and destruction of the naïve virgin sentiments of the soul.  Feeling himself solitary, excommu-
nicated, diffident, and embarrassed, he often regarded the
intrepidity and effrontery of more audacious though inferior men, his associates, with admiration, and almost
envy : they, on the other hand, recognized his rare gifts,
plied him with compliments, urged him forward, rallied
him with jests on his shrinking self-depreciation and fear.
He wrote in his journal, " To me it is insupportable to
appear other before men than one is before God.  My
severest punishment at this instant is the extravagant estimate formed of me by some beautiful souls."  Again :
" I lose half of my soul in losing solitude.  I enter the
world with a secret horror."  Going into Paris, "trembling and shivering as a scared deer," distrustful of himself, and afraid of men, he prays, " My God ! close my
eyes ; keep me from the sight of the multitude, the view
of whom raises in me thoughts so bitter, so discouraging.
Let me traverse the crowd, deaf to the noise, inaccessible to the impressions which crush me as I pass through
it.  Place before my eyes, instead, a vision of something
I love, — a field, a vale, a moor, Le Cayla, Le Val, an
image of some object of nature."

The isolation and unhappiness of this poor youth were
unspeakably piteous.  At eighteen, he speaks of being
"possessed by an inveterate melancholy, and fed on a
sad diet of regrets and miseries."  It is obvious that he
was never a misanthrope or an indifferentist, but painfully
concerned about his fellow-men.  He had an absorbing
ambition in combination with a haunting sense of a lack
of the organic strength and perseverance necessary to
sustain the tremendous labors which alone could ever
purchase the proud attainments he coveted.  This ambition, and this conviction of defect, kept him making
comparisons, — personal, artistic, critical ; and constantly

lowered him in his own eyes, distressed him and preyed
on him. In one of his moods of keen self-scrutiny, he
asks, "Why am I so depressed by the sight of mediocre
productions? Is it a dolorous pity for that saddest of all
spectacles, powerless vanity? Or is it conscience, and
return upon myself?" His own halting works and futile
efforts, set against the models of the great masters and
the standards of perfection which his imagination revealed, were a contrast too sharp for his peace ; and, in
the annihilation of complacency, he laid his face in the
dust with bitter sorrow. In a youthful letter to his con-
fessor, he writes, "Poverty and sorrow are hereditary in
my family : the most of my ancestors died in misfortune.
I believe this has had an influence on my character.
Why should not the sentiment of unhappiness communicate itself with the blood, when we see fathers transmitting
to their children physical deformities? My first years
were extremely sad and lonely. When I was only six
years old, my mother died ; and, brought up in scenes of
mourning, perhaps I acquired the habitude of melancholy." "Several causes belonging to my nature, interior and exterior, very early turned me back upon myself.
My soul was my first horizon. Ah, how long I gaze
there ! I see vapors rise from the bottom of my being as
out of a deep valley, and take form under the breath of
chance, — indescribable phantoms which ascend slowly
and without cessation. The powerful fascination which
this monotonous spectacle exerts on me makes it impossible for me to turn my eyes from it for a moment."
The two chains of physical pain and moral pain held
him fast in an over-acute consciousness of himself. He
says he often suffered inexpressibly " from sudden contraction of his being after an extreme dilatation." His
profound and strange misery is to a large extent susceptible of a physiological explanation. It resulted from the
possession of a faculty of life much greater than its sup
ply. His soul was a noble engine with insufficient fuel
and fire, and the incongruity produced agonizing want.
His spirit was effusively expansive : his nerves scantily
furnished. The former is seen in the close of a letter to

his friend Marzan : "I love you, and embrace you with all the strength of my arms and my heart, at the risk of suffocating you with the one, and inundating you with the other." The latter is betrayed in these dismal sentences : "The moral expanse which my life embraces is like a solitude covered by an iron heaven, motionless, without seasons." — "I project a shadow alone : every form is opaque, and struck with death. As in a march at night, I go forward with the isolated feeling of my existence, amidst the inert phantoms of all things." The flowers of his being, the brain and heart, exhausted the roots ; and this excessive spiritual vitality, based on a defective animal vitality, could not but manifest itself in misery. A deficiency of organic force fixes attention unduly on the experiences of loss, disappointment, decay. It sheds paleness and shadow over all things. It furnishes the ideas of sorrow, hollowness, evanescence, and death as the ever-ready links of association, to put their dominant stamps on all things. It gives our transitions of consciousness a downward movement. We settle from the puissant to the petty, from the magnificent to the mean, instead of soaring from the low and poor to the high and grand. With abundant organic force and health, the action of sympathy — our spiritual connection with all outside of ourselves — terminates in ascent and expansion, enhancing our life. But in morbid, drained conditions, the tendency of that action is towards descent and contraction, depressing and impoverishing us still more. Then the universe grows ashy, life becomes bitter, the leaden hue of death spreads over all, everywhere sounds the lugubrious salutation of the brothers of La Trappe, *Frère, il faut mourir.*

This was the case with the unhappy Maurice. The aching voids of defective vitality continually recalled his attention, and every meditation ended with vacancy and death. His exquisite taste and proud ambition, joined with his deep modesty and intense perception of the standards of perfection, ought to have solaced him with the joy of progressive attainments, but they stung him with wretchedness ; because, instead of rising to fasten with sympathetic appropriation on the higher ranks and wider ranges of things, and stay there, after admiringly regarding them, he sank in despair, to fasten on the examples of failure and thoughts of grief below. Every contemplation of the glorious models of the masters ended in mortification over his own defeats. "When I study history, or the works of a great man, my imagination and my desires burn; but a thought quickly follows which makes me bitterly feel the folly of my wild dreams; for no one has a lower opinion of me than I have myself." "High above my head, far, far away, I seem to hear the murmur of that world of thought and feeling to which I aspire so often, but where I can never attain." "I wear myself out in the most futile mental strainings, and make no progress. My head seems dying; and, when the wind blows, I fancy I feel it, as if I were a tree, blowing through a number of withered branches in my top. Study is quite out of my power. Mental work brings on, not drowsiness, but an irritable and nervous disgust." "I am one of those who, I know not by what strange malady of the soul, nourish a deep disgust for every social function save that of friendship. O, take me by the hand, my friend, for I shall suffocate in the crowd if you do not give me place."

Maurice affected solitude, simply as a protection from influences and claims he was not strong enough to grapple with. But "friendship," his sister Eugénie testifies, "was his sweetest and strongest feeling, the one he most thoroughly entered into, best liked to talk of, and took with him, I may truly say, unto the tomb." On first drawing near to Lamennais, he said he felt "that mysterious trembling of which we are conscious when we approach divine things and great men." Guérin suffered the same trouble in his relations with other men as in his relations with himself, the trouble which comes from fine affection with lack of confidence and complacency. He seems to have feared that his wretched inability to realize his own aspirations would either make him an object of hopeless unconcern to others, or else bring on him from them the same dislike and condemnation he visited on himself.

This is a very original and interesting trait. The class to whom it belongs is of the rarest. He looked not down seeking the uplifted eye of homage ; he looked up for the condescending eye of love. Most persons crave admiration ; he craved compassion. He had none of that indomitable haughtiness which owns not a master and longs to see the world at its feet ; his self-love rather delighted to feast on praises. Avid of celebrity, according to his own confession, he was more sensitive to scorn than to any other injury. With this softer kind of pride, he had the strongest feeling of his own wretched nothingness. With a humility extremely affecting under the circumstances, " he thought he could be loved only by a soul fond of stooping to an inferior, — a strong soul desirous of bending to a feeble one, not to adore, but to serve, to console, to protect, as one does for the sick, — a soul endowed with a sensibility as lowly as it is profound, which strips off the pride so natural even to love, in order to put on the shroud of an obscure affection which the world will not notice ; to consecrate itself to a creature all weak, languishing, and interior ; to concentrate all its rays on a flower without brilliance, — a tremulous and sorry flower, which returns to it indeed those perfumes whose sweetness charms and penetrates, but not those which intoxicate and exalt to the happy madness of rapture."

Guérin strove heroically to conquer his misery ; but there were fatal errors in his methods. He needed spiritual rest, that his organism might accumulate force ; but he kept up an incessant spiritual activity, an uninterrupted waste. " My thought is ever passing in review things present and things absent ; and, always carrying with itself the image of death, spreads a funeral veil over the world, and never brings any object before me on its smiling side." A wearing intellectual anxiety usurped the place of the leisurely and complacent assimilation of intellectual nourishment which he needed. Instead of sedulously cultivating every means of avoiding introspective and critical thought, to give room for repose and recuperation, the worse he suffered the more he analyzed and criticised, still adding to the already excessive exhaustion.

The style of his thinking was as much at fault as its persistency. If he could have made his transitions of thought, outward and upward, to rest on great objects of peace, permanent standards of duty and good, his ideas would have reacted wholesomely on his body, radiating a tonic refreshment through the nervous system. But, as the last direction of his prevailing modes of mental association was inward and downward, returning from the ideal to the actual, and stopping at last on personal defects and longings, his ideas were constantly shedding back irritating and melancholy influences on the body. He says, "The idea of existence necessarily evokes the idea of death. And, passing from the destiny and fragility of man to his interior miseries, to the eternal trouble of his heart, to the agonies which his passions cause, to that astonishing mixture of haughtiness and weakness, of grandeur and degradation, of complaints and hopes, of finite and infinite, of perishable and immortal, — who can say, after having thus studied and dissected man, — who can say, I am happy here?" Nothing can be more morbid and pernicious than this manner of thinking. It uses reason and imagination to aggravate the ills of our lot, by ideally multiplying, intensifying, and extending them ; whereas reason and imagination are properly used when they supplement the poverty and neutralize the hurts of the soul by taking our attention momentarily off from petty errors, defeats, and woes, and fixing it permanently on the grand truths, triumphs, and blessedness of existence. He thought too much : he should have trustfully fallen back on rest. He thought too brain-sickly, turning from what ought to be, to what is : he should in his thought have turned from what is to what ought to be, and gained serenity from the serene ideal.

Another injurious mistake he made was in the everlasting diagnosis of himself, the ceaseless fingering of his mental wounds. He knew and described his own case with such profound exactness that it is surprising he did not see the cure. That he did not perceive and practise the true treatment for his disease was, however, less his fault than the fault of the morbid theology and ethics in

which he was trained.    He followed the same course with
Petrarch, Pascal, and scores of other examples of un·
happy genius in modern times.    Their panacea was self-
contempt, detachment, denial, annihilation.    Our desires
torment us : let us renounce them, destroy them, die out
of ourselves into a patient waiting for God to redeem
us in eternity !    "My God !" exclaims Guérin, "what I
suffer from life ! not in its accidents, a little philosophy
suffices there ; but in its very substance.    As I go on in
age, my spirit drops a thousand spoils upon its path, ties
break, prejudices fall, I begin to show my head above the
flood ; but existence itself remains bound, — always the
same dolorous point marking the centre from the circumference.    O Stoicism ! founded to combat grief by firmness of soul, who only knewest to combat life by death, —
we have not yet gone a single step beyond thee."    But
surely death is not the *cure* for the ills of life : it is their
close.    The genuine remedy for the disturbances of the
soul is the healthy attunement of the discordant faculties
and forces of the soul.    Not denial, but fulfilment, is the
real key to content.    The genius of the Christian period
is characterized by an unprecedented development of
sensibility, — sensibility to finer and larger standards of
good.    Now, the keener, the more numerous, the wider
the ranks and ranges of obligation and desire of which
the soul is susceptible, so much the greater its exposures
to confusion, interior conflict, fermentation, — in a word,
unhappiness.    Sympathy is the crude material of our
moral nature.    All the standards of good which sympa-·
thy can recognize are elementary powers to be taken up
and organized into a firm and mature conscience.    Then
a stable self-consistency and concord will result.    Human
life is "the continuous adjustment of internal relations
with external relations," or the reflection of nature in
us.    The attempt to invert it, and make nature reflect
us, adjust her laws to our desires, must lead to misery.
The purpose of human life is the fruition of the functions
of our being in proper co-ordination.    Let any man
fulfil the functions of all his faculties in their due hie-
rarchical order, and he will be happy, because there will

be no war in him. Interior unison, self-respect, and complacency are the indispensable foundations of happiness, though they are not attainable while nebulous expanses of sympathy are floating, meteoric masses of passion darting about, in the soul.

When Maurice de Guérin strove to escape his misery by denying his ambition, scourging down his aspirations, and courting an apathetic resignation, he only made his state worse. His true refuge would have been harmony and fulfilment, with quiet submission to the inevitable. " I die secretly every day : my life escapes through invisible pricks. Some one told me that contempt for mankind would carry me far ; yes, and especially if sourness mingles with it. Every profession disgusts, every object fatigues me. I am irritated with the men who are still children. I hate myself in these miseries, which give me the most violent desires to leap on a free shore, and spurn the hateful boat that bears me. I laugh my pretensions to scorn. I scoff at my imagination, which, like the tortoise, would journey through the air. I ridicule the superb ego which vainly kicks against the goads of interior sarcasm. I bite myself, as the scorpion in the brazier, to end more quickly."

If he could have ceased to think upon himself so disatisfiedly, broken the gnawing bondage of self-consciousness, and rested calmly in a contemplation of the everlasting laws of beauty, goodness, and joy on which all creation reposes, he would have lost his misery. There was no other cure.

In such occasional passages as the following, he appears himself to have seen this : " My God ! how we distress ourselves with our isolation ! I was a long time possessed by this madness. It was because I lived wrongly, and established false relations between creatures and my soul, that I suffered so much, and that the creation repelled me from its joys. I wasted myself in a profound solitude : the earth seemed to me worse than a desert isle all naked in the bosom of a savage ocean. It was a silence to make one afraid. Madness, pure madness ! There is no isolation for him who knows how to take his place in the universal harmony, and to open his soul to all the impressions of this harmony. Then one comes to feel, almost physically, that one lives for God and in God."

One half the soul of Maurice de Guérin alone was partly plunged in evil ; the other half ever remained inaccessible to stains, high and calm, amassing drop by drop the poetry he hoped afterwards to shed on the world. The beauty of his descriptions of nature is almost unapproachable. The many paragraphs from his pen on friendship have a tone of penetrative sincerity and sweetness. The sufferings incident to his over-sensitive spirit plaintively reconcile us to the earliness of his death. One easily transfers to him the anecdote he has related of his master Lamennais. On a summer day the mournful prophet sat with Maurice under two Scotch firs, behind the chapel at La Chênaie. Drawing with his staff the form of a grave in the turf, he said, " It is there that I wish to rest ; but no sepulchral stone, only a bank of grass. O, how well I shall be there ! " He teaches us, both by what he has written and by what he was, many a striking lesson from which souls finely made and finely exposed may profit. He was one of those mentally impassioned persons, — not physically impassioned, — the victims of consumption, who appeal so profoundly to our sympathy ; whose lungs, material and spiritual, seem woven of a texture so gauzy that the common air of life works on it like a corrosive fire, who need the more distilled and aromatic breath of love to sustain and feed them, and who fade away into the one great good of eternity, with outstretched arms and vain longings after the many little goods of time.

### HEGEL.

THE great philosophers leading an absorbed inner life, with their metaphysical systems, bodies of thought hopelessly unintelligible to ordinary minds, form a class of lonely men. Such was Heraclitus, nicknamed the dark, declaring that nothing is, but that all flows; in other words, that being is not a station but a motion; a perpetual becoming: so that no one ever crosses the same stream, or sees the same picture, twice. Such was Pythagoras, with his esoteric mathematics, his secret society, his long novitiate of silence, occult instructions and signs. Such was Parmenides, with his unfathomable propounding of the One. Such were Plotinus and Proclus, with their super-refinement of bewildering speculations as to the phenomenal and the real, the transient and the eternal, multiplicity and unity. Such were the unknown founders of the oriental idealism, whose view of things was an intellectual alkahest, melting the universe into an idea. Such were the mystics, like Dschelaleddin Rumy, to whom the whole of things was an intoxicating dream, or a vision of self-identifying bliss.

It is clear enough that not one out of millions can enter appreciatingly into the mood expressed in the following lines : —

The Loved One bears the cup, and sells annihilation ;
Who buys his fire ecstatic, quaffs illumination.
He comes, — a flood of molten music round him gushing ;
He comes, — all veils are raised, the universe lies blushing.
I snatch the cup, and, lipless, quaff the Godhead's liquor,
And into unity of bliss the self-lights flicker.

It is probable that still fewer are capable of understanding the absolute idealism of Hegel. His learning is so vast, his analysis so remorseless, his abstractions so transcendental, his terminology so abstrusely knotty, his synthesis so all-comprehensive, that his system is the standing scandal of students, baffling all but the very best-equipped and toughest thinkers. He claims to have made metaphysics an exhaustive science, a closed circle of circles. He begins with Being as the absolute Affirmation, and Nothing as the absolute Negation, and shows their identity in Becoming. He proceeds, through a constant reconciliation of the contradictory pairs between which alone thought can exist, to the conclusion that the All is a thought, and that every genuine thought which penetrates to know itself is the All. It is, whether true or

false, in subtilty and comprehensiveness as tremendous a
piece of thinking as was ever performed by a human
head.   The popular inability to comprehend what he said
left him by himself.   He declared, "Only one man understands me, and he misunderstands me."

To master his system requires as special an intelligence
and training as to master the Fluxions of Newton, and in
a far higher degree.   Yet those least fitted to judge are
frequently the readiest to assume superiority, and to name
his industry charlatanry or folly.   No one who is yet lingering in rudimentary arithmetic will presume to call the
geometrical calculus an empty imposition ; he knows it to
be unmeaning only to his ignorance.   But it is quite customary for one who in philosophy has not finished simple
numeration to stigmatize the metaphysical calculus of
Hegel as little better than idiotic jargon.   Common
sense, which is the rule of mental averages, seeing how
far he varies from it, complacently considers him a fool.
No wonder his speech is cutting and caustic with irony
towards the intellectual pygmies who stumble at his out-
works, and fancy themselves stalking above him when
really dealing with their own dwarfed reflections of him.
His system may be illegitimate science, it is certainly fruitful gymnastics, a tremendous regimen of mental enlarge-
ment, mental emancipation, mental enrichment.   The
complacency of those who have neither taste nor faculty
for such studies, nor modesty to feel their failure, often
leads them to stigmatize him as an impostor and his product as emptiness.

But he, meanwhile, — where is he ?   Occupied with his
own indomitable effort to understand everything, to leave
absolutely no mystery uncleared, to know even God himself, he is "out in the void desert, separated from the
world of man by endless days and nights, and eternally
recurrent and repeating solitudes, lonely, mysterious, inexplicable, a giant dreamland, where the sense of Being
and the sense of Nothing, like two boundless vapors confronting each other, the infinite vaporous warp and the
infinite vaporous woof, melting, interpenetrating, wave
and weave together, waft and waver apart, to wave and

weave together again. He has wrested himself from the place of mere mortals, on the outside, groping into concrete delusions ; he sits in the centre of pure thought, and sees an immense magical hollow universe construct itself around. Does he not come out from the centre of that world, that secret chamber of his, begrimed with powder, smelling of sulphur, like some haggard conjurer, his voice sepulchral, his accent foreign, his laugh demoniacal ? Contrast him with the simple, pious soul, on the green earth, in the bright fresh air, industrious, loving, penitent, sure of a better world and a better life !"

To the ignorant eye of unsophisticated trust he would thus seem to be at a sore disadvantage. But the reverse is the truth that — according to his own view — appears to perfected insight. For the obvious little that he loses he gains an occult infinitude. By his rounded survey of human thinking and natural phenomena, penetrating from nadir to zenith, he knows his own incommensurate intellectual superiority to other men. He can understand them and their errors ; they cannot understand him or apprehend his certainties. He has drawn a circle around their outermost, sunk a shaft underneath their lowermost. The scores which to them are series of blind hieroglyphics he has the key to read off into music. The poor delusions in which they are enveloped, and to whose vulgar promises they cling, have scaled from him and left him to grasp the divine prizes which these had but degraded by their mocking simulations. Instead of remaining content, as they foolishly do, with the verbal phantoms of God, freedom, and immortality, he grasps the immediate substance of the thoughts themselves, seeing God in that process of universal Spirit which plays through the universe, realizing freedom in the consciousness of his own powers, and possessing immortality in the indestructibleness of all that really is. Thus where common men see hideousness and horror, he sees beauty and boon. Where they shrink from necessity and nihilism, he aspires to substantial liberty and infinity. What they interpret in their spurious mythological conceptions as the riot of the worm and the blackness of the grave, he recognizes

w

as the fruition of eternity and the splendor of the god-head. They cloak the omnipresent climax of contradictions with the sprawling sophistry of ignorance ; but he with logical dialectic probes the problem to its solution in reciprocal identity. Thinking his way irresistibly through, from the beginning to the end, he sees that the logic of thought is the logic of things, the ideality of man a reflection of the reality of nature, the reality of nature a reflection of the ideality of God, and all these at bottom identical, just as man sees in the mirror not a reflection of himself, but himself on the curving beams of light. This is the way it seemed to Hegel.

The builder, occupant, and master of such a system of thought is a spiritual hero and monarch of the world, a conqueror of destiny, who has put all things under his feet. But in the weird spaces of its amplitude, if not divinely companioned, he must be awfully alone. So far as any intelligent sympathy from the average order of men is concerned, he would be as much alone in the academy of Berlin as by the flaming crown of Hecla or at the frozen core of the Antarctic. To the ordinary mind, man is both a product and a beholder of the outward world. To the idealist, he is a seeing producer of the world. To Hegel, the mature philosopher is at once a product and a producer of the world ; he is both the seer and the sight ; his consciousness, unified with primal thought, begetting everything out of itself. In his Logic he constructs, *a priori*, the autobiography of God as He is in his eternal being ; the history of the Absolute Spirit, from the beginning, when He was alone, without any creation or finite spirit, to the end, when "He quaffs His own conscious infinity from the cup of the kingdom of finite minds." It is true that this is wonderful speculation, — wonderful in its power and in the magnitude of its concerns. It may justly be considered by every modest man as unbecoming presumption. But it cannot properly be treated as puerile chatter. Its master, as compared with common men, is not an audacious baby alone in his nursery with a toy, but an intellectual king alone on his mighty altitude with the universe.

The Hegelian metaphysics may be a baseless phantasy. Those who do not understand it nor know its historic development as the completion of foregone systems, may scout it as foolishness. But it was a reality to Hegel, and its inferences were realities to him. It is as strenuous an effort, as stupendous a construction, as is seen anywhere in the history of thought. And only by an incredible irony can one who appreciates it believe, that when its author had finished his work, and was identifyingly elevated to "absolute knowledge," the cholera grabbed this "infinite God" by the bowels and dragged him into the grave. Surely such was the fate of the body of Hegel. Surely the spirit, the force, the *begriff* that was Hegel, is Hegel still, and does not moulder under the sod of the Prussian churchyard.

> Life, like a dome of many-colored glass,
> Stains the white radiance of eternity,
> Until Death tramples it to fragments.

What journeys he travels, what toils he undergoes, what adventures he encounters, who makes mental pilgrimages to the spiritual shrines and landscapes of all the great thinkers of the world! Mastering the chief efforts of mankind to solve the problem of being, he leaves far away the multitude who do not so much as guess that there is any such problem to solve. He sails over the "water" of Thales; soars through the "air" of Anaximenes; reasons to the "mind" of Anaxagoras; sees the "atoms" of Leucippus flying into groups; studies the occult powers of the "numbers" of Pythagoras; contemplates the archetypal "ideas" of Plato; busies himself with the "entelechies" and "categories," the "matter" and "form" of Aristotle; comprehends the "artistic fire" of Chrysippus; plunges into the ontological "ecstasy" of Iamblichus; and gazes with tranced intuition into the "nirwána" of Gotama. With Des Cartes, sceptically stripping himself of all opinion and prejudice, all but the two notions of "thought" and "extension," he perceives that the essence of these is their reciprocal negation, the answer to What is matter? being, Never mind! to What is mind? No matter! and proceeds thence by successive steps to universal truth. With Spinoza, he converts thought and extension into attributes of one sole "substance," compared to the lair of a lion which many footsteps enter, but none leave, — individual entities blotted into the all-subsisting God. Passing on to Leibnitz, he conceives an infinity of "monads," each monad an obscure mirror of the creation, a little God, and with these builds a dynamic universe. Then come the intelligible "phenomena," and the forever inapprehensible "noumena" of Kant; the "reals," and the standards for "remodelling the conceptions of experience," of Herbert; the "I" and the "Not I," or the creation as a self-limitation of the ego, of Fichte; the various phases of the "identity-system," or objective transcendentalism, of Schelling; the "absolute idealism," or development of self-consciousness to a height where it logically constructs the universe, of Hegel; the wishful dynamism, or "world-as-conception-and-will," of Schopenhauer; the manifestation of "modes of force" in "forms of matter," or summing of all that can be known in a "law of evolution," of Spencer.

To travel in mental space to these mighty monuments of thought and aspiration, and a score of other kindred ones, traverse their labyrinthine apartments, and comprehend their contents by reproducingly entering into their genesis and development, is an achievement that few have the desire, the leisure, the patience, and the power to accomplish. What prouder ambition can any man cherish than the purpose to do this? to advance with assimilating docility along the biographic line of human thinking, through the schools of the Ionics, the Eleatics, the Socraticists, the Neoplatonists, the Scholastics, the Mystics, the Spiritualists, the Materialists, the Positivists, and, with the healthy mastership of it all, emerge at last under the blue empyrean of reality, where science and faith preside together with a cheerful acquiescence in each other's functions. It is the romance of the mind, the interior epic of humanity. Every master of it is thereby mentally isolated, since he lives in a world of thought which the ignorant can hardly enter; possesses for attachment to every object he regards and every state of consciousness he feels, a complicated mass of associations of which they never dream. He looks serenely down on the petty brawls of the selfish and the idle misgivings of the credulous. If he is a good man, as ripe in characteristic wisdom as in learned study, he occupies, in the words of Martineau, " an intellectual eminence above surprise, whence the great movements of humanity can be watched in the quiet air of piety and trust, and where the distant voices of its prayer and strife cannot reach him." The turmoils and uproars of the world are reflected to him as silent pictures. He has reached to the calm which lies at the centre of agitation, the heart of the hurricane of the forces of time. Let it not be said that the unsophisticated peasant, who simply trusts in the name of Jesus, is better off than this man. He too has a pious trust none the less firm in its blessedness. Recognizing all things " in the diamond net of one perfect law," he beholds himself as an intelligent emergence from the unknowable, dealing for a season with phenomenal relations, and re-entering the unknowable. He thrills with adoration before the spectacle, and rests peacefully in his own thought. It is his fault if he is not as much superior in sentiment and faith to the mere innocent believer as he is more favored in vision. His ideal sweeps of disinterested sympathy, of joyous and worshipping affection, as well as of intellectual contemplation, should be comparable with the experience of that bold aeronaut, who, while others were toiling in their low nooks or asleep in their beds, made in seventeen hours a balloon flight of five hundred miles, from London to Weilburg, in Nassau, — the passage over the dark sea, and the Belgian district of furnaces, the sea of mist below in the morning, with the rustling of forests coming up like the sound of waves on the beach, the paling of the stars, the gorgeous sunrise shedding its colors over the vast heavens and the earth far underneath, yielding him sensations inexpressibly solemn and beautiful.

### SCHOPENHAUER.

ONE of the most vigorous and piquant writers, bold thinkers, snappish and gloomy spirits of our century was Arthur Schopenhauer, the German philosopher, who died at Frankfort in 1860. Among the many strong and strange qualities of his character, loneliness was, perhaps, the most prominent, almost from the cradle to the grave. He was imaginatively suspicious and timid, proud and shy, with an astounding assurance of his own greatness and noble destiny, and at the same time with a furious moral irritability, and a morbid physical cowardice. He was a most singular being, interesting and odious, wise and absurd, endowed with a gigantic intellect which shrank from no problem or conclusion, and vehement affections discordant among themselves and awry towards the world. His tender need of sympathy and fierce craving for success balked and thrust back, made him feel deserted, a sort of outcast ; his subsequent curdling hate and scorn, and wilful hardening of his heart in haughty self-protection, made him feel doubly isolated. His biographer says : " Although remaining in the midst of society, never has a man felt more separated and alone than Schopenhauer. The Indian anchorite is a social being in comparison with him ; for the solitude of the former is accidental, or rests on practical motives ; with him it was essential and the result of knowledge. Therefore this feeling in his consciousness reached an intensive strength which admits of no comparison with mere retirement."

As soon as he began to think, he seems to have found an impassable chasm between himself and the world ; astronomic distances divided him from those whom he should live with and love. At first he feared the difference and opposition were his fault, and this often filled him with sadness. But his native pride and complacency, strengthened by a constant feeding on the ideas of kindred spirits in literature, — such as Machiavelli, Rochefoucauld, Chamfort, — caused the world to lose, his self-esteem to gain, something with each conflict. Up to his fortieth year he had felt frightfully lonesome, and had continually sighed, " Give me a friend." In vain ! He still remained solitary. But now, after such incessant disappointment, he concluded that humanity was infinitely more penurious than he had imagined, looked around on the earth as on a desert, made up his mind firmly that he was one of the intellectual rulers of the race, and that he must bear his royal solitude with dignity and patience. He said men shrunk from seclusion and sought association because they were so poor and empty. They and their society reminded him of Russian horn-music, wherein each horn can sound but one note, and a whole band is necessary to play a tune. The rich, many-toned wise man is a piano-forte, a little orchestra in himself. From this time he became systematically unsocial, and appeared deliberately to nourish by all means the worst possible views of life and men. Desiring fame with an eagerness proportioned to his estimate of his rank and of the value of his system of philosophy, he sought to cover and soothe his bitter chagrin at its long delay, by casting contempt on it, and expressing disgust at the mixed and unworthy throng who most easily gain it. " Fame is an existence in the heads of others, — a wretched theatre ; and its happiness is purely chimerical. What a rabble crowds into its temple, of soldiers, ministers, quacks, gymnasts, and millionnaires !"

It is a proof of his originally deep and high heart, that while the men around him were so empty and repulsive to him, he lived in delightful intimacy with the great minds of previous times. The thoughts left behind by those great men, who, like himself, were alone amidst their contemporaries, were his keenest enjoyment. Their writings came to him as letters from his home and kindred to one banished and wandering among islands destitute of men, but where all the trees are full of apes and parrots. That he should have used such an illustration as the foregoing, also proves how sorely wrenched and irritated his heart had become. In ethics a graduate and continuator of the school of Plato and Kant, he defined

the worst man as the one who makes the greatest distinction between himself and others ; the best man as
the one who makes the least distinction between himself
and others.   But his practice amazingly violated his in
sight.   In all his habitual modes of personal thought and
feeling, instead of minimizing he maximized the distinction of himself from othei men.

During the latter half of his life he considered every
contact with men a contamination.   He pretty faithfully
practised his own precepts wherein he said, " The world
is peopled with pitiful creatures, whom the wise man is
born not to fellowship with, but to instruct.   They are a
foreign species, with whom the wisest has the least to do,
regarding himself and deporting himself as a Brahmin
among Sudras and Pariahs."   He would not ordinarily
call his fellow-beings men, but contemptuously, with a
grim humor like that of Carlyle, whose Teufelsdröck describes the human creature as a " forked radish," characterizes them as " bipeds," the " two-footed."   He asserted that he was not a man-hater, but a man-despiser.
To despise the species as they deserve it was necessary
not to hate them.   Two classes of men, however, he did
hate with especial relish and virus.   First, the University
professors of philosophy ; ostensibly because they were
charlatans, dishonest smatterers ; really because they enjoyed the place and attention he coveted for himself, refused to give his works and genius the tribute he deemed
his due, and formed a conspiracy, as he fancied, to prevent all public recognition of him.   Second, the Opti-
mists ; because their system seemed a biting irony in
view of the facts of sin, sorrow, and death, — a shallow
mockery of the inexpressible wretchedness and emptiness
of existence as presented by his theory and emphasized
by his experience.   He was himself a Pessimist, one who
reverses the proposition that this is the best of all possible worlds.   The ingenious argument on which he based
his reversal of the scheme of Leibnitz was this : Life is
crowded with examples of discord, baseness, and misery ;
the whole system is so exactly interdependent, that if
the least feature of it were altered, made worse, all would

go to destruction : therefore, this is the worst possible world !

He esteemed himself an imperial mind, his contribution one of the richest the world had ever received. At nineteen he said he would become the philosopher of the nineteenth century. Forty years had passed, and his books were lumber, his name unknown. He rebelled with injured and wrathful arrogance against the injustice. He imagined it was the result of a malignant coalition. This mischievous conceit worked like vitriol in his blood, poisoned all his peace, aggravated his worse traits, filled even his philosophic works with savage invectives, and made him chuckle with ignoble delight over the flattering notices his books at last began to win. He exclaims : "I have dismal news to communicate to the professors. Their Caspar Hauser, whom for forty years they had so closely immured that no sound could betray his existence to the world — their Caspar Hauser is escaped. Some even think he is a prince. In plain prose, that which they feared above all things, and took every conceivable means to prevent, has befallen. Men begin to read me, and henceforth will not cease."

Ardently wishing the complacent sense of being admired and renowned, he turned angrily against those who withheld the boon. Cynically secluded in Frankfort, neglected, deprived of all the associations and sympathies he most desired, " a solitary thinker in a den of money-changers, he mused, and plodded, and nourished the grudges which disappointment had engendered in a nature predisposed by some radical vice or defect to misanthropic gloom." He sought to support himself by two artifices. First, by aggrandizing his own sense of his own merit and of the sure reward yet awaiting it. " I have lifted the veil of truth further than any mortal before me. But I should like to see the man who could ever boast of being begirt by worse contemporaries than I have had." " The world has learned many things from me which it will never forget." " Since the great soap-bubble blowing of the Fichte-Schelling-Hegel philosophy is done, there is greater need than ever of philosophy.

Now people will look about for solider nourishment, and this is to be found alone with me ; for I am the only one who has labored purely from an inner vocation." When asked where he would be buried, he proudly replied. "It matters not ; they will find me out."

The second artifice to which he had recourse was the sour-grapes principle. The prize is contemptible. The love and praise of such unworthy creatures as men are hateful. The world is a hideous place, existence a cursed burden. Absolute detachment is the supreme good. In this way the misery of his own experience infects and dis-colors all. He pours over the whole scene of life an inexhaustible tempest of execrations, contempt, and gloom. His pictures of the "Nothingness and Sorrows of Life," his eulogies of death and annihilation, are not surpassed in their energetic blackness and perverse gusto by the most disgusting portrayals of Oriental pessimism, the Buddhist catalogues of the evils of existence.

"In this world," says Schopenhauer, "there is very much that is very bad, but the worst thing in it is society." "The more I go among men the less of a man I come away." "Conversation with others leaves an unpleasant tang ; the employment of the soul in itself leaves an agreeable echo." "The jabber of companies of men is as profitless as the idle yelping of packs of hounds." Under the influence of such a doctrine of the penal character of life and the loathsomeness of man, society contracted to a unit, and solitude expanded to a boundless desolation. It could not otherwise than dilate and inten-sify the woe it was meant to antidote. Yet there is in the doctrine a weird horror that allures while it affrights. As Hedge says, "Nature shudders, but curiosity tempts. It is the fascination of the cavern and the catacomb. The world of this philosophy is a world of darkness which no sunshine or starshine irradiates, but whose only illumination is the phosphorescence of the animal matter contained in it."

The poor Titan took the wrong way. Instead of aggregating the topics and motives of unhappiness he should have aggregated the stimulants and materials of peace

and joy, moderating, soothing, attuning his faculties by a
sober and firm discipline with reference to the standard
of personal perfection. Maurice de Guérin felt himself
a personality more apart than above, select rather than
superior. He was consumptive ; physical lassitude and
superfluous sympathy and longing made him unhappy.
Schopenhauer felt himself a supreme man. He had the
best digestion, firm strength, and sound sleep. But his
superb complacency, with the restless exactions of its
social direction, kept him in irritating relations with the
world ; and he too was an unhappy sufferer. To look to
others, either with humble supplication, as Guérin, or with
irate command, as Schopenhauer, — to look to others
for the love or admiration they cannot, will not give, is
to be miserable. Then to cosset this misery, as a proof
of spiritual superiority — the unhappier, the greater and
worthier — is the mad sophistry of self-love. The wise
course is to try to do our duty, and perfect ourselves, and
harmonize our desires with the conditions of truth, in
comparative independence of the opinions of other people. Had our chief of modern Pessimists lowered the
denominator of his haughty desires from society, and
raised the numerator of his humble self-surrenders to it,
he would have solved the equation of his happiness. The
disappointment of his ambition gave him a chronic moral
nausea : he needed a constant contemplation of the
idea of the race in the individual, as a mental gargle to
enable him to relish men and be content in himself.
But unfortunately he reversed this, seeing in mankind
only the multiplied images of the individuals for whom he
had a particular aversion. The accuracy of the foregoing
diagnosis is confirmed by the mollifying influence which
the surprising renown he acquired in his last years had
on his character, the great happiness and new spirit it
gave him when the first rays of the morning sun of fame
gilded the evening of his life. "Time," he smilingly
said, referring to his blanched hair, "has brought me,
too, roses, but white ones." That under the bristling of
his bitter outside lay a soft heart of love is indicated by
his fondness for his poodle Putz, who slept on a black

bearskin at his feet, and invariably kept him company in his walks.

Left alone, by the momentary absence of his servant, a sudden rupture of the lungs snatched the sick Schopenhauer easily out of the world. The physician, entering the room immediately after, found him sitting in a corner of the sofa, with calm countenance, dead. His Indian Bible, the Oupnekhat, lay on the table. On the mantelpiece stood a gilded statuette of Gotama Buddha, the great leader to Nirwána. The strong, tart sage of Frankfort had followed the puissant thinker of the East to the city of peace. Thenceforth, for him, the poorness and sorrow of life, against which he had so long chafed, were no more. No more could the meannesses, impertinences, and vexations of the world of men rain on his weary head or beat against his sore heart. The insatiable search after the knowledge of reality ended, the distressful journey done, the pack and staff flung away, he had vanished into the night of eternal mystery, where friends and foes, victors and vanquished, equally go.

One of the small company that gathered around his grave, before they lowered the laurel-crowned head, stepped forth and spoke, among other sentences, these: "The coffin of this extraordinary man, who, after living among us so many years, remained as foreign to us as when he came, awakens strange emotions. No one united with him by the sweet bonds of blood stands here. Alone, as he lived, he has died. And yet, in this still presence, something tells us he has found satisfaction for his solitude." Yes, some true satisfaction. For, lonely as he was in character, thought, habit, death, we think not of him now as lonely. If the Christian heaven be a verity, he is there with the Saviour who revealed the God of the parable of the Prodigal Son. Pardon cannot be wanting for one whose ideal of man was so grand that his scorn burned against its degraded foils and forfeits; one whose loyalty to truth was so supreme that in all his life he never, in self-interest, swerved one step from his high, lonely way. If that heaven be the dream he thought it, why then he is where he aspired to be, with Kapila,

Sakya Muni, and the other conquering kings of mind, blent in the unknown destiny of the All, clasped in the fruition of Nirwána.

## EUGENIE DE GUÉRIN.

The name of Eugenie de Guérin has won unexpected celebrity through the posthumous publication of her journal and letters, which reveal a personality of singular depth and purity, and record an experience of rare interest, notwithstanding its monotony. Her life was clasped in by the first half of the present century; but the force and simplicity of her character are such as were more frequent in ages less complex and sophisticated than ours. The three central chords, constantly struck, and making the sympathetic music of her soul, are friendship, solitude, religion. Her love for her brother Maurice is one of the marvels in the history of affection : it deserves to be classed with the absorbing passion of Madame de Sevigné for her daughter. Placed in a lot extremely lonesome, bare, and regular, the keenness of her exuberant and tenacious consciousness made her doubly sensitive to the weary isolation in which she lived. Her rich and stainless feelings, denied sufficient lateral expansion in the social relations, forced a vertical vent, and broke upward in religious flame. But of the three refrains, love, loneliness, piety, the second is the one that recurs oftenest, and sounds with the most piercing tone. Her life is as plaintive as it is pure, as painfully stamped by the hunger of unsatisfied affections as it is divinely impressed with self-renunciation in faith and duty. To dwell on the pages of her writing is like entering a mountain chapel, where we breathe at once the charm of nature, the quiet of seclusion, and the peace of God. At the same time there is an unspeakable pathos in the pale face that looks out at us, and a strange sigh of human want and woe in the voice that speaks so calmly. Altogether the character and life of Eugenie de Guérin have profound lessons for those of her sex gifted with natures earnest enough to learn them ; a select number in that throng of women whose attention is frittered on trifles whose existence is a shallow distraction when it is not a tedious drudgery.

The retirement and sameness of the life at La Cayla were oppressive.  To have a visit or pay one, to write a letter, to receive a letter, were the great events.  Eugenie writes, " A poor stranger has passed by ; then a little child.  This is all that has shown itself to-day."  Again, " To-day nothing has come, nothing has stirred, nothing has got done in our solitude."  Once, when her sister was absent, she wrote, " I have passed the day in complete solitude, alone, — quite alone.  I am taking account of my thoughts by the light of a little lamp, now my only companion at night."  She said her days were as like each other as drops of water.  " Would that my arms were long enough to reach all those I love ! " — " Everything belonging to the *world* soon wearies me, since I always feel myself a stranger there." — " God be praised for this day, spent without any sadness !  Such are so rare in life.  A word, a memory, a tone of voice, a sad expression of face, a nameless nothing, will often disturb the serenity of my spirit, — small sky, that the lightest clouds can tarnish."  Her heart was great, and inconceivably susceptible even to the most delicate impressions.  Such hearts are indifferent to whatever does not give them life ; they shrink with powerful instinct from the careless slights and rude collisions that would bruise them and drain them.  Hence the inevitable solitariness of such characters as Eugenie.  Profoundly humble as she was, she could not help saying, " When mixing with the world, I feel that I am not like others." — " How much sadness in this isolation, this chill, this frost, of which the heart is conscious while surrounded by pleasures, and by those who partake of them ! "  Eugenie often expresses this forlorn and sorrowful desertedness. After the death of her idolized brother, she writes to him in her journal, " I feel a want to be alone, and not alone, — with God and thee.  I feel myself shut out in the midst of all others.  O living solitude ! how long wilt thou be ? "

But solitude, in spite of this bitter pain of unfulfilled desires, was her best refuge, filled with her truest joys. She abundantly celebrates its charms. "Nevers wearied me with its little society, its little women, its great dinners, dresses, visits, and other tiresome things without any compensation. Loneliness, calm, solitude, — recommencement of a life to my taste." While she was visiting Paris, a woman having said that, for her, "friendship was a velvet couch in a boudoir," she replied, "Let me be outside the boudoir, sitting on a lofty peak, high above the world. To sit apart from all, in this way, delights me in the same manner." This thoughtful, down-looking withdrawal removed the perturbations of her too susceptible soul, and soothed and nourished her self-respect. Yet, as we read over the confessing pages of her journal, how obvious it is that solitude is not the true destiny of the human heart ; but only a retreat wherein, when that destiny has been baffled, it comforts itself with consolatory substitutes ! The instinctive affections, extraordinarily strong in Eugenie, balked of their normal fulfilment, in still loneliness solaced themselves with other objects, with ideal activities, with heavenly aspirations. No pleasure rivalled that she knew in solitude, with God, books, and the thought of Maurice. "Though talking and loving each other much, two women alone find their solitude very blank, — great desert places in it : books, books, are the only recourse." "Verily a book is a priceless thing for me in this my desert and famine of the soul." It is impressive to notice how constantly, without her knowing it herself, the expressions of Eugenie show solitude to be, not the normal fruition of our being, but a retreat from a storm, a healing and compensatory covert from hurts and griefs. "A convent is the true home of sad spirits, of such as are strangers in the world, or who are timid and take shelter there as in a dovecote" "O, what enjoyment to be free from distraction, with God, and with one's self!" Had not her family held her back, she would gladly have adopted the vocation of a nun. "For a long while I have been saying, with St. Bernard, 'O blessed solitude ! O sole beatitude !'"

One of her most terrible trials was the frequent discovery of baseness in those she had trusted and admired.
She passionately loved to reverence and confide ; and
the knowledge of the treacherous and deformed side of
human nature cost her too much.   To see venerable and
beloved brows discrowned was an agony worse than
death to her.   Then the calm of her little chamber, the
starry solitude of night, were guardian sanctuaries into
which her soul fled.   " It is strange how much I enjoy
this being apart from everything."    " This has been
one of my happy days, of those days that begin and end
sweet as a cup of milk.   To be alone with God, O happiness supreme ! "   " When I am seated here alone, or
kneeling before my crucifix, I fancy myself Mary quietly
listening to the Saviour.   During the deep silence, when
God alone speaks to it, my soul is happy, and, as it were,
dead to all that is going on below or above."   A more
perfect picture of loneliness, a more convincing proof of
the genius of solitude in its author, than is afforded by
the following passage, can hardly be found : " My window is open.   How tranquil everything is !   All the
little sounds from without reach me.   I love the sound
of the brook.   The church clock is striking, and ours
answering it.   This sounding of the hours far away, and
in the hall, assumes by night a mysterious character.   I
think of the Trappists who awake to pray, of the sick
who count hour after hour in suffering, of the afflicted
who weep, of the dead who sleep frozen in their beds."
Eugenie was well aware of the moral dangers of too
great and constant a withdrawal.   She religiously strove
to neutralize them.   " I observe that I hardly make any
mention of others, and that my egotism always occupies
the stage."   "There is a weakness in this bias of thought
towards one's self and all that belongs to one.   It is
self-love."   The complaint does her injustice.   For it is
the peculiar property of a suffering nature to tint the
world with its grief.   And she only poured out her self-
burdened soul as a relief in her journal, never meant to
be seen.   In all associations with others she was self-
forgetfully devoted to her duties, abundant in disinterested

attentions. Her morbid quality was really, not thinking too much of herself separately, but too much thinking of others in herself, and of herself in others. It was sympathy that was tyrannical, not egotism. To accuse themselves of a blamable self-love is the painful fallacy of those humble souls who are too tender, and not strong enough, nor enough detached from their neighbors. To be accused of such an excessive self-reference is the cruel wrong such souls always suffer from conceited and impatient observers.

The threefold characteristic of genius in affection is the richness, the intensity, and the tenacity of its emotions. The emotions of a meagre nature are comparatively narrow, pallid, and evanescent. Whatever once entered the heart of Eugenie de Guérin became complicated with aggrandizing associations, royal or tragic; throbbed with her blood, and stayed as a fixed part of her life. " At the foot of the hill there is a cross, where, two years ago, having accompanied him so far, we parted with our dear Maurice. For a long time the ground retained the impress of a horse's hoof where Maurice stopped to reach out his hand to me. I never pass that way without looking for that effaced mark of a farewell beside a cross." The prints on the Cayla road were transient as strokes on the air, compared with the perdurable impressions on that soft, faithful heart. This peculiarity, joined with a retired and leisure life, has a good side and an evil side : for " an exclusive feeling grows to immensity in solitude." In a soul of ample health and strength, it leads, by successive conquests, through an accumulation of glorious associations, to the noblest greatness and happiness. Its power is seen in that story of La Picciola, where a simple flower became the light, the comrade, the angel, the paradise, of a poor prisoner in whose cell it grew. Eugenie affectingly illustrates it when she says, " I must record my happiness of yesterday : a very sweet, pure happiness, — a kiss from a poor creature to whom I was giving alms. That kiss seemed to my heart like a kiss given by God." Under such conditions, the littlest things are more than the greatest things are in a crowding and dissipated existence.

x

On the other hand, this accreting and embalming quality of genius, this incrusting of experience with associations, in a drooping, timid soul, defective in elastic energy, leads to the most melancholy results; it exaggerates every evil, confirms and preserves every depressing influence.  It fastens on the unfavorable aspects of things, heaps up sad experiences, emphasizes all dark omens, until society becomes odious, action penance, life a way of dolor, the earth a tomb, the rain tears, and the sun a funeral torch.  How profoundly Eugenie suffered from this evil, hundreds of passages in her writings reveal, like so many wails and sobs translated into articulate speech.  Thoughts of death and feelings of sorrow occupy that relative space, which, on any sound philosophy and estimate of our existence, ought to be occupied by thoughts of life and feelings of joy.  "At night, when I am alone, the faces of all my dead relatives and friends come before me.  I am not afraid; but all my meditations dress themselves in black, and the world seems to me dismal as a sepulchre."  Her moods of spiritual exhaustion appear from the grateful approval she gave to the word of Fenelon in relation to irksome prayer, "If God wearies you, tell him that he wearies you."  She says, referring to a former period, "I got deeper and deeper among tombs: for two years I thought of nothing but death and dying."  She calls "Inexorable dejection the groundwork of human life"; and adds, "To endure, and to endure one's self, is the height of wisdom."  Surely, poor is the office of the angel of religion, descending and ascending between God and men, if at the last he can only waft us this message of despair.  No, the highest wisdom is not, in sackcloth and ashes, to endure existence and ourselves.  The highest wisdom is, instead of submitting to the will of God as its penitential victims, to conform to it as its grateful executives and usufructuaries, appreciating all the goods of life in the just gradation of their values.  Instead of saying with Bossuet, "At the bottom of everything we find a blank, a nothingness," a healthy religious faith finds, at the top of everything, the bottom of something better.  The misery of Eugenie lay in her ungratiñed natural affections, whose disappointments held the germs of death against which she had not sufficient vitality to struggle into serene victory. *Lack of life* is the ground-tone of her grievous music, which would sweetly seduce the weak to death, but loudly warns the wise to a better way. Her betraying pen writes, " My soul lives in a coffin." Again, " I find myself alone, but half-alive, — as though I had only half a soul." And finally, with the anguished heroism of a total renunciation, so willing to perish as to be unwilling to leave a trace behind : " I am dying of a slow moral agony. Go, poor little book, into forgetfulness, with all the other things which vanish away !" Such an utterance proves the irritable feebleness of the centres of life to be so great that it is painful for them to re-act even upon the idea of posthumous remembrance. The fondled thought of extinction and oblivion is soothing then. Through its inner wounds, one may almost say, the very soul itself slowly bleeds to death.

There is a bird, the arawonda, that lives in the loneliest glens and the thickest woods of Brazil. Its notes are singularly like the distant and solemn tolling of a church-bell, as they boom on the still air, and plaintively die away. Sitting on the tops of the highest trees, in the deepest parts of the forest, it is rarely seen, though often heard. It is difficult to conceive anything of a more solitary and lonesome nature than the breaking of the profound silence of the woods by the mysterious toll of this invisible bird, the swelling strokes with their pathetic diminuendo coming from the air, and seeming to follow wherever you go. The tones of the character of Eugenie de Guérin are like the notes of the arawonda.

Her lot was thorny, yet not without roses. The world itself was a convent, in which she lived as a vestal, with bended knees, upraised eyes, a consecrated will, but an aching and bleeding heart. Poor, rich, unhappy, blessed maiden ! we cannot bid her farewell without deep emotion and a lingering memory. Her journal is a nunnery of sad, white thoughts, with here and there one among them revealing, as the snowy robe of style is lifted, a heart of

agonizing flame.  We pity her sufferings, admire her fortitude, revere her holiness, bow before her saintly faith
and patience.  What a thought of peace it is to think
that she is now in God!  There love is infinite, and repose perfect.  No ungenial society can vex, no weary
solitude burden, the freed inhabitants there.


## COMTE.

The character and life of Auguste Comte, author of the
Positive Philosophy, affords a forcible example of the loneliness of a mighty personality, of the trials it is subjected
to, of its temptations to misanthropy, and of the comparative neutralization of those temptations by sublime ideas,
personal purity, and devotion.  He believed himself born
to introduce a new and better faith in philosophy and in
religion.  The burdensome superstitions which had so
long darkened the minds, clogged the efforts, disturbed
the souls, and afflicted the lives of men, — these accumulated errors and evils he would teach the world to
throw off, and, by a complete organization of the hierarchy of the sciences, proceed more rapidly to fulfil and
enjoy their true destiny.  Instead of wasting their energies
in vain attempts to discover the unknowable ultimate
causes of things, they should limit their inquiries to the
grouping of facts and appearances, and to the discovery
of their laws.  He would instruct them how to outgrow
their selfish antagonisms and rivalries, in a disinterested
co-operation for the perfection of each other and the
whole.  They should no longer expend their devotional
sentiments in the worship of a metaphysical abstraction,
but should recognize, at last, with clear consciousness,
the true Supreme Being, namely, the collective Humanity,
made up of all the human beings who have lived, all who
now live, and all who are hereafter to live.  This imper-
sonated totality of mankind they should love with all their
mind, heart, soul, and strength, and worship with appropriate rites of good works, expansive sentiments, and
symbolic offices.

There was in Comte, undoubtedly, all his life, an un-balancing bias of egotism. Neither mental health and modesty, nor a wide range of careful comparisons and tests, enabled him to make a fit and sound estimate of his own relative importance. He had a prodigious idea of his own spiritual dimensions and rank. He had a most despotic will, a morbid unwillingness to take his cue from anybody else, no feminine abnegation in society, but a masculine necessity for dominating. He irascibly re-sented influence, and repelled commands. So, republi-can in spirit and rebellious in disposition, he offended his official superiors, and was expelled from one post after another. He sympathized easily and strongly with the great world of men below, whom he was to instruct and uplift; but to the nominal superiors, who overlooked or despised him, he gave a proud scorn. He used to sing the revolutionary Marseillaise with electric vehemence; and the prefaces to his different volumes express immeas-urable contempt for his opponents. His feeling is shown in his appeals to the two classes, women and proletaires, as ready for the acceptance of his catechism. But, for the great thinkers of earlier time he cherished a glowing admiration, and scrupulously acknowledged his obliga-tions to them. He felt allied to them as typical prede-cessors of himself.

Deeming himself intrusted with a transcendent mis-sion, he sought with heroic devotion to fulfil it by master-ing all foregone history, philosophy, and science, eliminat-ing the true from the false, supplementing the incomplete, and imparting their perfected lesson to the world. To this immense task he gave an immense toil. Shutting out from his life all that could distract him, supplying his humble wants by instructing private pupils, troubled by no wish for premature fame, year after year, with stern perseverance, he concentrated all his powers on his lofty undertaking. So deep was his withdrawal, that, for many years, he did not even read the newspapers. From the retirement of his study he sent out volume after volume, to gain few disciples, many assailants, and general neglect.

Comte says he received from a very tender mother certain interior chords eminently feminine in their character. He also speaks of writing, all in tears, some passages of his positive philosophy. It is clear, in the evidence of his whole life, that his capacity for loving was as much greater than that of common men as his intellect was more capacious. He was accustomed even to mystic and rapturous expansions. His early teachers and co-laborers had broken with him, he had lost public employment, he was forced to earn a precarious livelihood by private teaching and lecturing. He felt himself poor, solitary, and injured, in the great, brilliant, careless city; he who believed that his thought was the most advanced and inclusive any man had ever known, and that to his work coming ages would be the most deeply indebted. Regarding himself as a thinker for mankind, whose service was of incomparable importance to the world, when some of his admirers sent an annual contribution for his support he haughtily accepted it as his right, and angrily resented as an inexcusable wrong the subsequent withholding of it. When he had acquired celebrity, his baffled rivals and enemies, together with the theologians and metaphysicians whose views and interests his doctrines so scornfully swept aside, kept a pitiless storm of obloquy blowing around him. Yet through all these provoking conditions of bitterness and despair he held to his task with unswerving consecration, with indomitable energy, his mind calm at bottom, his heart sweet at the core. For, deep below his disappointments and wrongs lay certain authoritative assurances, far within his exasperated personal relations lived certain disinterested affections, which gave him inexhaustible support and comfort.

His angry feelings towards individuals were soothed, their misanthropic tendency neutralized, by his philanthropic theory of the whole, by the heroic purity and self-sacrifice of his life, and by his poetic and devotional meditations. He had, perhaps, the completest and vividest idea of Humanity, as a personified unit comprising all human beings from the beginning to the end, that any one has ever entertained. His religiousness of feeling towards this grand ideal existence, his effusive communion with it, worship of it, readiness to toil and suffer for it, were unique. They healed his soreness, fed his aspirations and strength, gave him rare joy, and have made a fresh contribution of very great and enduring value to the future development of the higher human feelings. When he thought of this "Supreme Being," when he worshippingly communed with the immortal thinkers of the past forever incorporated with it in the grateful reverence of mankind, he forgot his foes and his irritations, had a fruition of his own greatness, and was wrapt in wonder and love. The reflection, too, over his fellow-men, of his own character, marked by such self-denial and toil for truth and humanity, tended to ennoble and aggrandize them in his eyes. He indeed says in one place, "Many men remain in a parasitic state, swarms of creatures which are in truth burdens on the Great Being, reminding us of the energetic reprobation bestowed on them by Ariosto as 'born upon the earth merely to manure it.'" But his constant motto, the phrase in which he concentrated his entire system of morals, was, "Live for others." He said, "The greatest pleasures are the pleasures of devotedness to others." He repeatedly quoted with unction the admirable sentiment of Metastasio: "He deserved not to be born who thinks he was born for himself alone." He everywhere expresses boundless indignation and contempt for those who deny all disinterested sentiments to human nature, and himself enthusiastically enforces those sentiments. The last pupil he had testifies that his nature "was full of smothered kindliness," and that he was reminded in the sight of him of one of "those pictures of the Middle Ages representing St. Francis wedded to poverty." He offered three daily prayers; read every day in the sublime poem of Thomas à Kempis, the Imitation of Christ; was extremely fond of Dante; and never failed to devote his regular seasons to the study of the highest strains of poetic emotion, and to religious meditation on the Great Being and its worthiest representatives.

It is easy to sneer at the extraordinary egotism of Comte, more helpful to appreciate his rare powers and

services.  It is easy to throw ridicule on that remarkable
passage of affection, his sacred love for Madame Clotilde
de Vaux, a chaste passion which developed new faculties
in him and raised him from the style of a Priestley to
that of a Petrarch ; but ridicule is unseemly with reference to an experience so blameless in its conduct and so
profoundly instructive in the surprising ardor and tenacity of the purposes it inspired.  To stigmatize his philosophy as a shallow materialism, his religion as a puerile
and atheistic sentimentality, — as if that were all that justice required to be said of him, — is a cheap invective,
dishonoring less its object than its employer.

His name will live forever on the list of the illustrious
few who have imparted an original impulse to the intellectual progress of the race, the evolution of the science
and faith of humanity.  There is something imposing,
regal, in his self-sustained power of resolution, labor and
trust.  Differenced from the community by despotic idio
syncrasies, an individual creed, a separate mode of life,
and peculiar aims and sympathies, he walked the streets
of Paris unnoticed or scorned, emerged from his chamber
alone, re-entered it alone, sat by his midnight lamp alone,
without the slightest faltering of his aim, his faith and pride
supporting him like a rock amidst the ineffectual lappings
of the battalions of cold waves of indifference and hate.
Sweeping over the ages with his generalizing eye, gathering up their significance, lawgiver of history and science,
he felt, " I am the autocrat of mankind, the first intellectual potentate of the earth, the supreme pontiff of the
church of humanity : all future generations will bring their
homage to my grave ! "  Notwithstanding his defects, extravagances, and aberrations, he deserves, and will ever
hold, an honored place among the leading minds of the
world.  Prominent among the valuable lessons exemplified by his career is the illustration it gives of the antidotes to the germs of wretchedness and misanthropy
existing in a solitary lot.

## JESUS.

JESUS has probably contributed, more than any othei person who ever lived, to aggrandize the idea of man in the mind of the human race. No other has exerted so great an influence for the deepening of the spiritual life of the world and the production of the moral virtues. This he has done through the contagious working of his character, the commanding authority of his instructions, the persuasive beauty of his example, and the organiza- tion of the Church, a diffusive society whose explicit aim it is by all sorts of worthy motives and sanctions to cultivate goodness. So much the freest sceptic, who looks out over the picture of history with unprejudiced eye, will confess. But beyond this, in regard to the exact details of what Jesus actually was and did and said, there are insurmountable difficulties in the way of sure knowl- edge. In the narratives which furnish the only direct information we have about him, there are chasms, in- consistencies, incredibilities. The Christ of the Fourth Gospel appears not like a real being, but an imperson- ated theory, half-humanized and supplied with accordant speeches ; the incarnation of a philosophic and religious idea existing in the metaphysical speculations of that age, but moulded and colored by the peculiarities of the He- brew mind as well as by the genuine influences of the historic Jesus. His most extraordinary character and teachings must have transcended the comprehension of his companions even more than is shown by the avowed examples of it given in the New Testament. The idol- atrous affection and awe of later times wrought with a creative impulse upon everything pertaining to him, sur- rounding him with a halo of miraculous attributes and legends, through whose dazzling obscurity it is difficult to see his actual features. Furthermore, the absence of the spirit of scientific criticism in his contemporaries and biographers, the fragmentary meagreness of the records, the disguising perversions of nationalities, languages, and ages, through which his history has since passed, — all

agree to complicate the problem and make it virtually
impossible for us with entire accuracy to discriminate fact
from myth, truth from error, the meanings of the Teacher
from the interpretations of the reporters, and thus recover his authentic portrait.

One thing at least is certain.  The unspent regenerat-
ing force exerted on the world ever since, the redemptive
revolution working among men and distinctly traceable
to the time and place of the birth of Christianity, demonstrate that there then lived a man of unprecedented
originality and power of soul, divinely inspired in an
unprecedented degree.  The veritable Jesus of Nazareth,
whose blessed feet trod the fields of Galilee two thousand years ago, is the historic nucleus about which has
gradually gathered that supernatural nimbus which now
dazzles the imaginations of most of his followers with a
bewildered belief of his literal Godhead.  He whom Paul
called "the man Christ Jesus," is the highest historic
teacher, guide, and exemplar of our race ; not unlike
others in kind, however superior in degree.  Looked at
in this way he is no longer absolutely unique, but belongs
to a class, is the culminating flower of a type.  The great
prophets and founders of religions in other periods are
not to be contrasted with him as sheer impostors, casting
double mystery on him ; but their traits and doings, the
psychological phenomena they reveal, are to be studied
as helping us the better to understand him.  Abraham,
Moses, Isaiah, Zoroaster, Buddha, Socrates, Mohammed,
and scores more of the holiest and grandest spirits of
our race, have communed with God at first hand, been
inspired by Him, felt themselves intrusted with special
messages and a general mission.  These too have collected disciples, transmitted themselves, and, in their
various modes, formed churches, theologies, rituals.  To
such as these a superhuman birth, supernatural endowments and feats, have commonly been attributed after
they had gone and left their inexhaustible influence at
work, their immense echoes rolling behind them.  Most
students of the history of Jesus have singularly neglected
to avail themselves of this help, —a competent investiga-

tion of the characters and careers of such men as Samuel,
Elijah, David, Pythagoras, Apollonius, Francis of Assissi,
Bernard of Clairvaux, — the magnetic natures of the
world, the fascinating personalities of history, the mystic
souls of biography, the imperial wonder-workers of time.
The more thoroughly we enter into the experiences of
this stamp of men, and into the mythologizing action,
with reference to them, of the minds of subsequent and
inferior men, the better able we shall be to understand
Jesus and the vulgar theory of him ; though, after all this
aid, his overtopping authority, the overawing mystery of
his genius, may still baffle our measure and compel us to
say with the Centurion, " Truly this was the Son of God."

In distinction from the historic Jesus, there is the theological Christ, who is a theoretical personage, a speculative abstraction, a spectral dogma, a creation of scholas-
tic controversies. In distinction from both of these,
there is the practical Saviour of the heart, the working
Christ of the Church, the Master really revered and loved
by the world of true disciples. This Christ is partly historic and partly ideal, but wholly divine. He is the movable index of the conscience of mankind ; the reflex of
the world's sense of its duty ; the picture of perfection,
freshly shaded and tinted in every age, borne by the
marching human race. This transfusion of the ideal with
the historic in the image of its Saviour, is a necessity in
a life of humanity which is not a fixture but a process.
Every quality of beauty and of good developed in the
evolution of advancing history and man, with its new re-
finements, complexities, and expansions, is seen reflected
in this authoritative ideal, in order that it may be taken
up by the assimilating forces of reverence, obedience,
and love. Seen in this light the mythical and ideal elements in the popular Christ are not coincident with false-
hood or illusion, but are an inevitable factor in that historic process of revelation, or that revealed process of
history, through which God educates mankind, — are a
divine arrangement for leading men to redemption.

It is the delirium of historical scepticism to deny that
there was an authentic man who served as the centre for

this construction and glorification which has grown into the moral and religious head of the civilization of the Christian world.   Greek  philosophy,  Hebrew  tradition and hope, and Roman domination, furnished the ripe materials  and  conditions,  and Gibbon says, " Christianity was in the air."   Yes, and it would have remained in the air had not a crystallizing personality appeared to collect and draw it thence.   The elements of Christianity were held in solution in the world ; the character of Christ, moving through them, precipitated Christendom.

But historic actuality as Jesus was, no one without strong wilfulness or credulity can accept the present portrait, painted in the imagination of Christendom, as an exact transcript of the primitive original.   Each critical inquirer, who, unwilling to remain in confessed ignorance, or to accept with blind faith what is told, desires to get at the facts, must do his best to extricate the real image from the mingled darkness and radiance of history, myth, legend, and speculation enveloping it.   The Christian Consciousness, the collective sense of Christendom, is competent to determine what is congruous, what incongruous, with the true idea of Christ ; to cut off superfluities and supply defects in the transmitted form.   The purest and highest souls, who know the most of biography, history, and science, the most of the mysteries of human nature, who have been the most perfectly trained in the personal experience of the spiritual life, and who therefore have an ineffably quick tact to detect moral consonances, discrepancies, and requirements, are the authoritative representatives of this totality of Christian perception and feeling.   But the difference between fact and truth, history and spirit, the typical idea and the concrete reality, must always be borne in mind.

The present sketch comparatively limits itself to those aspects of the character of Jesus which have relation to solitude, loneliness, grief, and the various temptations incident to these experiences.   It does not attempt to present a full portraiture of him and his career.   He was a soul so pure as to be an organ of the Spirit of the Whole, that is, an inspired representative of God.   He was a gen-

ius so fine and strong as to master by spontaneous intuition moral and religious principles and sentiments which
the wisest philosophers and poets, aided by the richest
training of the schools, have apprehended only after a
lifetime of toil and aspiration. His organism was so in-
teriorly soft and deep, that fulfilling emotions of peace
and bliss, such as the rarest mystics in their highest moments have known, were his effortless acquisition. The
greatest and most original thoughts, the most direct perceptions of fundamental truths, the most beautiful and
persuasive images, the most entrancing expansions of
feeling, came to him so like instincts unawares, that he
could not claim them as his own, but only attribute them
to God, the Infinite Father, with whom the sweet simplicity of his self-renounced heart felt itself in unison
through all the loveliness and mystery of His works and
ways. No poet or moralist ever created fresher or more
charming apologues than he, or spoke in a richer dialect
of audacious insight and beauty than he in his speeches
of the lilies, of the birds, of the sun rising on the evil
and on the good, of the rain falling on the just and on
the unjust. No reformer ever scouted the hoary traditions of ages and reversed the rooted prescriptions of his
time with more fearless superiority than he. His receptive
and responsive capacity of genius brought him into unparalleled intimacy of fellowship with humanity, nature,
and God, made him independent of the teachings of
others, gave him a supreme authority, ingravidated his
utterance as with the weight of worlds. " Heaven and
earth may pass away, but my words shall not pass away."
It is the voice of a God.

Many earnest students of the character of Jesus are
perplexed, confounded, by what seems to be the astounding arrogance of his personal claims, the contradiction
between the sublime sincerity of his precepts and practice
of self-sacrifice on the one hand, and the unapproachable
egotism of many of his declarations on the other. Three
considerations go far to remove this difficulty. First,
there is great reason to believe that much of this self
assertory language was either not used by him at all, but

reflected back from the ideas subsequently entertained of him, or was employed by him in an official sense referring to his Messianic rank and functions, not in a personal sense. Secondly, in several instances it is clear that the ostentatious assumption is only apparently such. For example, when he says, "I am meek and lowly of heart," there is nothing like vanity or boasting; it is the simple truth, expressed by an innocence so naïve, an unsophisticatedness and sincerity so august, as to be wholly unconscious of self. He had no thought of awakening admiration, but aimed, through pure truth of example and word, to bless others by winning them also to meekness and lowliness. It is self-regardful vanity that in such a case would hesitate to speak the truth from fear of the effect, and be immodest in appearing modest. Thirdly, when Jesus made use of such expressions as, "He that eateth my body and drinketh my blood hath eternal life," he means not himself, but the divine quality shown in him, a gift of God. He likewise says, "He that eateth my flesh and drinketh my blood, dwelleth in me and I in him," that is, between him and the disciple animated by the same indestructible principles there is a community of spirit. Also, still more clearly, "The words I speak, they are spirit and they are life." The doctrine he taught, the faith he held, the spirit he was of, — it is this, and not his own personality, that he demands such astounding deference to. "I and my Father are one": here he does not sink God in himself, — boundless egotism, — but identifies himself with God, — boundless renunciation, — feels that God inspires him, lives and speaks in him, and does the works. He so surrenders and blends himself with the truth as to represent it, and say, "I am the truth." He found himself in possession of great moral and spiritual truths, — truths far in advance of the time. He did not know how they came, but felt that he did not himself achieve them. He supposed that God had given them to him and laid on him the mission of proclaiming them. He identified the revealing spirit with God; and justly so. It is God alone who can give to the finite and perishable individual the perception of the universal and

eternal, so that, an inspired prophet, he shall say, " I will
utter things which have been kept secret from the foundation of the world." It is not any personal ego, but the
voice of divine reality, that speaks then. "No man
cometh unto the Father except by me." That is to say,
No man can live in that communion with the Father in
which 1 live, except by means of the same faith in the
Father which I have, except by means of that idea of the
Father which I have declared. Correctly understood,
there is no egotism in these declarations ; they are natural and dignified expressions of the facts of the case :
they are the style proper to the seer.

Whatever cannot be explained by these considerations
is to be rejected as spurious ; because the evidence is
irresistible that Jesus was the most self-abnegated, sacri-
ficing, lowly, and loving, of the sons of men. The central
germ of his divine originality consisted in this very thing,
his utter superiority to all the hollow ambitions, pomps
and prides of the world, his unrivalled sympathy with the
poor, the sinful, the outcast, the lost. He came not to be
ministered unto, but to minister. Leaving to others the
uppermost rooms at feasts, and the chief seats in synagogues, he devoted himself to the cure of vice and the
relief of pain, and by the assiduous practice of his own
aphorism, "Let him that would be first among you be
your servant," inverted the scale of Pagan virtue, and instituted a new order of greatness on the earth. It is
because he has lived that we are now able to say, looking
down on the wretched with pity and up to the ransomed
with desire, The lower a man carries his love the loftier
he lifts his life. In every objectionable sense of the word
egotism, Jesus was one of the least egotistic souls that
ever appeared among men. Every word inconsistent with
this interpretation of his character is falsely ascribed to
him. The mighty *Ipse dixit* of Pythagoras reflects not
the personal assumption of the great Crotonian ; it reflects the impression made on his disciples by his inscrutable personality and genius. The more inscrutable soul
of Jesus would naturally work a deeper effect and secure
stronger expressions. This explanation is to be empha-

sized with the fact that in that period exceptionally impos·
ing and gifted men were often regarded as deities.  Peter
said to those who would have worshipped him as a god,
"See ye do it not ; I also am a man."  The priests and
people at Lystra would have sacrificed garlanded oxen to
Paul and Barnabas, believing them to be Hermes and
Zeus.  It is therefore clearly unnecessary to think that
Jesus is God because he has been believed to be God.
The mystery of the soul of Jesus, the strange authority
of his knowledge, the marvellous effects he wrought, are
to be ascribed to a special heightening of the ordinary
intercourse of God with human nature.  To undertake
to explain them by the notion of something superhuman
and preternatural, a unique incarnation of the Godhead,
is to leave the region of reason and law and enter the
region of fancy and chimera.

He who taught men that the path of the moral commandments was the path of salvation, that a life of philanthropic works was a title to redemption in the judgment,
who took an innocent child for the best image of heaven,
who sought no kingdom but truth, no honor but love, who
said "Of myself I can do nothing," and who in the farewell hour instituted a feast of love to keep his name in
remembrance, — could never have dreamed of the mediæval doctrine of the atonement, could never have expected to be deified, nor have wished to be personally
worshipped.  And all worship that, resting in him, stops
short of the Absolute Highest, simply makes him the
purest and sublimest of fetishes.  He is then the head
of that series of idols which sinks past the picture of the
Italian bandit, and the leaden image of the Portuguese
sailor, to the toad, tree, stick, and stone of the savage.

To merge the divine humanity of Jesus in a factitious
theory of his Deity is to lose more than can be gained.
For we can get no good from him except as we drink his
spirit.  He can benefit us only by influencing us to become
like himself.  The only redemptive relation to him is a
spiritual not an official one, an adoption of the quality of
his character, not any ceremonial attitude towards his
name or person.  The essential thing is not a formal

belief in the saving efficacy of his blood, but a willing·
ness, imbibed from him, to shed our own for the good of
mankind. We read that a diseased woman once pressed
through the crowd and touched the hem of his garment,
and by the power of her faith was immediately healed.
Is not still the loyal disciple, who gets near enough to
touch him in spirit and draw forth the inspiring virtue he
delivers, made spiritually whole? But to neglect the text,
" Be ye perfect as your Father in heaven is perfect," for the
text, " I will raise him up at the last day," to value the
groaning and the unloosed napkin by the grave of Lazarus more highly than the conversation by the well of
Jacob and the parable of the Good Samaritan, is to
vulgarize the wisdom of Jesus into clairvoyance, and
materialize his spiritual divinity into a physical thauma-
turgy. It is doubtless easier for the Confraternity of the
Sacred Heart of Jesus to worship the symbol of the visible
organ than vitally to appropriate the moral substance.
But the genuine heart of Jesus is to be seen in his sayings, Suffer little children to come unto me ; Go, and sin
no more ; It is my meat to do the will of Him that sent
me ; Love one another as I have loved you ; and is fitly
worshipped only by a personal assimilation of these
sentiments.

The whole mass of declarations and imagery, accordingly, in which Jesus is represented as virtually asserting
that no one can be saved without a direct and professed
relation to him in his Messiahship, arrogating to himself
personally a forensic position of inconceivable power and
grandeur, — this language, if regarded as authentic, and
taken in its literal sense, would force us to believe that he
labored under a gross delusion. It is impossible for any
mind fit to grapple with such a subject, to credit, as the
true account of the plan of God for the future history of
the earth and man, the mechanical hypothesis, the melo·
dramatic mythology, that on a fixed day a trumpet is to
sound, clouds of angels to fly down and reap the harvest
of the burning world, Jesus himself to appear in omnipotent array and to cause a resurrection of the dead from
their graves, and then sit in person in the awful assize,

and apportion their doom to the good and the bad. This is no tone from the infinite harmony of truth. It is a jarring figment of fancy. We cannot believe that he whose mind, with its matchless intuitive scope and penetration, was so soundly poised, ever taught any such thing. The genius of the religion he founded, the prominent congruities of his character, require us to think that such assertions are exaggerations thrown back by later theories, or misreports fastening on his expressions an exterior meaning foreign to his intention. His other thoughts are irreconcilable with this monstrous forensic and theatrical personal prominence. "The word that I have spoken, the same shall judge him at the last day ; for I have not spoken of myself." "To sit on my right hand and on my left, is not mine to give." "Why callest thou me good ? There is none good but One, that is God." "All the law and the prophets hang on these two commandments, Thou shalt love God with all thy heart, Thou shalt love thy neighbor as thyself." "Except ye be converted and become as little children ye cannot enter the kingdom of heaven." "If I honor myself, my honor is nothing." "Ye shall know the truth, and the truth shall make you free." "God is Spirit, and they that worship him must worship him in spirit and in truth." The man who was the author of these thoughts could not have believed himself destined to ride down space with a cherubic escort, as the conquering hero of the Universe, and to set up his judgment chair on an expiring world amidst the rising millions of the dead. Never. It is historically traceable as mixed Persian and Jewish fancy, and its authorship has been only erroneously attributed to Jesus.

The whole dominant style of character exemplified by Jesus, as summed up in his chief maxims, such as, — Resist not evil, Love your enemies, My kingdom is not of this world, Go forth as lambs among wolves, Humility and service are the true exaltation, — was startlingly new and strange in his age. This singularly original personality, so close to God, so harmonized to truth, so full of love, could not but set him far apart in spirit, and make him a baffling enigma to his contemporaries. Even those simple hearts who yielded to his charm, and followed him with loving reverence, could not pierce the mystery that surrounded him, but constantly "marvelled what manner of man he was." Whether we think of him as pacing the highways, in the still village synagogue, tossing in the midnight tempest on the lake, riding through hosannas, over strewn garments and palms, besought by the wondering multitude to become their King, seeking the sheltered shades of Gethsemane or the starry top of Olivet, — always he seems to us transcendently alone, wrapt in the solitude of his own magical originality.

In addition to the distinguishing effect of the wonderful impression he made on persons, and of the wonderful works of healing and renewal he wrought, — the matchless penetration of his genius, as shown in his parables and his beatitudes and his answers to puzzling questions, must have removed him from other men by making it impossible for them at once to comprehend his teachings. Repeatedly we read, "They understood not what things they were which he spake unto them." And once he said, as if sorrowfully forced back upon himself in a chill insulation, "I have many things to say unto you, but you cannot hear them yet," and then he invoked the Spirit of Truth to teach them afterwards what was at that time unintelligible to them. Not one man out of a million at this day can fathom by any direct perception the full meaning of his utterances, "Blessed are the poor in spirit, for theirs is the kingdom of heaven"; "Blessed are the pure in heart, for they shall see God." How few also are competent to appreciate his intense moral idealism, as shown both in his ethics and in his doctrine of prayer; a practical idealism far superior to the speculative idealism of Berkeley or Fichte. His interior realization was so entrancing as to make the inner consciousness all, the outer facts nothing. He removed his tribunal from the outer court of words and acts, where the rabbinical priests held theirs, and established it in the inner sanctuary of thoughts and affections. His appalling sentences are, "Whoever looketh on a woman to lust after her, hath already committed adultery with her in his heart"; "He that hateth his

brother is a murderer." It is true the latter text is not
from his lips ; but it belongs to his favorite disciple, and is
conceived perfectly in his spirit. If a man ask anything
of God, without doubting, he shall receive whatever he
asks. If, with absolute belief, he say to a mountain, Be
thou removed into yonder sea I it shall be done. This is
true in the ideal sphere, not in the material. A supreme
faith is omnipotent in its own realm, the world of con
sciousness. Whatever a man asks or orders, in entire
faith, with no opposing doubt, is subjectively granted.
It becomes real in his inner life, though not to the eyes
of others. If the unqualified language of Jesus be correctly assigned to his lips, it is explicable only in this
way ; and it ranks him in the same order of mind with the
supreme masters of thought who have held the universe
in solution in an idea. His expansiveness of intellectual
sensibility seems competent to any greatness. When
blamed for busying himself on the Sabbath, he said, God
ceases not his beneficence on this day : why should I ?
"My Father worketh hitherto, and I work." What an
unimaginable height of life such a level of thought implies !
Is it not indeed the speech of a Son of God? It was a
grand achievement to go into the depth of the sky, and
read the law of creation, — the attraction of matter according to the inverse square of the distance. It was a
grander achievement to go into the depth of the soul, and
read the law of salvation, — the free and conscious renunciation of self. "Whoever will save his life shall lose
it ; and whoever will lose his life for my sake shall find
it." In other words, Whoever, by an ignoble compromise,
would escape any hardship, shall incur a greater evil than
he avoids : Whoever, for a sacred cause, would sacrifice
any personal interest, shall take up a greater good than
he lays down. It is unquestionably the wisdom of the
inspiration of God.

Jesus was made lonely, furthermore, by the peculiarities
of his mission. God had laid on him a special work of
infinite importance ; and it absorbed him, he felt strait-
ened until it was accomplished. His divine call, and his
perfect devotedness to it, set him beyond the pale of the fel-

lowshiping sympathies of the crowd, remote from their interests and passions. It invested him with a sphere of strangeness which produced curiosity in some, hatred in others, awe in most, and a feeling of unlikeness and distance in nearly all. In fulfilling the Messianic office he was called to be a Messiah surprisingly different from the one his countrymen were expecting with such eager desire. He was not anointed to gratify their revengeful pride by overthrowing their enemies, and putting them at the head of the world in the administration of a visible theocracy; but to teach them humility, and love, and faith, and silently inaugurate an unseen kingdom of heaven by preaching the gospel of the enlightenment of the poor, the deliverance of the captive, the comforting of sorrow, the healing of disease, the removal of the sins and miseries of the world. The lowly circumstances of his origin, the contradiction of the purely spiritual functions he exercised to the pompous material functions of the anticipated Messiah, the fatal opposition of his living teachings to the system of dead traditions and rites in vogue, inevitably engendered in the established teachers a deadly feud against him. The persecuting hatred of all those classes who monopolized the offices of honor and power in the nation must have deepened his feeling of loneliness, and emphasized it with a dark sense of danger and suffering. This steep alienation, this irreconcilable antagonism, was steadily aggravated, on the part of the Scribes and Pharisees, as Jesus became more known and influential, and as the revolutionary character of his vital morality and religion grew more clearly pronounced; and on the side of Jesus, as he saw more fully the rank hypocrisy and tyranny of the Scribes and Pharisees, and the cruel burdensomeness and barrenness of their ceremonial system. He had come to set men free with the freedom of the living truth, to cleanse the augean bosom of the world by turning through it a river of pure enthusiasm. They were the opponents of his work. They had substituted in place of a renewing personal faith and love an oppressive and corrupting mass of formalities. They hid the key of knowledge, neither going in nor letting others in. He saw that before he could accomplish his mission of establishing the genuine religion of the love of God and man, the authority of these selfish and wicked fanatics must be destroyed. The battle made his lot that of an outcast. But he shrank not. Fired with holy indignation at the sight of the impious wrong and injury they were doing by their monstrous inversion of the moral law in their characters and of the religious law in their traditions, he flamed against them with the angelic wrath of the Lamb. He exposed them as sophists, blind guides, hypocrites, who would strain out a gnat and swallow a camel, who blew a trumpet before giving alms, made long prayers of ostentation, and would not stretch forth a finger to relieve the distresses of humanity. In return, with the malignity and terror of cowardice, they sought his life.

The spiritual solitude of Jesus, resulting from his transcendent personality, his inspired originality of genius, and the absorbing speciality of his mission in an alien world, acquired a culminating intensity from the series of cruelties and indignities he endured. He knew all the bleakness and hardship of a despised lot of poverty, and toil, and homeless wandering. Many a time, footsore and weary, he paused to refresh himself with a crust, and a draught from the wayside well. Many a time the stars looked between the branches of the olive-trees into his eyes, and the night-damps fell on his head by the shores of Gennesaret. And everything demeaning or odious that could be connected with his history was caught up by his envious neighbors or his public foes and flung against him in sneers and taunts. "Is not this the carpenter's son?" "Can any good come out of Nazareth?" His own kindred, unable to appreciate a soul so much above their own, turned against him, saying, "He is beside himself." His words were perverted, his actions misrepresented, his aims misinterpreted. They stigmatized him as a gluttonous man and a wine-bibber, because he was no dark ascetic. They accused him of a degraded preference for the society of publicans and sinners, because he divinely stooped in love to soothe the unhappy and save the lost. They called him a blasphemer, because he uttered the words breathed into his soul by the Spirit of God. They accused him of immorality and irreligion, because he had an incomparably finer perception of moral principles and deeper truth of devotion than they.

The worst sting in the injustice which the highest benefactors of the world have always suffered, is to have those immeasurably beneath and behind them assume to look down upon them and back upon them, with anger, hate, and scorn, and to see the rewards which ought to be theirs bestowed on persons utterly unworthy of them. The purest lovers of men and worshippers of God, the spiritual heroes who reject current dogmas and conventional feelings from allegiance to higher and better ones, are regarded as traitors to truth and violators of piety, on account of the very superiority of their virtue. The most royal souls of the race, who so truly love and honor their fellow-men as to sacrifice everything selfish for their good, who achieve wider ranges of knowledge and peace, giving men light in place of darkness, love in place of hate, trust in place of fear, — who win stores of bread of life for generations to come, — are either feared as dangerous innovators and persecuted as wicked heretics, or neglected to die of want and heart-break, while merely titular kings, without one attribute of merit beyond the place they accidentally occupy, selfish voluptuaries and tyrants, are boundlessly honored and pampered by the people they mislead, prey upon, and despise. This is the tragedy of history; and Jesus felt it in its darkest extremity. What imagination can reproduce his feelings when he saw the people choose Barabbas rather than him, ranking that brutal wretch above him, and heard the hoarse yell breaking on his ears : " Crucify him! crucify him !" Next he proved the lonely agony of treachery and desertion. One of his immediate disciples betrayed him for a price, and the rest fell away in the gathering gloom. He was left alone with his enemies and the blind fury of the mob. Then he sounded to its very bottom the deepest depth of loneliness and woe, — the tortures of mockery. All the

billows of injustice and ingratitude had gone over his soul.
And now the pitiless probe of sarcasm was to be applied.
Ah, how little it was dreamed, as the governor led him out,
bleeding from the degradation of the scourge, and said
to the multitude, "Behold the man!"—how little it was
dreamed that the voice of that silent sufferer would thrill
the world forever, his face melt the heart of all posterity!
They platted a crown of thorns and put it on his head,
and they put a purple robe on him, and a reed for a
sceptre in his hand, and they tauntingly bowed the knee
before him, and mocked him, saying, "Hail, thou king
of the Jews!"   It was the cruelest irony ever known on
earth, because the disparity was the vastest between what
he deserved and what he received.   His merit was Godlike, his treatment fiendish.   Him, who never spurned the
lowliest thing that wore the shape of man, but unweariedly
went about doing good, they nailed upon the cross.   Him,
whose kingdom was the truth, whose royal function was
succoring the needy, they charged with traitorous usurpa-
tion, and put to death.   Was there ever so tremendous a
jibe as the descent from his idea of a moral throne of
beneficence and love subduing all souls in universal goodness, to their estimate of him as desiring to wear the Hebrew crown and be joined with the vulgar despots of
history?   And he had to endure this.

Unmistakable indications of his sufferings from loneliness, neglect, and abuse, are scattered through the narratives of his life.   It could not be otherwise.   He must
have been as extraordinarily susceptible of pain from lack
of sympathy, from injustice and unkindness, as his interior softness, richness, and fire were extraordinary.   His
want of the usual domestic ties, his frequent withdrawals
from the crowds who gathered to listen to him, his constant habit of wandering by himself for meditation and
prayer in the grove, by the lake, and on the mountain,
the account of his solitary temptation in the desert,—
throw light on this sad and interesting phase of his character.   What a revelation of his yearning for affection is
made in his words to Simon, "Thou gavest me no kiss
when I came in," and in his deep satisfaction from the

love of the sinful woman who washed his feet with her
tears and wiped them with her hair! The picture of him
with John, the beloved disciple, leaning on his bosom at
the feast, will never fade before the eyes of the world.
We catch glimpses of his hunger for sympathy, of the
sorrows of his wronged affectionateness, in many of his
utterances. "Simon, son of Jonas, lovest thou me?"
"Could ye not watch with me one hour?" "Will ye also
go away from me?" "They hated me without a cause."

This last expression, with other kindred ones, enables
us to trace something of his reactions towards those who
repulsed his words and his person. There is a blending
of a grieved feeling of personal injury and an indignant
feeling of public injury in several of his speeches concerning those who rejected his mission and persecuted him
because he aimed to supersede their traditions and ceremonies by a living religion. He saw at once the malignant style of character out of which their antipathy sprang,
and the pernicious corruption which subordinated righteousness, mercy, and faith to tythings of mint, anise, and
cummin, — and he unsparingly condemned them in the
name of God. He did not refrain from invective and irony:
and in this some personal feeling always mingles. Satire
is curdled poetry. Satire is the very recoil of stung sensibility. Is not something of this perceptible in such texts
as the following? "It cannot be that a prophet perish
out of Jerusalem." "Many good works have I showed
you from my Father ; for which of these works do ye stone
me?" "I am come in my Father's name, and ye reject
me ; if another come in his own name, ye will receive
him." "No prophet is accepted in his own country."
"When the blind lead the blind, both fall into the ditch."
But how divinely the disinterested feeling rose over every
merely personal feeling is sublimely shown by his bearing
under the greatest moral outrage he ever suffered. They
said, "He casteth out devils by Beelzebub, the Prince of
Devils." Then Jesus said, after a silencing dialectic refu-
tation of their statement, "Whosoever speaketh against
the Son of Man, it shall be forgiven him ; but whosoever
speaketh against the Holy Spirit, it shall not be forgiven

17 *

him." There is one text, however, fit to be the motto of all truly aristocratic souls who writhe back from the stinging wrong and scorn of a misappreciating world, — a proverb, which, if he be really the creator, or even the quoter of it, more than any other utterance, betrays at least a temporary soreness in his mind. "Cast not your pearls before swine, lest they trample them under their feet, and turn again and rend you." He who breathed such divinity of tenderness, such inexhaustible magnanimity of forbearing pity and love towards all men, who would give the pearl of great price purchased with his own blood to the lowest child of humanity, who in the agony of death yearned over the broken malefactor by his side with the promise of Paradise, — what pain, what unutterable revulsions of feeling, he must have undergone before he could have said *that!*

But if Jesus had sharp temptations to misanthropic pride and despair, his helps for neutralizing them, and overcoming the world, were also great. They proved sufficient to make him the most glorious victor among all who have ever fought the bitter battle of life; inspiring model, great Captain of Salvation, to all subsequent fighters of the good fight. He kept himself constantly employed in fulfilling his mission, relieving the mental distresses and bodily infirmities he encountered, sowing broadcast the seeds of his kingdom. And no sweet spirit thus busied in disinterested works of philanthropy and religion ever curdles. When overtried by the multitude, he found solacing restoration in the beautiful retreats of nature, in communion with God, by the brook Cedron, in the vale of Siloam, and other dear haunts of his feet. "And in the morning, rising up a great while before day, he went out, and departed into a solitary place, and there prayed." He likewise knew the sweets of friendship: for, besides those "who followed after him for the loaves and fishes," there were some who devotedly loved him for his own sake. There was a humble home in Bethany, where, with Lazarus and the two sisters, he who was homeless often delayed. If of the ten lepers whom he cleansed he was forced to ask, Where are the nine? one gratefully returned and clove to him. He had, too, the unfaltering approval and support of his own conscience, a clear, energetic assurance from the inspiring spirit of God within him. When all were scattered, and he was left alone, he could firmly say, "And yet I am not alone, for the Father is with me."

He was sustained and animated by two ideas, of the sublimest import and of unprecedented novelty in his time. First, the idea of one God who is an Omnipresent Spirit, the Universal Father, who is to be worshipped by loyal openness to truth, purity of heart, righteousness and beneficence ; who sees in secret and appropriately rewards every hidden act. Secondly, the idea of Humanity as one great unit or family of brothers covering the earth, to be saved and brought into co-operating affection and blessedness by one law of love. He was the Son of Man, the child of collective Humanity, as well as the Son of God, — the earliest in history to bear those conjoined titles. "Whosoever doeth the will of my Father in heaven, the same is my brother and my sister and my mother." The law of undefiled morality and religion, the universal will of God, is the fine consanguinity which constitutes the Brotherhood of Man. It is the ineffaceable glory of Jesus to be the first in history to affix the full significance of the name Father to the unity of the unknown Godhead, and to derive the legitimate consequences. Whether or not science shall ever supersede this conception with another, it was a step of progress, of immense historic and moral importance, which was inevitable, sooner or later ; and the name of Jesus is identified with it. The Greeks and Romans had spoken of father Zeus, omnipotent father Jove, parent of gods and men ; but it was a pale philosophic glimmer, an ineffectual poetic image. So the infrequent theoretic perception of the unity of Humanity played as a cold light in the head of antique philosophy, with no power to overcome the jealousies and hates of families, classes, tribes and nations. But Jesus, by his exemplification of the doctrine of the Fatherhood of God, gave that feeling of the Family, so common and so powerful in antiquity, that intense sentiment of one blood from one parentage, with its affiliating obligations, new life and expansive energy, and turned it through the world in a warm and voluminous flood of humanity. It was a practical discovery in morals as important as the invention in mechanics of reverse motion by the cross-band.

He comforted himself with the sympathetic idea and forefeeling of fame, honorable and affectionate remembrance according to his deserts. Instituting the Eucharist, he said, " This do in remembrance of me." He sent his disciples forth to convert the earth, saying, " Lo, I am with you to the end of the world." Looking forward through many nations and ages, he saw little companies gathered together in his name, and felt himself in the midst of them. Suffering the loneliness of a leader who is out of sight of his followers, the idea of invisible millions behind imaginatively brought the inspiring solace of their companionship already into his heart. How deep his grateful feeling, how true his bold prophecy, with regard to the woman who poured the alabaster box of spikenard on him ! " Verily I say unto you, wheresoever this gospel shall be preached throughout the whole world, this also that she hath done shall be spoken of for a memorial of her." Inspired word of love fulfilled this day !

Entire devotion to the accomplishment of his mission, strengthening communion with the peaceful solitudes of nature, inspired consciousness of the presence of God, unconquerable love of humanity and assurance of a benign dominion in the appreciating future, were the supports which, in connection with his own holy genius, enabled Jesus to rise victoriously above the severe trials that beset him, and leave, in unapproached pre-eminence, a blameless example of heroism, nobleness, and beauty. From his first clear perception and assumption of the providential part assigned him he knew not the distress and waste of internal conflict, but was in interior unity with himself. This steady oneness of will and conscience is the supreme condition of strength and peace. He whose first recorded words, a strayed boy in the temple, were, " Wist ye not that I must be about my Father's busi-

ness ? " could say when the shadow of the cross was falling athwart his steps, " I have finished the work Thou
gavest me to do." In all the lone passages and human-
ly unfriended hours of his life he nourished his soul with
the angels' food of the sense of duty performed, love
exercised, self sacrificed, divine favor vouchsafed. And
the most essential lesson of that Gospel which he is
rather than preaches, declares that whoever, of all the
faltering strugglers with the world, will use the same
helps in the same spirit, shall win a kindred victory.

O what a victory that was! The wrong he received
was the cruelest, the return he made the divinest, within the compass of history. With godlike benignity he
stooped to pour out on all forms and conditions of men a
pitying and redeeming love never equalled in purity and
measure before or since ; and they left him to wander, for
the most part neglected, friendless, shelterless, in sorrow
and pain. Imaginatively extending his individuality to
the limits of the race and the earth, he identified himself
with all the outcasts, prisoners, sick and destitute of all
ages, all unhappy victims bleeding under the miseries of
humanity, and invoked for them the same tender treatment that he thought he deserved himself; and they
hoisted him between two thieves in the place of infamy,
to die the most ignominious and torturing of deaths.
And when, with magnanimity unmatched in the annals of
humanity, he preferred, as a ground for their forgiveness,
their ignorance of the real nature of their own deeds, they
wagged their heads at him, and reviled him, and mocked
him. This was his last sight below, an upturned sea of revengeful and sarcastic visages. Then came a moment, —
moment of most awful loneliness ever felt by man, — when
with the ebbing strength of the body the spirit too shrunk
from the encroaching darkness, and he cried, " My God,
my God, why hast thou forsaken me ? " It seems as if
that cry might have pierced immensity, shaken the
farthest stars, and wrung a response from the inscrutable
lips of Fate. Instantly the eclipse passed, — eternal
light broke, — " Father, into thy hands I commit my
spirit," — and the earthly tragedy subsided into stillness

forever.    It was inevitable that the impression on the minds of those who afterwards learned to appreciate the infinite contradiction between the worth of the august sufferer and the doom he bore, between the spirit he showed and the treatment he took, should express itself in stories of preternatural portents, the veiling sun, the shuddering earth, opening graves, and rending temple.

There are doctrines connected with theoretical Christianity which may never command universal assent. There are speculative disputes on points relating to the person and biography of its founder which may never be satisfactorily settled.    But, practically considered, in the authoritative beauty of his character and example, which carried the high-water mark of human nature so far above all rival instances, no purer expression of the divine in humanity is to be expected.    And good men can cherish no worthier ambition than to make the whole world a Christopolis, whose central dome shall lift the lowly form of Jesus in solitary pre-eminence to draw all men unto the discipleship of his spirit, while, with ever-progressing intelligence and liberty, they co-operate in the mazy industries of the sciences and arts of human life below.

---

## SUMMARY OF THE SUBJECT.

THE foregoing pages have furnished abundant proof that persons of extraordinary sensibility are likely to experience the loneliness and unhappiness of human life in an extraordinary degree.    Probably no previous age was so rife as the present in interior discords, baffled longings, vast and vague sentiments whose indeterminateness is a generating source of misery.    Probably there were never before so many restless and weary aspirants, out of tune with their neighbors, dissatisfied with their lot, unsettled in their faith, morbidly sensitive, sad, and solitary.    To make a true estimate of what the trouble is with these victims of self-love and the social struggle, to give them

sound sanitary directions, explaining the causes of their
wounds, and the best curative treatment, we cannot but
think will be a service of especial timeliness.  To these
innumerable sufferers, writhing under distressful relations
with themselves and others, would it not be an invaluable
boon to be guided to a tranquil oblivion of their injuries
and resentments, their uneasy desires and woes, in remote
retreats of thought, in cool and sweet sanctuaries of sentiment, in undisturbed temples and glens of faith and love?
If the studies of the preceding chapters, and the personal
experience which first led to those studies, furnish any
qualification for this office, it may be in some degree
discharged by summarizing for the reader the practical
results of the whole investigation.  He to whom a hundred veiled wounds of his own have given an accurate
knowledge of the wounds of other people should know
how to impart therapeutical instructions, and also how
to soothe the unhappy souls about him with soft magnetic strokes of sympathy.  Blessed art, why do so few
practise thee?

There is inexhaustible help for the suffering man in an
adequate knowledge of the sufferings of others ; how they
originated, and to what issues they led ; the warnings of
those who were defeated by their trials and ignominiously
perished under them ; the examples of those who vanquished theirs and came out in victorious cheer.  Nothing
can be more stimulative and fruitful for the unambitious
recluse than sympathetic contact with the experience of
the noble spiritual heroes who have spotlessly worn their
crowns, throned on the summits of society.  On the other
hand, nothing can be more blessedly solacing and sed-
ative for the overwrought champion of the arena than
contemplation of the inner drama of those delicate and
listening minds, those deep and dreamy hearts, who pass
their days in an ideal sphere detached from the intoxicating prizes of outward life, far from the bewildering roar
of the world.  This is indeed the choicest value of literature, the deepest art of life, — to supplement the defects
of our own experience by appropriating from the experience of others what we stand in want of.

Beyond a question, the welfare of society and the hap
piness of the multitude have increased from the palmy
days of Egypt, or those of Sparta, to the time when the
serfs of Russia were emancipated, and when the telegraphic cable girdled the world.   Beyond a question, on
the other hand, when we turn from Cyrus to Napoleon,
from Pericles to Pitt, from Socrates to Schopenhauer, from
Pindar to Lamartine, we must see that the moral discontent of individuals, the difficulties in the way of inward
unity, the mental fatigue and soreness of superior persons,
have been increasing.   This is owing to the greater complexity of elements and stimuli in modern life ; also to
the greater development of conscience, alliances with
impersonal interests, obscure connections of dependence
and responsibility with huge masses of public good and
evil.   The greater the number of the interests a man
carries, and the greater the number of external relations
he sustains, the more delicate and arduous becomes the
problem of harmonizing them, fulfilling his duties, and
satisfying his desires.   The sympathetic ties of the individual were far less numerous and extended formerly than
at present.   Consciousness spreads over a wider surface
and along more lines ; every breast is a telegraphic office
throbbing with the vibrations of the communicating web
of civilization.   Christianity, the historic moral progress
of the race, has also introduced quicker and larger standards of right and wrong, developed an intenser sense of
divine authority and human brotherhood, and made men
feel themselves amenable to a much more diffused and
exacting spiritual tribunal than was known to the careless
children of the early world.   All this increases the difficulty of any chronic self-complacency ; and, as Aristotle
says, " happiness is the attribute of the self-complacent."
It is natural that as extension and complication remove
narrowness and simplicity from the life of the individual,
he should with diminishing frequency attain the happiness
of a contented unity with himself.   This must be especially true when a profound sense of the presence and
perfection of God, of the rebuking examples of the saints,
of the infinite nature of duty, gives him a constant feeling
of his own unworthiness, vanity, and transitoriness.

In antiquity the individual was sunk in the mass as a political tool. Now he has a keen feeling of a separate personality, freedom, and responsibility; yet, at the same time, and as a consequence of the same causes which have produced this, he has an acute feeling of his moral relations with the mass. The deep sense of God, humanity, duty, eternity, which adds so much to our dignity and joy when it is healthily co-ordinated with our nature, often makes us so much more susceptible to self-reproach, grief, and fear. In every age an earnest experience of religion has segregated men from the world; but Christianity did this in an unprecedented degree when it filled the deserts and valleys and mountains of Christendom with hermits. One great consequence of the modern enhancement of self-consciousness, and enhancement of the consciousness of the external relations of self, has been the feeling of individual loneliness in the crowd, a melancholy shrinking and sinking of the heart from the miscellaneous public, a sad or fond courting of solitude for the application there of ideal solaces to the soul. There is in the following lines by Sterling a tone of sentiment marking them so distinctly as a product of our modern Christian epoch that no one could suppose them written by any poet of antiquity.

> Lonely pilgrim through a sphere
>   Where thou only art alone,
> Still thou hast thyself to fear,
>   And canst hope for help from none.

Andrew Marvell, the friend of Milton, and quite his mate in soul, thus describes a noble character withdrawn into his garden and musing there; a character rich in mind and heart, and avid of a quiet retreat aside from the busy littlenesses of life:

> Nor he the hills, without the groves,
> Nor height, but with retirement, loves.
> Meanwhile the mind from pleasure less
> Withdraws into its happiness,
> Annihilating all that's made
> To a green thought in a green shade.

Z

Marvell, depicting the glory of Adam in Eden, thinks there was one drawback to his bliss, namely, that he had a comrade.

> But 't was beyond a mortal's share
> To wander solitary there :
> Two paradises are in one
> To live in paradise alone.

It is a touch of sentiment tinged with humorous satire impossible to any writer of India, Egypt, Persia, Greece, or Rome.   There is an ineffable charm for the modern heart in the picture of Paul and Virginia alone together on their island.   But the Philoctetes of Sophocles, the classic Crusoe, ten years alone on desolate Lemnos, listening to the lonely dash of the breakers, himself his only neighbor, the cliffs echoing his groans, reveals the horror the social Athenians had of solitude, and shows by contrast the joy the sunny-hearted Greek took in the society of his fellow-men.   There is something grand in the words of Gotama Buddha, as reported in the Dham-mapadam :—" If you can find no peer to travel with you, then walk cheerfully on alone, your goal before, the world behind : better alone with your own heart than with a crowd of babblers."   But how clear the difference between the temper of ancient Buddhist isolation and the temper of modern Christian isolation is when we compare with the above sentence the following one by Martineau! " Leave yourself awhile in utter solitude ; shut out all thoughts of other men, yield up whatever intervenes, though it be the thinnest film, between yourself and God ; and in this absolute loneliness, the germ of a holy society will of itself appear ; a temper of sympathy, trust-ful and gentle, suffuses itself through the whole mind ; though you have seen no one, you have met all, and are girt for any errand of service that love may find."

The same reasons that make the feeling of loneliness and moral wretchedness more frequent and strong in modern times than it was before the Christian era, like-wise make the achievement of a steady concord and happiness more arduous to the man of exceptional sensi-bility and ambition than it is to average men.   Unhappiness results when the imagination outruns the heart; when great faculties have no correspondent desires to animate and use them; also when great energies have no adequate motives and guides. A man with tremendous oars, but no rudder, will hardly reach the port : and one with a tremendous rudder, but neither sail nor paddle, promises as poorly. Glorious talents and affections, placed in a lot of harsh adversity, may prove little better than a gold saddle on a galled back. The vaster one's perceptions and emotions are, the harder it is to adjust them to one another. Genius wants a life as prolonged in time and space as its own ideas and feelings of itself are in its own imagination. Destiny says, " Why dost thou build the hall, son of the winged days ? Thou lookest from thy towers to-day. Yet a few years, and the blast of the desert comes ; it howls in thy empty court, and whistles round thy half-worn shield." And with pride of mournful resignation genius replies, " Let the blast of the desert come ! We shall be renowned in our day." In the light of the collective biography of the finer members of our race we are almost tempted to say that it is as hard for a man of ambitious and sensitive genius to be happy, as it is, according to the Scripture, for a rich man to enter the kingdom of heaven. Schopenhauer says, with affecting eloquence, " The great thoughts and beautiful works given to the genius by nature, and which he gives to the world, lead his name through open halls into the temple of fame ; but his heart goes bleeding through the narrow gate of self-denial into the eternal realm of peace." In many nations, for many generations, hundreds of the choicest spirits have been kept wretched by the despotism of their rulers, the slavery of their country, the dire necessity of suppressing their noblest energies, silencing their divinest inspirations. The history of Italy, Spain, France, Austria, Poland, Russia, teems with touching examples of this misery in the lives of artists, poets, historians, musicians, republican patriots and nobles, who have been gagged, imprisoned, executed, or banished. This was ever so in a degree. Firdousi, after heartless persecution at court, died in poverty and in a foreign land. Ovid knew a kindred

fate.  But it has been more common since the facilities of
the printing-press and of travel have made the communication of ideas and sentiments swifter, and more danger·
ous to tyranny.   Genius, while it is the most resentful of
despotic interference, is the most likely to suffer it.   For
conventional establishment looks on genius as its foe.
Of the few famous names in Slavonic literature, the
greater part have chafed bitterly under their censorship,
or openly rebelled and paid the penalty.   Pushkin was
exiled, and fell in a duel at thirty-seven.   Lermontoff,
who, at thirty, also fell in a duel, before he went into
exile wrote to a friend, " Heaven taught me to love ; men
teach me to hate."   A cloud hung over his soul, like a
fog over the blue sea.   He said, " My whole life has been
a series of gloomy and miserable contradictions to my
mind and my heart."

Senancour, in his Obermann, that melancholy and beautiful epic of the heart, that breathing psychological picture
of the nineteenth century, has portrayed the spiritual sorrow and pain of our time with inimitable courage and fulness, and has prescribed the cure.   The recital of his
experience of misery, and the account of the process by
which he achieved peace, are marvellously truthful, and
should be medicinal for similarly afflicted spirits.   Senancour chanted this mysterious monody of the unhappiness
of his soul unheard by his contemporaries.   At that very
time the kindred sufferings of Rousseau, Goethe, Byron,
and Chateaubriand, had won an immense popularity.   By
none of these was the tragedy of the soul treated with
such depth of sentiment and wisdom as in Obermann.
Yet the author scarcely won a hearing.

> Some secrets may the poet tell,
>   For the world loves new ways ;
> To tell too deep ones is not well,
>   It knows not what he says.

Matthew Arnold, in one of the worthiest of poems, and
George Sand, in one of the weightiest of prefaces, have
effectively called attention to this neglected but fascinating book.   Its final lesson of self-renouncing adjust-

ment in the love of nature and the love of man, contains a benign medicament for the wounds of thousands of suffering hearts.

George Sand says that " the present period is signalized by a multitude of moral maladies, unobserved before, contagious and mortal now." The great centres of civilization abound with souls of febrile intensity, overtasked by enormous toils. " The springs of personal interest, the powers of egoism, stretched beyond measure, have given birth to monstrous vices and torments to which psycholo· gy has as yet assigned no place in its annals." There are the sufferings of desire deprived of power, the sufferings of power deprived of desire, the sufferings of disappointed passion baffled in its aims, and the sufferings of disenchanted passion finding nothing worthy of its efforts. Different types of these unhappy experiences are set forth in Saint Preux, Faust, Manfred and Childe Harold, — solemn figures marked by a complete individuality, profoundly discontented and imposingly solitary. George Sand refers the chief examples of constitutional unhappiness to these three causes. First, passion opposed in its development ; that is, the struggle of man with circumstances. Second, the feeling of superior faculties, but without force to make them available. Third, the confessed feeling of incomplete and insufficient faculties. These three orders of wretchedness are exemplified in Goethe's Werther, Chateaubriand's René, and Senancour's Obermann. Werther illustrates the indignant reaction of lofty faculties, irritated and injured, revenging his wrongs both on the world and on himself. René illustrates genius without will. Obermann illustrates moral superiority without genius, a morbid sensibility without commensurate intellectual energy. Werther, whose violent passion has sombrely divorced him from the hopes of human life, says, I have nothing to live for ; cruel world, farewell ! René says, If I could wish, I could do ! Obermann says, Why should I wish ? I could not do ! " Werther is the captive who would die suffocated in his cage : René is the wounded eagle who reattempts his flight ; Obermann is that bird of the cliffs to

whom nature has denied wings, and who sings his mild
and melancholy lay on the strands where ships depart and
wrecks return." The first finds a barrier everywhere; the
second finds satiety everywhere; the third finds vacuity
everywhere.   An excessive chafing against fate, a reverted
pride, and  habitual apathy, and  ineffectual aspiration,
respectively make them unhappy.   Instead of using faith
and imagination to embellish life and aggrandize its aims,
they suffer experience to disenchant the world, strip society of every charm, and throw them back upon the
revolving of their own thoughts and emotions in a solitude full of pain.

Obermann is a romance of the soul, tracing in firm and
tender lines the evolution of an entire destiny, — the
destiny of a nature extraordinarily pensive, expansive,
and susceptible, but feeble and indeterminate.   The successive phases through which this soul passes in the increase and decrease of its pain, are dismay in presence
of the overwhelming claims of a society all whose parts
are too rude for it, idleness, nullity, confusion, sourness,
anger, doubt, enervation, fatigue, tranquillization, benevolence, material labor, repose, forgetfulness, sweet and
peaceful friendship.   It is the course of a soul of powerless reverie, of desires merely sketched in pale outline.
It is the frank confession of a soul avowing the incom-
pleteness of its faculties ; the touching and noble exhibition of a weakness which becomes serene and happy by
renunciation of the ambitions too mighty for it, and resignation to the humble conditions fitted for it.   Ceasing
to groan over the infinity between what he was and what
he longed to be, he resigned himself to be only what he
was.   Formerly he had cried, "I wish no more desires;
they only deceive me.  If hope flings a glimmer into the
surrounding night, it only announces the abyss in which
it fades: it only illumines the vastness of the void
in which I seek, and where I find nothing."   Now, in
the last and well-contented phase of his experience, he
effaces all egoism, and, in the stillness of the Swiss
valleys, in the peaceful cares of pastoral life, in the satisfactions of a reciprocated friendship, his days glide away,

until finite illusions are lost in the infinite reality. Feeling keenly the invisible grandeur of his soul, but knowing
perfectly his inability to reveal or assert that grandeur, he
wisely renounces his exactions and gives himself quietly
up to a modest existence in the love of nature and the
love of his friend. This is the great lesson for the innumerable sore and restless spirits of our age, whose too
much ambition and too little wisdom, excessive sensibility and defective will, fill the air with secret sighs.

The true destiny of man is the fruition of the functions
of his being, the purest and fullest exercise of his faculties, in their due order, in internal unity and in external
harmony. He should therefore seek to perfect himself in
the light of the great standards of truth, virtue, beauty,
humanity, and God ; and to be contented with himself as
reflected by these standards. To seek, instead of this, to
see himself flatteringly reflected in the estimation of other
people, in whose judgments these standards are often re-
fracted in broken distortions, is the sure way to wretchedness. He who aims at perfection, going out and up in
thought and feeling from his defects to its standards, will
be happy. He who aims at fame, coming down in
thought and feeling from his rich desires to the poor
facts, will be miserable. Happiness is the successful
pursuit of an aim. Perfection is the grandest of aims,
and the only one in which a continuous success is morally possible for all. The happiest of men are the saints
and mystics, in whom the social exactions of self are
lost in a fruition of the sublimest standard ; each wave
of force goes out and dies in ecstasy on a shoreless good.
But the selfish plotter feels each wave of force rise and
move inward to die, with egotistic disgust, of extinction in
the centre. Whoever would live contented and die happy
must not pursue public applause, but must give more
than is given him, and love without asking a return.

The chief cause of failure to lead a blessed life is
the immodesty of our demands, and their fitfulness.
Happiness cannot consist of orgasms. Few can expect
to win either the heart or the eye of the world. And the
constant effort to gratify exorbitant desires exhausts the

soul into a chronic state of self-nauseated weariness in-capable of enjoyment.   Here the finger touches the very disease of modern genius, the reason why a Teian Anac-reon was so much happier than an Alfred de Musset. The vast sphere opened for emotion, the thousand excit-ing interests concentrated in a man of cosmopolitan intel-ligence, wear his nerves to a feverish feebleness.   His divineness makes him dainty.   The glare and stare of noisy society become odious to him while they enslave him.    Monotony clothes the world and tedium fills the day.  Nothing is worth anything; an eternal spiritlessness in the breast creates a universal tastelessness in life.   He falls into the melancholy habit of valuing the live hopes of other people by his own dead ones.   His soul frets and pines in a sour solitude : for, when disgusted with ourselves, we have most need, but are least able, to de-light in others.   This condition breeds a feeling of being wronged, a bitter mood of complaint, a general depres-sion and discord.   Giusti says, " The habit of believing ourselves to be unhappy leads us to accuse the order of nature of injustice, makes us think ourselves solitary on the earth, and ends by reducing us to a state of apathy degrading to a man."   Men of genius are more exposed to this than others, because they are more likely to over-exert their powers, and thus disturb the balance of the nervous system.  The incessant spin of activity in their brains drains their force.   Unwise as this course is it is hard for them to avoid it ; and many a son of glory, by losing his health in the acquisition of knowledge and fame, is

> Like one who doth on armor spend
> The sums that armor should defend.

The one prescription for him who would be happy is, Keep yourself in generous health.   Nerves glowing with vigor shed the miseries which irritable nerves invite. Power easily surmounts obstacles to which feebleness as easily succumbs.   The king strides over hecatombs ; the beggar stumbles at a crumb.    A brain well maga-zined with energy by a good digestion, will make be-leaguring trials raise their siege.   An undertoned state of the nervous centres is the greatest predisposing cause of unhappiness. It is often an important relief for one to know that his wretchedness has this physical cause, and is not the shadow of some ominous calamity. For the preservation of a victorious health faithful care must be taken to secure nutrition and rest. No one can stand an uninterrupted drain, especially if it come in disturbing shocks or in a chronic harassment. The more over-worked and unstrung one is the harder it is for him to get rest and nourishing refreshment, and the more likely he is to neglect their claims. Poor Lenau, one of the finest of the German poets, who makes his solitary Faust climb a mountain in a dripping fog, and sigh, " Ah, that my doubts might melt and run off as these mists !"— poor Lenau, whose extreme mental labor and worry reduced him to a dyspeptic and wretched moral state, could not stop his over-action ; would eat nothing but tidbits, cake, and candy : and the cerebral impoverish-ment that resulted filled him with weeping agony, and he ended his days, a most pitiable object, in an asylum for the insane. Proper rest and nutriment would have avert-ed his misery and saved his life.

Pleasing thoughts, faith, affection, serene self-sur render, are helps for the assimilation of strength. Painful thoughts, doubt, fear, hate, pride, are a great source of spiritual waste. To fine souls emotions are as costly as deeds. A feeling may draw off as much as a con-vulsion. Hardly any one appreciates the effect which his *modes of thought* have on his health and strength. Pierce a butterfly with a pin and fasten him to the wall, and he will flutter till his ganglia are emptied of force and he is dead. Every dissatisfying fixed idea is such a disastrous pin. The idea that truth is unattaina-ble, the idea that you are wronged and undervalued, the idea that the world is worthless and full of misery, the idea that human nature is false and contemptible, the idea that history is no benignant plan, but a frightful chaos of chances, — every such painful fixed idea is a probe, pinning the soul against the wall of self-consciousness, and keeping up the wasteful flutter of its forces. These

18

ideas, all thoughts with better thoughts at strife, are therefore to be avoided or neutralized.   Every thought that
chronically impoverishes or lowers consciousness, is a
waste-gate of life: every thought that chronically enriches
or heightens consciousness, is a supply-gate of life.   A warm
and close communication must be kept open between the
heart and those vast masses of authoritative good de-
noted by the love of truth, the love of nature, the love of
men, the love of virtue, and the love of God.   The great
standards of justice, beauty, perfection, eternity, divinity,
are ideal goods meant to serve as ideal supplements to
the actual goods of life; they should uplift the actual
with their reinforcements, and not be allowed to degrade
it by hostile contrast.   The mighty ideal bodies of good
signified by such words as wealth, rank, glory, fame,
society, duty, kosmos, Deity, are either invaluable allies
or fatal foes to the placidity and power of the soul : an-
tagonized by alienated sympathies, all their stimuli act to
exhaust our spiritual reservoirs ; appropriated by friendly
sympathies, their stimuli act to feed and replenish us.
This sub-conscious action of the individual soul in its relations with the goods and authorities of the universe,
either keeping up a nutritious supply of force or a depressing leakage of force, is of unspeakable importance
both to mental health and to bodily health.   Grace is as
instinctive in the symmetrical as awkwardness in the un-
gainly ; the waddle of the duck is no more natural than
the sail of the swan.   So happiness is as properly rooted
in an affectionate and assimilating habit of thought as
unhappiness is in the opposite habit.

He who collects for contemplation all dark enigmas,
discords, failures, crimes, and sorrows, if not already a
weeping philosopher is likely to become one ; while he
whose thoughts collect all clear knowledges, concords, triumphs, virtues, and joys, may easily be a smiling philosopher.   Both classes of views are real; it would be partial
to omit either.   But which should give the gazer his dominant bias ?   On which should his prevailing thought go
out ?   On neither, in itself; but on the idea of the whole,
the eternal laws, the steady tendencies, whose historic

increments will finally take up all exceptions and impedi-
ments in their accumulated swell and sweep. What
is the authoritative ideal of things? What are the deepest
and slowly winning tendencies of society? It is either
disease or wilful perversity that denies these to be good.
It is our duty, then, as we pause apart and look out over
life, to see these, and let them pacify and bless us. Sitting in this higher unity, which synthetises the contradictions below, we may now dip with a smile towards Democ-
ritus, again with a tear towards Heraclitus, but we should
always end by overlooking them both in a complacent
surrender to the Universal Providence.

Make sure that thou shalt have no fault to find with thyself, and thou art inaccessible to unhappiness. Such is the
maxim of Fichte, which stirs us like a blast from the clarion
of an angel. It is the half-truth, on the side of individuality,
recommending the cultivation of self-accord, self-respect,
self-sufficingness. But it needs to be complemented by
the other half-truth, on the side of society, recommending
the cultivation of self-renouncement at the voice of duty,
telling us to live, by the sympathy of love, in the blessings
of others, and not, by the contrast of envy, in their misfortunes, nor, by the isolation of selfishness, in a frozen
indifference alike to their weal and their woe. Every
heart should entertain, in addition to its own affairs, the
great concerns of humanity at large ; as the little lake
among the highlands not only holds the subtances that
form its bed, and the rocks and shrubs that fringe its
borders, but also embraces in its transparent breast the
surrounding landscape of mountains, the endless exhibition of passing clouds, and the nightly pomp of stars.
Fuithermore, the generous believer who traces in the laws
of history the evolution of evil to good, and good to better, may derive a comfort never known to morose plotters
or torpid earthlings, by forecasting the destinies of men
in the happier times to come, when tyranny shall be over-
thrown, and wisdom and love be general.

And a yet surer resource remains for a good man. He
may turn from the petty agitations about him to the sublime peace of God. Why fret and rage? Why sadden

and droop? Enemies pass, and obstacles subside. Accept as your own the Eternal Will that must be done. Let trust sink into peace beneath the struggling vex of mortality, and vision soar into peace over it, as the sky and the deep slumber above and below while tempest and sea hold their terrible dialogue between.

Whoso follows these directions, however afflicted he may be, will never be without great sources of consolation at his command; however warred on, will never be conquered; however solitary, will never be desolately alone. With others, or by himself, the exacting man is discontented, the abnegating man is satisfied. Which is the easier, to cover the world with leather? or to put your foot in a shoe? That is to secure happiness by public conquest; this is to secure it by private renunciation.

Heavenly blessings follow the creature that bears a gentle mind. Solitude is the ravishment and the torture of the soul: love can make it the former; hate or indifference can make it the latter. There are painful external relations of the soul, which act as rasping frictions to wear away strength; and there are painful internal relations of the soul, which act as fretful corrosives to devour peace. That *welt-schmerz*, or world-sorrow, and this *selbst-schmerz*, or self-sorrow, easily create each other, are transmuted into each other, aggravate each other. And it is the saddest of truths that the soul naturally most high and affectionate is in greatest danger of suffering these griefs. Hundreds of gifted men in every generation, like the noble Börne, from mere lack of deferential and loving treatment, become cynics, and die in savage agony and despair. It is milk or wine that sours; water only putrefies.

Reconcile the various counter-claims of thought and passion, adjust your desires to the inevitable conditions of your lot, cultivate some genial occupation, cherish a disinterested affection for your race, a sublime enthusiasm in contemplation of the universe, — and you shall find no hour in life without a glad inspiration, no spot on earth an unwelcome solitude.